T0199239

RECENT TRENDS IN COMPUTATIONAL SCIENCES

This book is a compilation of research papers and presentations from the Fourth Annual International Conference on Data Science, Machine Learning and Blockchain Technology (AICDMB 2023, Mysuru, India, 16–17 March 2023). The book covers a wide range of topics, including data mining, natural language processing, deep learning, computer vision, big data analytics, cryptography, smart contracts, decentralized applications, and blockchain-based solutions for various industries such as healthcare, finance, and supply chain management.

The research papers presented in this book highlight the latest advancements and practical applications in data science, machine learning, and blockchain technology, and provide insights into the future direction of these fields. The book serves as a valuable resource for researchers, students, and professionals in the areas of data science, machine learning, and blockchain technology.

PROCEEDINGS OF THE FOURTH ANNUAL INTERNATIONAL CONFERENCE ON DATA SCIENCE, MACHINE LEARNING AND BLOCKCHAIN TECHNOLOGY (AICDMB 2023), MYSURU, INDIA, 16–17 MARCH 2023

Recent Trends in Computational Sciences

Edited by

Gururaj H L
Manipal Institute of Technology, Bengaluru, India

Pooja M R
Vidyavardhaka College of Engineering, Mysuru, Karnataka, India

Francesco Flammini
University of Applied Science and Arts of Southern Switzerland (CH)

CRC Press
Taylor & Francis Group
Boca Raton London New York Leiden

CRC Press is an imprint of the
Taylor & Francis Group, an **informa** business

A BALKEMA BOOK

First published 2023
by CRC Press/Balkema
4 Park Square, Milton Park, Abingdon, Oxon, OX14 4RN

and by CRC Press/Balkema
2385 NW Executive Center Drive, Suite 320, Boca Raton FL 33431

CRC Press/Balkema is an imprint of the Taylor & Francis Group, an informa business

© 2024 selection and editorial matter Dr. Gururaj H L, Dr. Pooja M R & Prof. Francesco Flammini; individual chapters, the contributors

British Library Cataloguing-in-Publication Data
A catalogue record for this book is available from the British Library

Library of Congress Cataloging-in-Publication Data
A catalog record has been requested for this book

ISBN: 978-1-032-42685-3 (hbk)
ISBN: 978-1-032-42687-7 (pbk)
ISBN: 978-1-003-36378-1 (ebk)

DOI: 10.1201/9781003363781

Typeset in Times New Roman
by MPS Limited, Chennai, India

Table of Contents

Current trends in cyber security

IoT technologies and applications

Recent trends in image processing

Preface

Data Science, Machine Learning and Blockchain Technology have emerged as the most dynamic and rapidly evolving fields in the current era of technological advancements. The Fourth Annual International Conference on Data Science, Machine Learning & Blockchain Technology brings together a global community of experts, researchers, practitioners, and industry professionals to share and discuss cutting-edge research, innovations, and practical applications in these domains.

The conference serves as a platform for knowledge exchange, collaboration, and networking among academicians, researchers, and industry practitioners. This book presents a compilation of research papers and presentations from the Fourth Annual International Conference on Data Science, Machine Learning & Blockchain Technology, held on 16th & 17th March 2023.

The book encompasses a wide range of topics related to data science, machine learning, and blockchain technology, including but not limited to data mining, natural language processing, deep learning, computer vision, big data analytics, cryptography, smart contracts, decentralized applications, and blockchain-based solutions for various industries such as healthcare, finance, and supply chain management.

The research papers presented in this book highlight the latest research trends, challenges, and solutions in data science, machine learning, and blockchain technology. The book serves as an invaluable reference for researchers, students, and professionals seeking to explore the latest advancements and applications in these fields.

We would like to express our gratitude to all the authors, reviewers, and organizers who contributed to the success of the Fourth Annual International Conference on Data Science, Machine Learning & Blockchain Technology. We hope that this book will be an inspiration and a valuable resource for all those interested in these exciting fields.

Foreword

It is our pleasure to write the foreword for this book, which presents a collection of research papers and presentations from the Fourth Annual International Conference on Data Science, Machine Learning & Blockchain Technology. As the fields of data science, machine learning, and blockchain technology continue to evolve and transform various industries, it is important to stay updated on the latest advancements, challenges, and solutions in these domains.

The Fourth Annual International Conference on Data Science, Machine Learning & Blockchain Technology brought together experts, researchers, and practitioners from around the world to share their knowledge, ideas, and experiences. This book represents a valuable compilation of research papers and presentations that were delivered at the conference, providing a comprehensive overview of the latest research trends, applications, and challenges in these fields.

The topics covered in this book are diverse and range from data mining and natural language processing to blockchain-based solutions for healthcare, finance, and supply chain management. The research papers presented in this book highlight the latest advancements in machine learning, deep learning, computer vision, big data analytics, cryptography, smart contracts, and decentralized applications, among others.

This book is a must-read for researchers, students, and professionals seeking to explore the latest developments and practical applications of data science, machine learning, and blockchain technology. We commend the editors for their excellent work in compiling this volume and for providing a valuable resource for the global community.

We would like to extend our gratitude to all the authors, reviewers, and organizers who contributed to the success of the Fourth Annual International Conference on Data Science, Machine Learning & Blockchain Technology. I hope that this book will inspire further research and innovation in these fields and help to advance the state of the art.

Committee Members

Scientific/ Technical Program Committee
Prof. Yu-Chen Hu, Professor, *Providence University, Taiwan*
Dr. Arun Solanki, *Gautam Buddha University Greater Noida, India*
Dr. Hanaa Hachimi, *Tofail University-National School of Applied Sciences of Kenitra, Morocco*
Dr. Prathosh A P, Assistant Professor, *Indian Institute of Technology Delhi*
Dr. Deepu Vijayasenan, Assistant Professor, *Department of ECE, NIT Surathkal, Karnataka*
Dr. Ramkumar Krishnamoorthy, *Department of Information & Comm. Technology Villa College Male' Maldives*

Advisory Committee
Dr. Hong Lin, Ph.D., *Professor at University of Downtown, Houston*
Dr. Heena Rathore, *ACM Distinguished Speaker, Assistant Professor at University of Texas, USA*
Dr. Paolo Trunfio, Ph.D., *Associate Professor of Computer Engineering, University of Calabria, Italy*
Dr. Fernando Koch, *IBM GTS Innovation Senior Technical Solutions Manager; Eisenhower Fellow; ACM Distinguished Speaker*
Prof. Amlan Chakrabarti, *Dean Engg. and Tech. & Director School of IT, University of Calcutta & Dist. Speaker at ACM, Sr. Member ACM & IEEE*
Dr. Varun Menon, *Distinguished Speaker at ACM, Associate Professor and Head of International Partnerships at SCMS Group*
Dr. Ullas Nambiar, *Principal Director AI at Accenture, Innovation Strategist, Startup Mentor Bengaluru, Karnataka*
Dr. Ramasuri Narayanam, *Senior Research Scientist at IBM Research, India Bengaluru, Karnataka, India*
Dr. G R Sinha, *ACM Distinguished Speaker of ACM New York & Professor at Myanmar Institute of Information Tech. (MIIT) Mandalay Myanmar, Myanmar*
Dr. Anand Nayyar, *ACM Distinguished Speaker, Professor, Researcher, Scientist, Author, Innovator, Inventor, Orator, Senior Member-ACM/IEEE*
Dr. Pradeep Kumar TS, *Associate Professor in School of Computing Science and Engineering at Vellore Institute of Technology (VIT) Chennai campus, Chennai, Tamil Nadu, India*
Dr. Prasanna Ranjith Christodoss, *Faculty of Information Technology. Shinas College of Technology Sultanate of Oman*
Dr. Neha Sharma, *Founder Secretary, Society for Data Science Senior IEEE Member, Execom Member, IEEE Pune Section*

Organizing Committee
Dr. Janhavi V, *Associate Professor, Department of CSE, VVCE*
Dr. Paramesha K, *Professor, Department of CSE,VVCE*
Dr. Balarengadurai C, *Professor, Department of CSE,VVCE*
Dr. Ramakrishna Hegde, *Associate Professor, Department of CSE,VVCE*
Dr. Aditya C R, *Associate Professor, Department of CSE,VVCE*
Dr. Natesh M, *Associate Professor, Department of CSE,VVCE*
Dr. Ayesha Taranum, *Associate Professor, Department of CSE,VVCE*

Dr. Prasad M R, *Associate Professor, Department of CSE, VVCE*
Dr. H S Madhusudhan, *Associate Professor, Department of CSE, VVCE*
Mrs. Shraddha C, *Assistant Professor, Department of CSE, VVCE*
Mrs. Divya C D, *Assistant Professor, Department of CSE, VVCE*
Mrs. Tanuja Kayarga, *Assistant Professor, Department of CSE, VVCE*
Mrs. Swathi B H, *Assistant Professor, Department of CSE, VVCE*

Acknowledgment

We would like to express our sincere thanks and gratitude to all those who contributed to the success of the Fourth Annual International Conference on Data Science, Machine Learning & Blockchain Technology, and the production of this book.

First and foremost, we would like to thank all the authors for submitting their research papers and presentations, which made this book possible. We also extend our heartfelt thanks to the reviewers who provided valuable feedback and constructive criticism, which helped to improve the quality of the papers included in this book.

We are grateful to the Management of VVCE, Principal, keynote speakers and session chairs who shared their expertise, insights, and experiences with the conference participants, and contributed to the intellectual richness and diversity of the conference.

We would like to acknowledge the efforts of the organizing committee members and volunteers who worked tirelessly to plan and coordinate the conference and ensured its smooth execution. Their dedication and hard work were instrumental in making the conference a great success.

We would like to express our appreciation to the conference sponsors AICTE-Grant for Conference for their generous support and contributions, which enabled us to organize and host the conference.

Lastly, we would like to thank the editorial and production team who worked diligently to produce this book, from reviewing and editing the papers to designing the layout and formatting. Their hard work and professionalism have contributed to the quality and readability of this book.

Once again, we express our sincere thanks to all those who contributed to the success of the Fourth Annual International Conference on Data Science, Machine Learning & Blockchain Technology, and the production of this book.

Machine learning and applications

Recent Trends in Computational Sciences – Gururaj, Pooja & Flammini (Eds)
© 2024 The Author(s), ISBN 978-1-032-42685-3

Fruits fresh and rotten detection using CNN and transfer learning

G. Anitha & P. Thiruvannamalai Sivasankar
Jakkasandra Post, Kanakapura Taluk, Ramanagara District, Bengaluru, Karnataka, India

ABSTRACT: The economic development of our nation is significantly influenced by agriculture. Fruit production with high yields and productive growth are crucial to the agriculture sector. 30 to 50 percent of the gathered fruit is wasted because there aren't enough qualified workers. Additionally, fruit identification, classification, and grading are not done accurately due to human perception subjectivity. Therefore, the fruit sector must impose an automation system. In order to save labour, production costs, and production time, this research suggests a method based on recognizing fruit flaws in the agriculture sector. These flawed fruits can infect healthy fruits if we are unaware of them. As a result, we suggested a methodology to stop corruption from spreading. From the input fruit photos, the suggested model distinguishes between fresh and rotting fruits. Apples, bananas, and oranges are the three types of fruits I used in this project. Softmax is used to categories the input fruit image into fresh and rotten fruits, and a convolutional neural network (CNN) is utilized to extract features from the input fruit image. Utilizing the Kaggle dataset, the suggested model's performance is assessed. This results in 81% accuracy. The findings demonstrate that the suggested CNN model can successfully classify both fresh and rotting apples. The suggested study investigated strategies for classifying fresh and rotting fruits using transfer learning models. The suggested CNN model outperforms transfer learning models and prior art methods in terms of performance.

Keywords: CNN, Agriculture, Transfer-Learning models

1 INTRODUCTION

One of the topics currently being researched in the agricultural sector is fruit classification. This fruits classification utilizing image processing techniques was created as a part of the present study topic. This fruit classification can be used to locate a fruit in a store or super-market and automatically calculate its price. Three fruits were classified as the first step in the suggested approach. If automated machine vision is used to categorize various types of fruits and vegetables in the agricultural industry, even farmers will gain. Fruits play a crucial role in our daily lives as a food. It delivers nutrients that are essential for our health and bodily upkeep. More fruit consumption as part of a healthy diet is likely to lower the risk of developing various chronic diseases. But not all fruits are treated similarly, and it is troubling that not everyone is knowledgeable about each fruit. This study may create an automatic fruit classification system using a convolutional neural network (CNN) and deep learning (DL), together with a dataset containing information about each fruit. This approach can guide us in choosing fruit that is right for us and instruct us on the traits of that specific fruit. These kinds of programmes can aid in educating kids and introducing them to fruits. Additionally, these algorithms can be used to train a robot to find the right fruit for its user, which is crucial for robots that are employed in tasks linked to fruit harvesting. Smart refrigerators are a significant application for fruit identification and recognition. Modern refrigerators with sensors can determine how fresh a fruit is, how many of each type of fruit are still available, and which fruits are in short supply and should be added to the shopping list. It is frequently seen that recommendations of nutritious foods are quite important as people's access to health information increases. An automated method for classifying fruits that is linked to a database of information can assist the shopper in making healthier fruit selections while also providing nutritional information. These kinds of technologies are also used by super stores nowadays to teach customers about each type of fruit, to keep track of what is sold and what is still in stock, and to determine which fruit items are the most popular. Such an automated system can be used quite readily even by online shopping companies. A reliable fruit detection and recognition system is necessary for all these tasks.Over the years, academics have tried a variety of solutions to the classification issue for fresh and rotten fruit. As will be mentioned below, numerous technologies have been used as well as expensive experimental models.With an emphasis on the advancement of state-of-the-art, Behera Santi *et al.* [1], Provide a succinct review of the methodologies put forward in the research articles from the years 2010 to 2019. The associated researches are contrasted with various methods for classifying, identifying, and rating fruits. This essay also discusses the current research's successes, constraints, and recommendations for additional study. By using a novel way to assess data, Dr. Chandy Abraham *et al.* [2] al proposed a method assists in obtaining the standard maturity level that is appropriate for importing the fruits. Hallur, V., Atharga, B., Hosur, A., Binjawadagi, B., Bhat, K. [4], gave this study which created a portable gadget for banana fake ripening detection. They used a stepper motor, banana holder, IR camera, image processing unit, microprocessor, and display, among other things. Positioned in a holder. The micro switch creates an interrupt when the banana is placed in the holder so that the microcontroller can detect it. When the request is accepted, the stepper motor begins to operate and the camera turns on. A picture of the banana is taken. The image processing module processes the collected images, which are then compared to the reference image.Nitin Kothari., and Sunil Joshi [5], They designed and developed a portable sensor-based prototype for real-time fruit ripeness monitoring in crop fields and storage are discussed in this work. The fruit they used are Musa acuminate (Banana- "Kela"), Psidium guajava (Gauva-"Amrood").

1.1 *Methodology*

The creation of a system based on transfer learning Resnet 50, a Convolution Neural Network, which, in response to input from the user via the user interfaces, accurately predicts when fruit is fresh and when it is rotten.

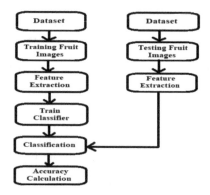

Figure 1. Methodology.

Figure 1 depicts the system's projected flow of our model for differentiating between fresh and rotten fruits. It is processed using the Transfer Learning Technique. The model for fruits classification according to their categories is trained and tested using the suggested dataset.

1.2 Data analysis

Three different fruit types—Apple, Banana, and Orange—make up the dataset. The CNN algorithm knowledge is mentioned in the Implementation Section with the suitable figure of flow (Figures 5 & 6). The approach is consistently used to train a specific model to determine if a fruit is fresh or rotten. There are 10901 images in the training set and 2698 images in the test set.
There are 6 image categories in the dataset.

Fresh Apple   Rotten Apple

Fresh Banana   Rotten Banana

Fresh Orange   Rotten Orange

Figure 2.   Six categories of fruits.

2   RESULTS

Using model fit, we are going to train our model on the training set and test set. The model is built for 80 epochs after that we got 81% of accuracy.

Table 1.

Epochs	Accuracy	Loss Value	Val Accuracy	Val Loss
10	0.4133	1.9018	0.5964	2.3018
20	0.6839	1.9920	0.6839	5.0413
30	0.7191	1.8101	0.6101	3.1813
40	0.7299	1.7753	0.7150	2.1511
50	0.7438	1.7542	0.6597	3.8025
60	0.7564	1.6502	0.7005	2.5516
70	0.7659	1.6433	0.7153	2.5270
80	0.7686	1.6964	0.8136	1.3915

Loss Function is defined as a function that compares the target and predicted output values. During training our objective is to minimize this loss.

Accuracy is defined as the number of classifications, a model correctly predicts divided by the total number of predictions made.

Accuracy = No of correct predictions / Total No of Correct predictions.

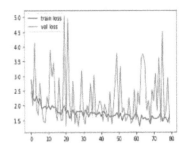
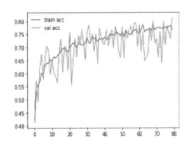

Figure 3.   Accuracy and loss over 80 epochs.

5

Classification   Classification

Result got as Fresh Apple   Result got as Rotten Banana

Figure 4.   Fresh apple and rotten banana.

3   DISCUSSION

3.1   *Convolutional Neural Network implementation*

CNN is a feed-forward deep learning system that can collect an input image, discriminate between objects, and give importance, i.e. biases and learnable weights. The development of CNN was based on the scientific idea that human brain neurons can learn highly abstract properties, classify objects effectively, and have a remarkable capacity for generalization. The capacity of CNN to share weights, which in turn Minimizes the number of parameters required for training, is one of the key arguments for choosing it. This encourages fluid

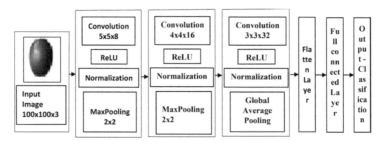

Figure 5.   CNN architecture.

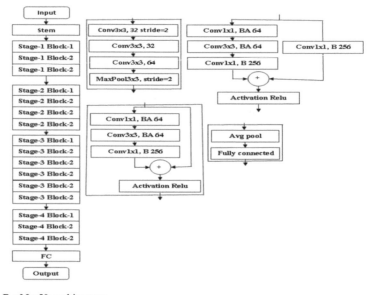

Figure 6.   ResNet50 architecture.

training and solves the over fitting problem. Multiple blocks of convolutional layers, an activation function, pooling layers, and a fully linked layer make up the basic CNN architecture.

One of the well-liked deep learning models, ResNet (also known as Residual Networks), also took first place in the ImageNet challenge. This model, which has 50 layers as its name says, is far deeper than the Prevision 2 models utilised in this project. Convolutional, ReLu, and batch normalisation layers make up a typical residual network block. Utilizing ResNet50 has the advantages of promoting feature usage, enhancing feature propagation, and drastically reducing the amount of parameters. The identity connection, a feature of ResNet, skips connections between layers and adds the output from the layer it is connected to, resulting in a richer model.

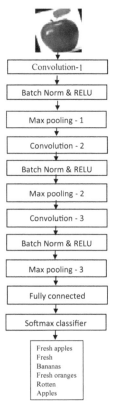

Figure 7.   Implementation architecture.

4   CONCLUSION

Agriculture places a lot of importance on the distinction between fresh and rotting fruits. In this research, we developed a transfer learning model for the classification of fresh and rotting fruits using a CNN-based model. In this case, the accuracy-proposed CNN model is compared to the transfer learning model ResNet50. We will examine the effects of different hyper parameters in this task, including batch size, number of epochs, optimizer, and learning rate. The findings demonstrate that the suggested CNN model, which has higher accuracy than the transfer learning model, can distinguish between fresh and rotting fruits with clarity. As a result, the proposed convolutional neural network model is used by someone to automate the process of 81% accuracy.

ABBREVATION

CNN, Convolutional Neural Network; ResNet, Residual Network; DL, Deep learning (DL)

COMPETING INTERESTS

The Authors declare that they have no Competing interests.

AVAILABILITY OF DATA AND MATERIALS

Plant Village Dataset: https://www.kaggle.com/datasets/emmarex/plantdisease

AUTHORS CONTRIBUTIONS

AG designed, carried out research and drafted the manuscript. PTS participated in research coordination.

ACKNOWLEDGMENT

AG would like to thank Dr. Shivashankar sir for guiding me to make up this paper.

REFERENCES

[1] Santi Behera, Amiya Rath, Abhijeet Mahapatra, and Prabira Sethy. "Review of Fruit Classification, Grading, and Identification Utilising Machine Learning & Artificial Intelligence". *Journal of Ambient Intelligence and Humanized Computing.* Doi:10.1007/s12652-020-01865-8,March 2020

[2] Chandy, Abraham, "In Farming, RGBD Analysis is Used to Identify the Various Fruit Maturity Stages". *Journal of Innovative Image Processing.* jiip.2019.2.006, 2019

[3] R. Thakur, G. Suryawanshi, H. Patel, and J. Sangoi. "An Innovative Approach For Fruit Ripeness Classification," *4th International Conference on Intelligent Computing and Control Systems (ICICCS),* Doi: 10.1109/ICICCS48265.2020.9121045,pp. 550–554, 2020

[4] Hallur, V., Atharga, B., Hosur, A., Binjawadagi, B., Bhat, K., "Design and Development of a Portable Instrument for the Detection of Artificial Ripening of Banana Fruit, in: *International Conference on Circuits", Communication, Control and Computing,* Doi:10.1109/CIMCA.2014.7057776. pp. 139–140, 2014.

[5] Nitin Kothari and Sunil Joshi. "Design and Implementation of IoT based Sensor Module for Real Time Monitoring of Fruit Maturity in Crop Field and in Storage". *Int. J. Curr. Microbiol. App. Sci.* 8 (03):582588. https://doi.org/10.20546/ijcmas.2019.803, 2019

Recent Trends in Computational Sciences – Gururaj, Pooja & Flammini (Eds)
© 2024 The Author(s), ISBN 978-1-032-42685-3

An unsupervised approach to creating a restaurant recommendation system

Metta Venkata Srujan, Rohit Viswam, S. Raghavendra & Ramyashree
Department of Information and Communication Technology Manipal Institute of Technology Manipal Academy of Higher Education, Manipal, India

ABSTRACT: Over the last few years, recommender systems have become increasingly popular due to the technological advancements occurring in the fields of data mining, predictive analysis, and machine learning. In this, we try to use an unsupervised learning approach to cluster restaurants that are similar and recommend restaurants to users according to the ones that they have ordered from in the past. The models used in this project gave us average results given that clustering techniques have their problems and are not the most novel architectures available. However, working on a real-life data set lets us take a deep dive into the nuances of mining data and making predictions in big projects taken up by companies. DBSCAN seemed to work better than K-Means given the latter is very susceptible to outliers. Motivation to use DBSCAN is, this method is able to represent clusters of arbitrary shape and better to handle noise. However, the models that we have built still need a lot of work such as hyperparameter tuning, cross-validation, etc. to increase their accuracy and to be used in real life.

Keywords: Data Mining, Clustering, K-Means. DBSCAN, Datasets.9

1 INTRODUCTION

Over the last few years, recommender systems have become increasingly popular due to the technological advancements occurring in the fields of data mining, predictive analysis, and machine learning. Massive corporations such as YouTube, Spotify, Netflix, Amazon, etc use recommender systems to recommend their products to their customers The methods used for this are usually either machine learning based, involving techniques such as collaborative filtering or clustering, two popular approaches to unsupervised learning, or they are algorithmic, that is just based on relevant attributes associated with their products [1].

In this, we try to use an unsupervised learning approach to cluster restaurants that are similar and recommend restaurants to users according to the ones that they have ordered from in the past. Using attributes such as the location, type, and cuisines offered by the restaurant, we have used two clustering algorithms [2] to recommend similar restaurants to users. We also adopt the strategy of frequent itemset generation and deriving association rules to recommend similar dishes and cuisines, based on market basket analysis.

The remaining part is structured as follows. The Survey about this work is described in Section 2. Section 3 present the Working Methodologies. The Results are found in Section 4. The Conclusion and future direction of this paper is finally covered in Section 5.

2 LITERATURE SURVEY

Sathya *et al.* [1]. Done the analysis the based on the matrix density and the distribution of customer evaluations for food and services, the study's authors provide a technique for recommending restaurants. A key aspect of restaurant recommendation would be the selection of attributes to analyze, and consequently base recommendations on. Owing to the widespread use of Internet-enabled restaurant aggregators and online databases, there is an abundance of data to process and train models for this task.

Yifan Gao et al. [2], Numerical ratings, star-based reviews are considered by the models presented in the paper 'A Restaurant Recommendation System by Analyzing Ratings and Aspects in Reviews'. User-submitted text reviews are another useful attribute in the task of recommendation.

Asani et al. [3], the authors present a model trained on user comments on restaurant pages. User-submitted text reviews provide insight into the experiences of individual users and can be semantically analyzed for the purpose of training a recommendation model. In the paper 'Restaurant recommender system based on sentiment analysis' to ascertain how well they were liked, the names of meals that were gleaned from user comments are sorted using the semantic method.

Alif Azhar Fakhri et al. [4], address some of the limitations of content-based filtering, collaborative filtering uses similarities between users and items both at the same time to provide recommendations. We see this applied to restaurant recommendations in the paper 'Restaurant Recommender System Using user-based Collaborative Filtering Approach: A Case Study in Bandung Raya Region'. The authors employ a user-based collaborative filtering technique to propose restaurants based on user reviews. To measure the similarity between users, they additionally employ two similarities: user rating similarity and user attribute similarity. They measure the precision of rating prediction using the Mean Absolute Error (MAE) statistic.

Anusha Jayasimhan et al. [5], 'A Preference-Based Restaurant Recommendation System for Individuals and Groups' restaurant recommendation system is created for both individuals and groups using data from Yelp. Create a ranking SVM model with variables that take dietary restrictions and culinary preferences into account for each of the 4.9k unique Yelp users, such as cuisine type, services ordered, environment, noise level, average rating, etc. 72% of predictions were correct. The model was trained on skewed data, and there was a significant class imbalance.

In the paper 'Food and Restaurant Recommendation System Using Hybrid Filtering Mechanism' by Amanuel Melese et al. [6], a hybrid filtering method is used to present the restaurant and food recommendation system. A hybrid filtering approach can be considered as a combination of the content and collaborative filtering methods. To suggest restaurants and meals to clients, a variety of filtering algorithms were applied to datasets. Customers are given product recommendations using hybrid filtering based on a variety of factors, including user reviews, the day's best deal, the season, sales, customer age, gender, and mood, browsing history, etc. To remove words linked with eating and dining establishments for some clients, content-based filtering uses the TF/IDF approach.

Kuppani Sathish, et al. [7], understanding consumer demands, gathering feedback, and analyzing it to determine customer satisfaction with the services provided by enterprises is the core goal of the paper. Clustering is a useful approach to solve the problem of recommendation as it uses a dissimilarity/similarity metric to cluster similar items together, allowing us to provide. In the work 'Restaurant Recommendation System Using Clustering techniques'

In the paper 'Restaurant Recommendation System in Dhaka City using Machine Learning Approach' by Taufiq Ahmed et al. [8], a clustering approach was used. This study created a machine learning algorithm-based model that can recommend a good restaurant based on consumer preferences. Based on the restaurants the customer has chosen, this recommendation system would also offer comparable restaurant suggestions.

3 METHODOLOGY

Used an unsupervised learning approach to cluster restaurants that are similar and recommend restaurants to users according to the ones that they have ordered from in the past.

3.1 *Data*

Zomato's catalog of restaurants has been used in this project. Zomato is one of India's biggest restaurant delivery services. The dataset has been restricted only to restaurants within Bangalore as the volume of data obtained from pan India is enormous. We obtained the dataset from Kaggle, a popular platform for hosting and exploring datasets as well as building models, collaborating with other users, publishing interesting findings, and learning about Machine Learning. A link to the dataset has been listed in the references [3]. It has been given a usability score of 8.24 which implies that it has been documented well enough to be used for a project of this scale.

3.2 *Data cleaning*

There are 51,717 data entries and 17 attributes for each restaurant. Redundant attributes have been removed. We found the most relevant attributes in the dataset to be name, online_order, book_table, rate, votes, location, rest_type, cuisines, approx_cost(for two people), listed_in(type), and listed_in(city).After removing duplicate entries and entries with null values, the number of entries left are 51,042.

  The attributes were also renamed to name, online_order, book_table, rate, votes location, rest_type, cuisines, Cost2Plates, and Type for convenience. For categorical data such as cuisines, and rest_type, labels with very few entries have been modified to a separate label called "others".

3.3 *Data analysis*

Extensive data analysis has been done to observe [4] a few trends and patterns in Zomato customers who live in Bangalore. A few plots of the same are shown in Figure 1.

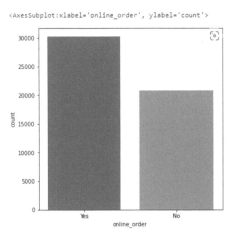
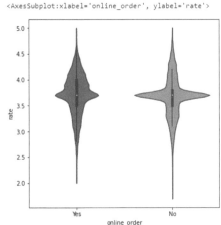

Offering online order: Yes (Blue), No (Yellow)   Violin plot of ratings (on 5) vs online_order

Figure 1.   Data analysis.

3.4 Data preprocessing

All categorical attributes were encoded using the ScikitLearn categorical encoder. After this, standard scaling was performed on all attributes, again using ScikitLearn to ensure no numerical biases exist in the dataset [5]. Since location, Restaurant Type, and cuisines seemed to be the most important attributes, these attributes were weighted to the powers of 3, 2, and 3 respectively.

3.5 Data modelling

This project uses the data mining algorithm called Apriori to extract frequent itemsets. We also used two machine learning algorithms, both of which are clustering algorithms to generate models [6].

3.6 Apriori algorithm

Apriori is an algorithm used to find frequent itemsets and learn association rules over a transactional database. Since the Apriori algorithm is intended for use on transactional data, we selected two candidate attributes of the given dataset, 'dish_liked' and 'cuisines', to apply the algorithm. The 'dish_liked' column includes a comma-separated list of popular dishes for each restaurant, aggregated based on user reviews on Zomato. The 'cuisines' column lists the multiple cuisines of food served at each restaurant, such as Chinese, South Indian, and Italian [7].We also extracted the last word of each dish, (eg: 'Pasta' from 'Alfredo Pasta') to prune the number of possible items for easier computation. This resulted in a transactional database with 747 different dishes. For cuisines, we didn't need to limit the number of items as there were only 20 different cuisine labels. The item lists were stacked and transformed into a one-hot encoded matrix of items, separately for both the dishes and cuisines.

3.6.1 Frequent itemset generation
We experimented with different values of min_sup (minimum support) for the dishes attribute, finally choosing min_sup = 0.05 (or 5% of the transactions) to get a representative sample of 30 frequent 'dishes_liked' itemsets. For 'cuisines' we chose a min_sup of 0.01 and got 15 frequent itemsets, including ten 1-sets, four 2-sets, and one 3-set (North Indian, Chinese, South Indian) are represented in Figure 2.

	support	itemsets	length
0	0.013263	(Biryani)	1
1	0.017139	(Cafe)	1
2	0.074972	(Chinese)	1
3	0.041598	(Desserts)	1
4	0.016708	(Fast Food)	1
5	0.014211	(Ice Cream)	1
6	0.108948	(North Indian)	1
7	0.029972	(South Indian)	1
8	0.010378	(Street Food)	1
9	0.759323	(others)	1
10	0.053914	(North Indian, Chinese)	2
11	0.014469	(Chinese, South Indian)	2
12	0.014211	(Ice Cream, Desserts)	2
13	0.014469	(North Indian, South Indian)	2
14	0.014469	(North Indian, Chinese, South Indian)	3

Figure 2.   Frequent itemsets generated for 'cuisines'.

3.6.2 *Association rule mining*

We then mine for association rules, a rule-based machine learning method for discovering relationships between items in transactional databases [8]. We used the metric of 'confidence' to filter out our association rules.

Rule: $X \Rightarrow Y$ .... (i)

Support: (Frequency (X, Y))/N .... (ii)

Confidence: (Frequency (X, Y))/(Frequency(X)) .... (iii)

The highest confidence rule was (Pasta) -> (Pizza), with a confidence of 0.51, closely followed by (Pizza) -> (Pasta) with a confidence of 0.483.

The association rules of (North Indian, South Indian) -> (Chinese) and (Chinese, South Indian) -> (North Indian) signify a common trend in many restaurants in Bangalore, to serve snacks/meals from all three cuisines simultaneously. Therefore, if two of the three cuisines are present, there is a very high chance that dishes classified under the third cuisine will also be served which is shown in Figure 3.

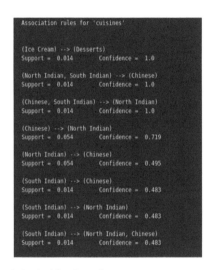

Figure 3.   Association rules of the 'cuisines' attribute.

3.6.3 *Clustering*

Clustering algorithms are popular unsupervised learning techniques to find patterns in datasets. It works on distance or density-based methods which imply that the similarity in the data entries is the deciding factor for which cluster an entry belongs to. Data points are grouped together based on similarity. In this project, we try to cluster restaurants based on the said attributes. Two popular clustering techniques, namely K means and DBSCAN have been implemented in this project. Many big companies use similar methods for building recommender systems for their services.

3.6.3.1 *k-Means clustering*

k-means is a distance-based clustering algorithm. Euclidean distance is used as the metric to measure distances between data points. K refers to the number of clusters into which the data will be clustered. First, k random data points are taken as the centroids for clustering the data. Then all data points are segregated according to the euclidean distance from each centroid [9]. Given that there are 51,104 data points in our dataset, we have used a k value of 1000 to get clusters whose sizes are approximately 51. A random state of 0 has been used.

3.6.3.2 DBSCAN

DBSCAN stands for Density-based spatial clustering of applications with noise. We use a density-based method to cluster as opposed to the earlier approach where Euclidean distance was the metric used. Points that are densely or closely packed together are clustered and those lying-in low-density regions are marked as outliers. The advantage of using a Density-Based approach [10] as against a distance-based approach is outlier detection. Outliers are more clearly demarcated in this approach and clusters have a lesser chance of being influenced by the same.Interestingly, like the metric that we had given to the K-means algorithm [11], the number of clusters found by the DBSCAN on its own is 1049, very close to 1000.

4   RESULTS AND DISCUSSION

To ease loading and reusing the data set, we decided to use Jupyter Notebooks which we ran on the Kaggle Platform for this project. They provide us with 4 cores of CPU and 16 GB of RAM for the same. We did all preprocessing, and data visualization and ran machine learning on the same platform. The notebooks took around 5 to 10 minutes for each model to run. For data preprocessing and creating the machine learning models, a popular machine learning framework called scikit-learn was used. It was chosen as it is an industry-standard and is very well documented.

The metrics for unsupervised learning problems are different from those used for supervised learning problems as there are no labels. Silhouette score is a metric that is used to measure the distance between the clusters formed. Interestingly, the DBSCAN algorithm did not give the best results but while we increased the number of clusters for K means, the score continued to increase which is shown in Table 1.

Table 1.   Silhouette score for different number of clusters.

The number of Clusters	Silhouette Score
1000	0.27
5000	0.34
10000	0.58
15000	0.67
18000	**0.82**

5   CONCLUSION AND FUTURE WORK

The models used in this project gave us average results given that clustering techniques have their problems and are not the most novel architectures available. However, working on a real-life data set lets us take a deep dive into the nuances of mining data and making predictions in big projects taken up by companies. The Apriori algorithm is an excellent tool to generate frequent itemsets, something that is extremely important in the field of data mining. Clustering algorithms while having their problems with generalizing the data provided are still extremely useful tools for preliminary data analysis.

DBSCAN seemed to work better than K-Means given the latter is very susceptible to outliers. However, the models that we have built still need a lot of work such as hyperparameter tuning, cross-validation, etc. to increase their accuracy and to be used in real life. The data, while very useful, still could be used to make data points more representative than in the form that they have been given in this dataset.

REFERENCES

[1]   Krishnaraj, Sathya Seelan. (2021). Restaurant Recommendation System Using Machine Learning Algorithms. *RIET-IJSET International Journal of Science Engineering and Technology*. 9.

[2]   Gao, Yifan & Yu, Wenzhe & Chao, Pingfu & Zhang, Rong & Zhou, Aoying & Yang, Xiaoyan. (2015). *A Restaurant Recommendation System by Analyzing Ratings and Aspects in Reviews*. 526–530. 10.1007/978-3-319-18123-3_33.

[3]   Elham Asani, Hamed Vahdat-Nejad, Javad Sadri, Restaurant Recommender System Based on Sentiment Analysis, *Machine Learning with Applications*,Volume 6, 2021, 100114, ISSN 2666–8270, https://doi.org/10.1016/j.mlwa.2021.100114.

[4]   Fakhri, Alif & Baizal, Abdurahman & Setiawan, Erwin. (2019). Restaurant Recommender System Using User-Based Collaborative Filtering Approach: A Case Study at Bandung Raya Region. *Journal of Physics: Conference Series*. 1192. 012023. 10.1088/1742-6596/1192/1/012023.

[5]   Jayasimhan, A., Rai, P., Parekh, Y., & Patwardhan, O. (2017). Recommendation System for Restaurants. *International Journal of Computer Applications*, 167, 23–25.

[6]   Amanuel Melese, Food and Restaurant Recommendation System Using Hybrid Filtering Mechanism, *North American Academic Research (NAAR) Journal 2021 APRIL*, VOLUME 4, ISSUE 4, PAGES 268–281 https://doi.org/10.5281/zenodo.4712849

[7]   Kuppani Sathish, Somula Ramasubbareddy. K. Govinda, E. Swetha, Restaurant Recommendation System Using Clustering Techniques, *International Journal of Recent Technology and Engineering (IJRTE)* ISSN: 2277–3878, Volume-7 Issue-6S2, April 2019

[8]   Ahmed T., Akhter L., Talukder F. R., Hasan-Al-Monsur, Rahman H. and Sattar A., "Restaurant Recommendation System in Dhaka City using Machine Learning Approach," *2021 10th International Conference on System Modeling & Advancement in Research Trends (SMART)*, 2021, pp. 59–63, doi: 10.1109/SMART52563.2021.9676197.

[9]   Pan, J., & Zhao, Z. (2022). Research on Restaurant Recommendation using Machine Learning. *arXiv*. https://doi.org/10.48550/arXiv.2208.05113

[10] Joachims T., Cosley D., Konstan J. A., Riedl J., Sajnani H., Saini V., Kumar K., Gabrielova E., Choudary P., opes C.,Joachims T., Ihler A." A Preference-Based Restaurant Recommendation System for Individuals and Groups" *International Conference on Machine Learning (ICML)*.

[11] Gomathi R. M., Ajitha P., Krishna G. H. S. and Pranay I. H., "Restaurant Recommendation System for User Preference and Services Based on Rating and Amenities," *2019 International Conference on Computational Intelligence in Data Science (ICCIDS)*, 2019, pp. 1–6, doi: 10.1109/ICCIDS.2019.8862048.

Recent Trends in Computational Sciences – Gururaj, Pooja & Flammini (Eds)
© 2024 The Author(s), ISBN 978-1-032-42685-3

Classification of alzheimer's disease using D-DEMNET framework

P.U. Neetha, S. Simran, G. Sunilkumar, C.N. Pushpa & J. Thriveni
University of Visvesvaraya College of Engineering, Bengaluru, India
ORCID ID: 0000-0003-4073-9611, 0009-0008-6870-975X, 0000-0001-7737-5271, 0000-0002-4307-6318

K.R. Venugopal
Former Vice-Chancellor, Bangalore University, Bengaluru, India

ABSTRACT: A neurological condition called Alzheimer's Disease (AD) causes the brain cells to gradually shrink and die. It is regarded as one of the common causes of Dementia. In this paper, we introduced a new framework that is based on the DenseNet-121 architecture. It is named as D-DEMNET since, it is an extension of DEMNET model which is specified in related work. The Synthetic Minority Over Sampling Technique (SMOTE) approach is used to address the issue of class imbalance. The testing accuracy of the framework is found to be greater than the base model and the accuracy is 95.16%.

Keywords: Alzheimer's Disease, Dementia, DEMNET, DenseNet-121, Neurological Disorder

1 INTRODUCTION

Alzheimer's Disease (AD) is a condition with respect to brain that slowly affect cognitive skills, memory, as well as abilities to perform basic tasks. It is one of the leading death causes in US [1]. AD is one of the type commonly found in most Dementia cases [2]. Due to advancements in brain imaging technologies, researchers can find the changes happening in the brain structure as well as function with the development and enhancement of Tau and Amyloid proteins in the alive brain [1]. Magnetic Resonance Imaging (MRI) and Positron Emission Tomography (PET) are some of the modern neuroimaging techniques which were used for structural as well as molecular bio-marker identifications that are linked to AD.

The researchers of computer-based are involved in the development of machine-based disease detection/diagnoses. Machine Learning (ML) methods have recently been paying good attention in those aspects. But, in spite of a good improvements resulting by these algorithms, there are still drawbacks persist that could be found in traditional ML algorithms which need to be further addressed. To overcome some of these limitations, the new field of ML which is called as Deep Learning (DL) has paid a lot of interest, especially in large-scale, high-dimensional medical image analysis [3].

So, in this paper, we have tried to use one of the DL algorithms called DenseNet-121 (Densely Connected Neural Network) which is considered the most promising Convolutional Neural Network (ConvNet/CNN) type architecture to detect AD. Similarly, to overcome the problem of class imbalance found in the dataset, we have used Synthetic Minority Over Sampling Technique (SMOTE) [4]. The dataset retrieved from the database called Alzheimer's Disease Neuroimaging Initiative (ADNI) [5].

Along with good performance in terms of accuracy, the DenseNet-121 architecture avoids vanishing gradient problem. Some contributions of the proposed framework D-DEMNET are listed here:

- The framework has a compelling benefit in resolving the vanishing gradient problem. This may be due to the use of DenseNet-121 architecture [6].

DOI: 10.1201/9781003363781-3

- Compared to traditional convolutional networks, the DenseNet connectivity pattern requires fewer parameters [6].
- SMOTE is in use to avoid overfitting problem [4].
- Adaptive Moment (Adam) optimizer is basically considered as an optimal optimizer for CNN models which is used here [7].

The paper is organized into Sections. Next section gives a brief literature survey about the similar works. Section III discuss in detail about the proposed framework. The results and discussions regarding it are done in Section IV along with the dataset description. Finally, Section V gives a conclusion about the work. The references are listed further.

2 RELATED WORK

In this particular section, we put forth some of the technical papers that utilize DL algorithms in diagnosing AD. Table 1 gives the summary of some related papers.

Table 1. Summary of some existing works.

Authors	Method	Dataset	Class	Accuracy	Drawback
Murugan et al. [4]	DEMNET	ADNI	Multi	84.83%	Comparatively, it is less effective in five class classification.
Ieracitano et al. [8]	PSD-based DL approach	IRCCS	Binary Multi	89.8% 83.3%	Limited to binary and three class classifications.
Liu et al. [9]	Siamese networks	ADNI	Binary	92.72%	Limited to binary classification.
Basher et al. [10]	Aggregation of CNN and DNN	GARD	Multi	94.82% /94.02%	Method is evaluated using relatively smaller private dataset.
Mehmood et al. [11]	Transfer Learning	ADNI	Binary AD/ NC eMCI/ lMCI	−98.73% 83.72%	Limited to binary classification.

3 PROPOSED WORK

(A) *Problem Statement:* To categorize MRI images of five classes with good performances using an architecture trained through a dataset collected from ADNI database, in order to early diagnose AD.

(B) *D-DEMNET Architecture:* The Figure 1. shows the architecture of the proposed framework. After the data pre-processing step, the input is forwarded to SMOTE technique if class imbalance found. Later, the entire dataset is divided into three sets namely test data, train data, and validate data. Once after splitting, the model is trained using DenseNet architecture. The steps which are used within the proposed framework after splitting the dataset are Densenet121 function, Dropout layer, Flatten layer, Batch Normalization layer, Dense layer, Batch Normalization_1 layer, Activation layer, Dropout_1 layer, Dense_1 layer, Batch normalization_2 layer, Activation_1 layer, Dropout_2 layer, and Dense_2 layer. Finally, as a classification layer, the softmax layer is utilized. The ReLU is used as an Activation Function [12].

The Input layer works similar to DEMNET model which is specified in the related work and which is considered as a comparison model for the proposed framework. In DenseNet-121 function encompass total 117-conv, 3 traditional, and 1 classification layers [13]. Dropout layers are very important because they keep the data in becoming overfit while training the methods [14]. The dimension is reduced to single vector by Flatten layer. The full connectivity between the neurons is found in Dense layer [15]. The inputs are normalized in

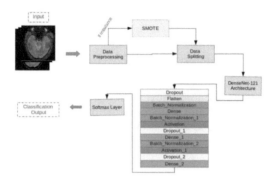

Figure 1. D-DEMNET framework.

Batch Normalization layer [16]. We evaluated the proposed framework using the metrics: accuracy, F1-score, recall, and precision [17–19].

4 RESULTS AND DISCUSSIONS

(A) *Dataset Description:* We collected a dataset from the ADNI database which has the objective of developing clinical, imaging, genetic, as well as biochemical biomarkers to early detect and follow AD. The total number of MRI images collected from this standard database is 1296.

Later, the images were resized into 176*176 size. There are five classes in the dataset, namely: AD with 145, lMCI with 61, MCI with 198, eMCI with 204, and NC with 493 images.

(B) *Experimental setup:* The proposed architecture is been tested in edition-Windows 10 following a device configuration of an Intel(R) Core(TM) i5-6200U CPU @ 2.30GHz to 2.40GHz with 8.00 GB CPU. The proposed architecture is trained with a default setting of 50 epochs and 0.001 initial learning rate value. The Tensorflow version-2.7 is used along with Numpy, Panda, Keras, and Matplotlib libraries. Adam optimizer is used as an optimization algorithm in training this framework [20].

(C) *Performance Analysis:* The proposed framework uses ADNI dataset as specified in the dataset description to analyze the performance. We compare the performance including SMOTE technique of DEMNET [4] architecture with the proposed framework for ADNI dataset.

In both DEMENT and proposed framework, the 1296 images of ADNI dataset were resized into 176*176. Later, the dataset was checked for class imbalance. If an imbalance found then the dataset was passed to SMOTE technique to avoid overfitting problem. After balancing dataset in the proposed framework, we pass the input to DenseNet-121 model for the results. The Figure 2. shows the training and validation curves of accuracy obtained from DEMNET and D-DEMNET frameworks during training process. When we compare both the graphs, it is observed that the training accuracies as well as validation accuracies obtained from proposed framework are much higher than the DEMNET model. This is because of DenseNet architecture usage in which we find a close connectivity between the neurons. The validation curve obtained from the proposed framework also specifies that the framework is less responsive to the noises.

Table 2 gives the performance indices of each classes in DEMNET and proposed framework. On an average, the precision, recall, and F1-score of DEMNET model are 0.76, 0.6, and 0.578 whereas, the precision, recall, and F1-score of the proposed framework are 0.806, 0.806, and 0.804. The Figure 3(a) and (b) gives a performance comparison between DEMNET and D-DEMNET frameworks in terms of average precision, average recall, average F1-score, and testing accuracy.

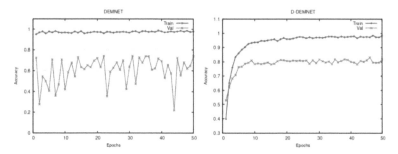

Figure 2. Training and validation curves of accuracy obtained from DEMNET and D DEMNET frameworks.

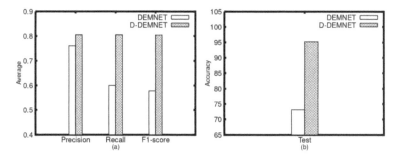

Figure 3. (a) Average performance comparison between DEMNET and D-DEMNET frameworks. (b) Accuracy comparison between DEMNET and D-DEMNET frameworks.

In total, the training accuracy, validation accuracy, and testing accuracy of the DEMNET model is 97.89%, 73.74%, and 73.12%, respectively. But, the training accuracy, validation accuracy, and testing accuracy of the proposed framework is 98.49%, 80.39%, and 95.16%, respectively. Hence, the proposed framework has much higher learning ability than the base model.

Table 2. Performance indices of individual classes in both DEMNET and D-DEMNET model.

	Precision		Recall		F1-score		
	DEMNET	*D-DEMNET*	*DEMNET*	*D-DEMNET*	*DEMNET*	*D-DEMNET*	Support
AD	0.81	0.88	0.78	0.85	0.79	0.86	112
CN	0.38	0.65	0.89	0.64	0.53	0.65	125
eMCI	0.84	0.78	0.27	0.81	0.41	0.79	117
lMCI	0.87	0.93	0.91	0.97	0.89	0.95	116
MCI	0.94	0.79	0.15	0.76	0.27	0.77	110

5 CONCLUSION

In this paper, a new framework called as D-DEMNET which is based on CNN architecture is proposed for improving the performance in classifying the AD stages to detect the progression of the disease. Here, we have done a comparison between the proposed framework and the base model to evaluate the performance. For class imbalance, we have used the same technique as explained in the base model. The accuracy obtained by the proposed framework is 95.16% which is better than the base model. Despite having good results, there are still limitations that can be highlighted. One among them is the "optimizer used". The optimizer considered in this paper has an unexpected action in certain incidents as specified in [21].

REFERENCES

[1] *Alzheimer's Disease Fact Sheet.* Available from: https://www.nia.nih.gov/health/alzheimers-disease-fact-sheet. 2021.

[2] *What is Dementia?* Available from: https://www.alz.org/alzheimers-dementia/what-is-dementia. 2020.

[3] Jo T., Nho K., Saykin A. Deep Learning in Alzheimer's Disease: Diagnostic Classification and Prognostic Prediction Using Neuroimaging Data. *Frontiers in Aging Neuroscience.* 2019;11. Available from: 10.3389/fnagi.2019.00220.

[4] Murugan S., Venkatesan C., Sumithra M.G., et al. DEMNET: A Deep Learning Model for Early Diagnosis of Alzheimer Diseases and Dementia From MR Images. *IEEE Access.* 2021;9:90319–90329. Available from: 10.1109/ACCESS.2021.3090474.

[5] *ADNI Dataset. 2017.* Available from: https://adni.loni.usc.edu/.

[6] Huang G., Liu Z., van der Maaten L., et al. *Densely Connected Convolutional Networks. 2016.* Available from: 10.48550/ARXIV.1608.06993.

[7] Yaqub M., Feng J., Zia M.S., et al. *State-of-the-Art CNN Optimizer for Brain Tumor Segmentation in Magnetic Resonance Images. Brain Science.* 2020;10(7). Available from: 10.3390/brainsci10070427.

[8] Ieracitano C., Mammone N., Bramanti A., et al. A Convolutional Neural Network Approach for Classification of Dementia Stages based on 2D-Spectral Representation of EEG Recordings. *Neurocomputing.* 2019;323:96–107. Available from: https://doi.org/10. 1016/j.neucom.2018.09.071.

[9] Liu C.F., Padhy S., Ramachandran S., et al. Using Deep Siamese Neural Networks for Detection of Brain Asymmetries Associated with Alzheimer's Disease and Mild Cognitive Impairment. *Magnetic Resonance Imaging.* 2019;64:190–199. Available from: https://doi. org/10.1016/j.mri.2019.07.003.

[10] Basher A., Kim B.C., Lee K.H., et al. Volumetric Feature-Based Alzheimer's Disease Diagnosis From sMRI Data Using a Convolutional Neural Network and a Deep Neural Network. *IEEE Access.* 2021;9:29870–29882. Available from: 10.1109/ACCESS.2021.3059658.

[11] Mehmood A., yang S., Feng Z., et al. A Transfer Learning Approach for Early Diagnosis of Alzheimer's Disease on MRI Images. *Neuroscience.* 2021 01;460. Available from: 10.1016/ j. neuroscience.2021.01.002.

[12] Brownlee J. *A Gentle Introduction to the Rectified Linear Unit (ReLU). 2019.* Available from: https:// machinelearningmastery.com/.

[13] Li X., Shen X., Zhou Y., et al. Classification of Breast Cancer Histopathological Images Using Interleaved DenseNet with SENet (IDSNet). *PLOS ONE.* 2020 05;15:e0232127. Available from: 10.1371/journal.pone.0232127.

[14] Baeldung. *How ReLU and Dropout Layers Work in CNNs. 2022.* Available from: https://www.baeldung.com/cs/ml-relu-dropout-layers.

[15] Verma Y. *A Complete Understanding of Dense Layers in Neural Networks. 2021.* Available from: https://analyticsindiamag.com/.

[16] *Batch Normalization layer. 2022.* 2022-08-10; Available from: https://keras.io/api/layers/.

[17] Pushpa C.N., Patil A., Thriveni J., *et al.* Web Page Recommendations using Radial Basis Neural Network Technique. In: *2013 IEEE 8th International Conference on Industrial and Information Systems; 2013.* p. 501–506. Available from: 10.1109/ICIInfS.2013.6732035.

[18] Pushpa C.N., Thriveni J., Venugopal K.R., *et al.* Web Search Engine Based Semantic Similarity Measure Between Words Using Pattern Retrieval Algorithm. In: *International Conference on Computer Science & Information Technology*; Vol. 02; 2013. p. 01–11.

[19] Pushpa C.N., Thriveni J., Venugopal K.R., *et al.* Web Page Recommendation System using Self Organizing Map Technique. *International Journal of Current Engineering and Technology.* 2014;44:3270–3277.

[20] Brownlee J. *Gentle Introduction to the Adam Optimization Algorithm for Deep Learning. 2017.* Available from: https://machinelearningmastery.com/.

[21] Bushaev V. Adam — *Latest Trends in Deep Learning Optimization. 2018.* Available from: https:// towardsdatascience.com/.

Recent Trends in Computational Sciences – Gururaj, Pooja & Flammini (Eds)
© 2024 The Author(s), ISBN 978-1-032-42685-3

Comparison of machine learning and deep learning methods for detection of liver abnormality

P. Vaidehi Nayantara & Surekha Kamath
Department of Instrumentation and Control Engineering, Manipal Institute of Technology, Manipal Academy of Higher Education, Manipal
ORCID ID: 0000-0001-5142-8304, 0000-0001-5567-166X

K.N. Manjunath
Department of Computer Science and Engineering, Manipal Institute of Technology, Manipal Academy of Higher Education, Manipal
ORCID ID: 0000-0001-8239-4047

Rajagopal Kadavigere
Kasturba Medical College, Manipal Academy of Higher Education, Manipal

ABSTRACT: Liver is a vital organ that performs various essential functions. Liver cancer can have severe implications on human health, including death. To reduce the mortality rate, early detection of hepatic cancer is imperative. For rapid diagnosis and treatment planning, developing a computerized system that can aid in the diagnosis is essential. In this paper, machine learning and deep learning methods were compared for identifying abnormal liver from computed tomography images. The liver was segmented using DeepLabv3 + network. The machine learning method used, histogram and gray level co-occurrence matrix for feature extraction and support vector machine for classification. The deep learning methods investigated were ResNet, ShuffleNet and EfficientNet. Accuracy, sensitivity and specificity were used for performance evaluation. The results of our experiments showed that the deep learning methods outperformed the machine learning methods. Among the deep learning models, EfficientNet gave the highest accuracy of 87%. As future work, the abnormal liver can be further classified into benign and malignant classes.

Keywords: Support Vector Machine, ResNet, ShuffleNet, EfficientNet, Liver disease, Computed Tomography

1 INTRODUCTION

Hepatic cancer is the third major cause of death due to cancer globally. According to GLOBOCAN, there were 830,180 deaths due to hepatic cancer in 2020, accounting for 8.3% of the total cancer deaths [1]. Early detection can lower the mortality rate and improve the quality of life of the patient. However, there is an imbalance between the demand and availability of expert radiologists worldwide [2]. The Computer Aided detection and Diagnosis (CAD_x) systems based on Artificial Intelligence (AI) can reduce the current crisis by supporting radiologists with a fast and accurate diagnosis. Computed Tomography (CT) is the commonly used medical imaging technique for hepatic cancer diagnosis due to its cost-effectiveness, short study time and high-quality images.

The Machine Learning (ML) based Hepatic CAD_x systems consist of three main stages: segmentation, feature extraction and classification. In the Deep Learning (DL) based CAD_x systems, the DL model performs the functions of the last two stages. Researchers have

DOI: 10.1201/9781003363781-4

explored image processing, ML and DL techniques to develop CAD_x systems. Segmentation is a very crucial part of these systems as the remaining stages depend on their output. Liver segmentation is difficult due to the related intensity of other abdominal organs like stomach, heart etc. that most often get segmented along with the liver and giant peripheral tumors that alter the morphology of the liver. Many conventional algorithms like region growing, Fuzzy C-Means and the like were used for liver segmentation in the past. These methods are based on pixel intensities and are mostly semiautomatic. Recent research has incorporated DL networks for liver segmentation. In this work, we have analyzed the ML and DL methods for classifying the abdominal CT images into normal and abnormal categories. The liver segmentation was performed automatically using a DeepLabv3 + network.

2 RELATED WORK

Over the years, researchers have proposed several CAD_x systems for liver disease diagnosis. Sreeja and Hariharan [3] detected liver with Hepatocellular Carcinoma (HCC) using Gray Level Co-occurrence Matrix (GLCM) features and Naïve Bayes classifier. Although they reported good accuracy, their method required expert radiologists to delineate the Region Of Interest (ROI) manually. Manual contouring is considered the gold standard in medical image segmentation but is very time-consuming and defeats the very purpose of a CAD_x system. Nayak *et al.* [4] used modified region growing to segment the liver interactively. The logistic regression classifier trained on histogram features was used for liver classification. Muthuswamy [5] extracted the liver through Fuzzy C-Means (FCM) clustering and used GLCM features and Support Vector Machine (SVM) classifier for liver abnormality detection. Krishna *et al.* [6] used Segmentation Based Fractal Texture Analysis (SFTA) and SVM for liver classification. Li *et al.* [7] proposed a CAD system that used Convolutional Neural Network (CNN) for classifying Hepatocellular Carcinoma (HCC). The liver and tumor segmentations were done through Fully Convolutional Network (FCN). Brunetti *et al.* [8] classified HCC using Google Inception v3 network. Sun *et al.* [9] investigated the significance of histogram and GLCM based features in identifying colorectal liver metastasis in CT images. The ROI was, however, manually delineated.

3 MATERIALS AND METHODS

The block diagrams of the CAD_x systems based on the ML and DL methods are shown in Figure 1. The details of the different stages are given in the following subsections.

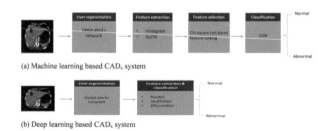

(a) Machine learning based CAD_x system

(b) Deep learning based CAD_x system

Figure 1.   Block diagrams of the artificial intelligence based CAD_x systems.

3.1 *Dataset description*

The CT data used in the work were mainly obtained from Kasturba Medical College (KMC), Manipal after approval from the institutional ethics committee. Some CT images for semantic segmentation were taken from a public database, 3D-IRCADb [10]. For training the DL model for segmentation, 2428 CT images were used, out of which 2242 images were from the public database and 186 images were from KMC, Manipal. The training and test sets for classification comprised 400 and 100 CT images, respectively. The training dataset had 200 images each in normal and abnormal categories. The test set had 100 CT images with 50 images from each category. The abnormal images had liver with different lesions like cysts, hepatocellular carcinoma, hemangioma etc.

3.2 *Liver segmentation*

Liver segmentation was done using DeepLabv3 + network. Its architecture comprises an encoder and a decoder as depicted in Figure 2. The encoder employs ResNet-50 as the backbone network and an Atrous Spatial Pyramid Pooling (ASPP) block. ResNet-50 performs feature extraction and the ASPP block applies atrous convolutions at multiple scales to captures multiscale contextual information. The decoder reconstructs the output with dimension same as the input [11]. ResNet-50 is a CNN with fifty layers (Figure 3). It has skip connections to overcome the problem of vanishing gradients.

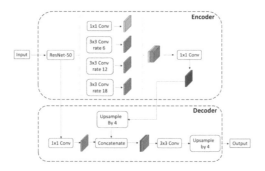

Figure 2.   Architecture of the DeepLabv3 + network.

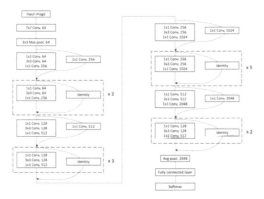

Figure 3.   Structure of ResNet-50 network.

For training the DL model 2428 CT images and their ground truths were considered. ITK-SNAP tool [12] was used for ground truth delineation under the supervision of an expert radiologist. The model was trained with Stochastic Gradient Descent with Momentum (SGDM) optimizer for 100 epochs. The momentum chosen was 0.9 and a constant learning rate of 0.001 was considered.

3.3 Feature extraction and classification

The ML and DL methods used for classifying the liver images into normal and abnormal classes are described in the following subsections.

3.3.1 Machine learning method

The ML approaches relied on feature extraction, feature selection and classification methods for liver categorization. In this work, we have used histogram and GLCM based features for characterizing the liver pixels. Chi-square test-based feature ranking was incorporated for feature selection and support vector machine was employed for classification. The methods are detailed below:

3.3.1.1 Features extracted

- *Histogram features*
  The histogram of an image is the gray-level intensity distribution of the pixels in the image that displays the frequency of occurrence of each gray value. The histogram features extracted were mean, variance, skewness, kurtosis, energy and entropy [13]. These features describe the magnitude, dispersion, asymmetry, flatness, uniformity and randomness of the histogram.
- *GLCM features*
  GLCM is a matrix that describes frequency of occurrence of pairs of pixels with specific intensities and in a given spatial relationship appear in an image. We characterized the texture of the liver image using contrast, dissimilarity, homogeneity, angular second moment and entropy [14].

3.3.1.2 Feature selection

Feature selection is the process of retaining only the important features from the original feature set to reduce overfitting, enhance generalization and improve interpretability of the ML model. We have used Chi-Square test for selecting the relevant features. This method chooses the features that are highly dependent on the response. Out of the 11 features only 8 relevant features were selected using this method. The selected features are mean, energy, entropy, variance, skewness (five features from histogram); and contrast, homogeneity and angular second moment (three features from GLCM).

3.3.1.3 Classification

Linear Support Vector Machine (SVM) was used for liver classification based on the extracted features. The algorithm determines an optimal hyperplane in the N-dimensional space that distinctly classifies the observations, where N is the number of features. It maximizes the margin between the support vectors of both classes [15].

3.3.2 Deep learning method

The DL models investigated in the work are ResNet-18, EfficientNet-B0 and ShuffleNet. All the DL models were trained for 30 epochs using SGDM optimizer with a learning rate of 0.01. The images were resampled to $224 \times 224 \times 3$ pixels. Since the size of the dataset is small, we augmented the images on-the-fly using random scaling (between 80–120%) and random rotation ($-10°$ to $+10°$) to increase the diversity in the dataset.

ResNet-18 is an 18-layer deep CNN with residual blocks to handle the problem of vanishing or exploding gradients. They employ skip connections to connect layer activations to subsequent layers by bypassing certain layers in between. These residual blocks are stacked together to form ResNets [16]. ShuffleNets are CNN that use pointwise group convolution and channel shuffling to decrease computational cost while retaining accuracy [17]. EfficientNetB0 is a CNN architecture and scaling approach that uses a compound coefficient to equally scale all depth/width/resolution dimensions. It has 237 layers and is the baseline network [18].

		Predicted Class	
		POSITIVE (Abnormal liver)	**NEGATIVE** (Normal liver)
Actual Class	**POSITIVE** (Abnormal liver)	True Positive (TP)	False Negative (FN)
	NEGATIVE (Normal liver)	False Positive (FP)	True Negative (TN)

Figure 4. Confusion matrix showing actual and predicted positive and negative classes.

3.4 *Performance evaluation metrics*

The performance of the different ML and DL methods were evaluated using three metrics namely, accuracy, sensitivity and specificity. These metrics are computed from the confusion matrix (Figure 4) using the True Positive (TP), False Negative (FN), False Positive (FP) and True Negative (TN) values using Eqns (1)-(3).

$$Accuracy\ (\%)\ =\ \frac{TP + TN}{TP\ +\ FN + TN + FP}\ \times\ 100 \qquad (1)$$

$$Sensitivity\ (\%)\ =\ \frac{TP}{TP\ +\ FN}\ \times\ 100 \qquad (2)$$

$$Specificity\ (\%)\ =\ \frac{TN}{TN\ +\ FP}\ \times\ 100 \qquad (3)$$

4 RESULTS AND DISCUSSION

The liver segmentation results obtained with the DeepLabv3 + network for some of the cases are depicted in Figure 5. The first column shows the input abdominal CT image and the second column shows the segmented liver image. We can see that pathological liver with multiple tumors within or close to the border of the liver were segmented well (rows 1-4). In row 5, liver and heart have similar intensity, despite that the liver has been segmented well. In the last row, both lobes of the liver have been correctly identified although there are other structures with comparable intensity around the liver. These results illustrate that the DeepLabv3 + network is very effective and robust in delineating the liver.

The results obtained with the ML methods for the test set are summarized in Table 1. Out of the two feature extraction methods, GLCM based method has given higher accuracy of 74%. With the histogram based method, the accuracy is reduced by 5%, this may be because it considers only the intensity distribution in the image. The GLCM features, on the other hand, capture the spatial relationship among the pixels. Hence, the texture features are more effective than the histogram features in classifying the normal and abnormal liver images. A higher specificity of 71.43% was achieved with the GLCM features. However, the sensitivity or true positive rate achieved was higher with the histogram features (80.65%) indicating that these features were better at identifying the abnormal liver images. Hence, we see that the overall capacity of the GLCM features was higher in classifying the two cases. However, the histogram features were effective at identifying the abnormal cases and the GLCM features were good for the normal cases.

25

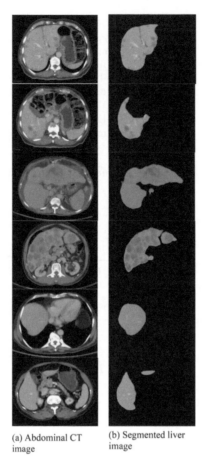

(a) Abdominal CT image

(b) Segmented liver image

Figure 5.   Liver segmentation results with DeepLabv3 + model.

Table 1.   Performance evaluation using machine learning method.

	Features		
Performance metrics	*Histogram*	*GLCM*	*Selected features*
Accuracy	69%	74%	**76%**
Specificity	63.77%	**71.43%**	68.06%
Sensitivity	80.65%	77.27%	**96.43%**

After applying feature selection, only eight relevant features were retained from the eleven features. As mentioned before, it was found that five histogram features mean, energy, entropy, variance, skewness; and three GLCM features viz. contrast, homogeneity and angular second moment were crucial for classification. The classification accuracy improved by 2% (to 76%). However, the specificity reduced to 68.06% and sensitivity increased to 96.43%. Hence, the reduced feature set has been able to successfully identify more of the abnormal images than the normal images.

The results obtained with the three DL networks namely ResNet, ShuffleNet and EfficientNet are shown in Table 2. The best performance was obtained with EfficientNet.

26

The classification accuracy achieved was 87% and the specificity was 79.37%. Sensitivity of 100% was obtained indicating that all the abnormal cases were correctly classified. ShuffleNet also gave good results: accuracy = 78%, specificity = 69.44% and sensitivity = 100%. The lowest accuracy achieved was 74% with ResNet.

Table 2. Performance evaluation using deep learning method.

Performance metrics	Deep learning model		
	ResNet	*ShuffleNet*	*EfficientNet*
Accuracy	74%	78%	**87%**
Specificity	70%	69.44%	**79.37%**
Sensitivity	80%	**100%**	**100%**

Comparing the two methods (Tables 1 and 2), it can be seen that the DL method has given better results. With the ML method, the highest accuracy achieved was 76% (with the reduced feature set), whereas with EfficientNet there was an increase in accuracy by 11%. There was considerable improvement in sensitivity and specificity as well. Around 80% of the normal cases and 100% of the abnormal cases were correctly predicted by EfficientNet. In addition, unlike the ML methods, the DL methods do not require the user to identify the features that are required for classification. Feature extraction as well as classification are done automatically by the CNN. The convolutional layers extract useful low level and high-level features from the CT images that are required for distinguishing between normal and abnormal liver.

5   CONCLUSION

In this study, a comparison of the machine learning methods (histogram and GLCM features with SVM classifier) was done with the deep learning methods viz. ResNet, ShuffleNet and EfficientNet. It was found that EfficientNet gave the best results. For liver segmentation, the semantic segmentation model, deeplabv3 + was used. Based on our experiments, we can conclude that deep learning is more effective in diagnosing abnormal liver from CT images than the machine learning methods. Moreover, the former method does not require the user to perform the tedious task of feature engineering. Hence, it is more effective and efficient in liver abnormality diagnosis. Our work is a preliminary study, in future, it can be extended to further classify the abnormal liver into subclasses like benign and malignant tumors.

ACKNOWLEDGMENT

We acknowledge the financial assistance provided by KStePS, DST, Government of Karnataka, India. We also thank Manipal Institute of Technology, MAHE, Manipal for providing the research facilities and KMC, Manipal for providing the images.

REFERENCES

[1] Sung H. *et al.*, "Global Cancer Statistics 2020: GLOBOCAN Estimates of Incidence and Mortality Worldwide for 36 Cancers in 185 Countries," *CA. Cancer J. Clin.*, vol. 71, no. 3, pp. 209–249, May 2021, doi: https://doi.org/10.3322/caac.21660.

[2]   Burute N. and Jankharia B., "Teleradiology: The Indian Perspective," *Indian J. Radiol. Imaging*, vol. 19, no. 1, pp. 16–18, Feb. 2009, doi: 10.4103/0971-3026.45337.

[3]   Sreeja P. and Hariharan S., "Image Analysis for the Detection and Diagnosis of Hepatocellular Carcinoma from Abdominal CT Images," *Lect. Notes Networks Syst.*, vol. 19, pp. 107–117, 2018, doi: 10.1007/978-981-10-5523-2_11.

[4]   Nayak A. *et al.*, "Computer-aided Diagnosis of Cirrhosis and Hepatocellular Carcinoma Using Multi-phase Abdomen CT," *Int. J. Comput. Assist. Radiol. Surg.*, vol. 14, no. 8, pp. 1341–1352, 2019, doi: 10.1007/s11548-019-01991-5.

[5]   Muthuswamy J., "Extraction and Classification of Liver Abnormality Based on Neutrosophic and SVM Classifier," in *Progress in Advanced Computing and Intelligent Engineering*, 2019, pp. 269–279.

[6]   Krishna A., Edwin D., and Hariharan S., "Classification of Liver Tumor Using SFTA Based Naïve Bayes Classifier and Support Vector Machine," *2017 Int. Conf. Intell. Comput. Instrum. Control Technol. ICICICT 2017*, vol. 2018-Janua, pp. 1066–1070, 2018, doi: 10.1109/ICICICT1.2017.8342716.

[7]   Li J. *et al.*, "A Fully Automatic Computer-aided Diagnosis System for Hepatocellular Carcinoma Using Convolutional Neural Networks," *Biocybern. Biomed. Eng.*, vol. 40, no. 1, pp. 238–248, 2020, doi: 10.1016/j.bbe.2019.05.008.

[8]   Brunetti A., Carnimeo L., Trotta G. F., and Bevilacqua V., "Computer-assisted Frameworks for Classification of Liver, Breast and Blood Neoplasias Via Neural Networks: A Survey Based on Medical Images," *Neurocomputing*, vol. 335, pp. 274–298, 2019, doi: https://doi.org/10.1016/j.neucom.2018.06.080.

[9]   Sun D., Dong J., Mu Y., and Li F., "Texture Features of Computed Tomography Image Under the Artificial Intelligence Algorithm and Its Predictive Value for Colorectal Liver Metastasis," *Contrast Media Mol. Imaging*, vol. 2022, 2022.

[10]   Soler L, Hostettler A, Agnus V, *et al. 3D Image Reconstruction for Comparison of Algorithm Database: a Patient-Specific Anatomical and Medical Image Database.* IRCAD, Tech. Rep; 2010.

[11]   Chen L.-C., Zhu Y., Papandreou G., Schroff F., and Adam H., "Encoder-Decoder with Atrous Separable Convolution for Semantic Image Segmentation BT – *Computer Vision – ECCV 2018*," 2018, pp. 833–851.

[12]   Yushkevich P. A., Gao Y., and Gerig G., "ITK-SNAP: An Interactive Tool for Semi-automatic Segmentation of Multi-modality Biomedical Images," in *2016 38th Annual International Conference of the IEEE Engineering in Medicine and Biology Society (EMBC)*, 2016, pp. 3342–3345.

[13]   Moldovanu S., Moraru L., and Bibicu D., "Characterization of Myocardium Muscle Biostructure Using First Order Features," *Dig J Nanomater Bios*, vol. 6, no. 3, pp. 1357–1365, 2011.

[14]   Haralick R. M., Shanmugam K., and Dinstein I. H., "Textural Features for Image Classification," *IEEE Trans. Syst. Man. Cybern.*, no. 6, pp. 610–621, 1973

[15]   Cristianini N. and Ricci E., *"Support Vector Machines BT – Encyclopedia of Algorithms,"* M.-Y. Kao, Ed. Boston, MA: Springer US, 2008, pp. 928–932.

[16]   He K., Zhang X., Ren S. and Sun J., "Deep Residual Learning for Image Recognition", *Proc. IEEE Conf. Comput. Vis. Pattern Recognit.*, pp. 770–778, Jun. 2016.

[17]   Zhang, Xiangyu, *et al.* "Shufflenet: An Extremely Efficient Convolutional Neural Network for Mobile Devices." *Proceedings of the IEEE Conference on Computer Vision and Pattern Recognition.* 2018.

[18]   Tan, Mingxing, and Quoc Le. "Efficientnet: Rethinking Model Scaling for Convolutional Neural Networks." *International Conference on Machine Learning.* PMLR, 2019

Recent Trends in Computational Sciences – Gururaj, Pooja & Flammini (Eds)
© 2024 The Author(s), ISBN 978-1-032-42685-3

Soil micronutrient detection using machine learning

Sonal Patreena Dalmeida & Surekha Kamath
Manipal Institute of Technology – Manipal Academy of Higher Education (MAHE), Udupi, India
ORCID ID: 0009-0001-3872-0463, 0000-0001-5567-166X

ABSTRACT: This paper presents an algorithm for detecting micronutrients present in the soil using Image Processing and Convolutional Neural Network technique. Soil test is necessary because excessive or lacking use of manure will destroy the crop plant. Huge quantity and less quantity of fertilizer can affect the crop yield. Earlier, farmer used traditional methods to grow the crops to prepare their own farm in the traditional method which they had learnt from their intimates without noticing the micronutrients present in the soil that can change from time to time. This lead to low productivity because the farmer's did not have the proper understanding of the soil. Nowadays, there are several methods to analyze the soil sample through soil test laboratories but these methods are time consuming and tidous. However, Image Processing and Convolutional Neural Network is used efficiently to determine the micronutrients and pH level of soil i.e., Zinc(Zn), Iron(Fe), pH, Manganese(Mn) and Copper(Cu). The soil images were captured manually from Udupi Soil Health Lab. The system is divided into five sections i.e., Image capturing, Image Pre-Processing, Image Segmentation, Training System and Result. Soil images are captured using Smartphone and are provided as an input to the system, from the input images HSV values (Hue Saturation Value) are extracted and these HSV values(Hue Saturation Value) are given as an input to Convolutional Neural Network and as a result soil micronutrients value of the input image is displayed. Based on the result, the soil micronutrients and Accuracy of the model is calculated and displayed. The dataset is split into training and testing wherein, 80% of data is used to train the model and 20% of data was used for testing the model. This project is implemented on MATLAB. The maximum accuracy of the model obtained was 95%.

Keywords: Image Processing, Convolutional Neural Network (CNN), Soil micronutrients, Hue Saturation Value (HSV), Image capturing, Image pre-processing, Image segmentation, MATLAB

1 INTRODUCTION

The agriculture sector is recognized as the pillar of India. With the increase in the modern agriculture technology, it is very important to educate the rural farmers with different farming techniques. To resolve the problem of low crop yield, soil examination is required before planting crop to ensure that the plants are getting right and equal amount of nutrients or manure from the soil. Thus, resulting of higher crop yield [1]. The pH level of soil indicates whether it is acidic, basic or neutral. Due to lack of knowledge, the farmers prepared their own farm in the traditional method without noticing the micronutrients present in the soil that can change from time to time [2]. From the recent few years, many methods were developed for the Prediction of soil micronutrient using Image Processing but those methods are time-consuming, tidous and more expensive [3]. The digital image based system accurately predicts the micronutrients of the soil with the help of Convolutional Neural Network technique. The soil images are captured using Smartphone or digital camera's and the image is fed into Image Pre-Processing, to reduce the unwanted noise, adjust the brightness and resize the image

DOI: 10.1201/9781003363781-5

into fixed dimensions. Once the Pre-Processing is done, the images are segmented, the features of the image are extracted using HSV(Hue Saturation Value) techique [4]. The HSV values are given as an input to the model and then using the Convolutional Neural Network, the convolutional layer is added as a first layer to extract features from the input image. Next, between two convolutional layers, the pooling layer is added to decrease the number of framework or parameters of a image, when images are large. Then, at the end Fully-Connected layer(FC) is added where all the features of the image are combined together to create a model and the desired result is predicted and model's accuracy is predicted [5]. Finally, the pH level of soil and soil micronutrient value of an input image is displayed.

2   LITERATURE REVIEW

The systematic study carried out in this work is summarized with respect to the various parameters and are briefed correspondingly as follows. In the paper [6], authors developed Convolutional Neural Network model to analyze the situation of plants depending on the plant image. In this experiment they made use of Inception-Resnet architecture. Okra plants images were used to train the model. They had used about one hundred and eighty four images for training and about forty seven images for testing process. Image augmentation method was applied to enlarge size of dataset.

In the paper [7], Redmi 3S prime Smartphone was used to capture the soil images. The captured images were used as datasets. The HSV(Hue, Saturation and Value) of the soil images are calculated and it stores Saturation and Hue. The Analysis of soil prediction was done using Artificial Neural Network(ANN), Linear Regression and KNN Regression. It was found that ANN had better result.

In the paper [8], the authors discussed about the soil in the farmland of Philippines, Image processing and computer vision techniques was used to to classify the soil Macronutrients i.e., Nitrogen,Phosphorus, Potassium and soil pH. Soil experiment were carried out to decide the soil's primary nutrients. Artificial neural network (ANN) was used for fast and accurate performance. There were about four hundred and forty eighty captured image data. chemical soil measurements were used to categorize fertility of the soil [9]. EC, P,Fe, OC, Zn, B, S, Mn, K, Cu, pH were used to categorize the soil as high, medium, low fertile. Random Forests (RF), logistic regression, Naive bayes, decision tree bagging, Support Vector Machine (SVM), Boosted Regression Tree (BRT) were used to identify the soil as high, medium, low. The RF classifier obtained best results. Experiment was conducted using Karnataka state soil profile data with 11 attributes and 92832 instances.

In the paper [10], soil images are captured using Smartphone and then images are segmented using K-Mean Algorithm and features of the soil image are determined using Hue, Saturation and Value (HSV) color image processing. Finally, features of the soil image are regressed with the actual pH soil value which is observed in soil pH laboratory. The soil samples were taken from the different region of west Guwahati zone.

In this paper [11], The datasets of soil were taken from TNAU website which included 32 districts of Tamilnadu. The algorithms that were included for the experiment are Bayes Net, IbK, and Naïve bayes to predict variety of crop suitable for the soil. The accuracy was determined using false positive value, true positive value, recall, f-measure, MCC and precision.

In this paper [12], The research was based on secondary data concerning to Kumaun region of Uttarakhand. The data is used to develop and train FIS (subtractive clustering based fuzzy inference system) and RBF NN (radial basis function neural network) on MATLAB, for predicting the soil pH based on minerals concentration i.e., Fe, P, Mn, K, Cu. Thus, the prediction performance was compared using mean square prediction error.

In this paper [13], The Arduino-based prototype was used for testing macronutrients and soil pH from soil testing upto fertilizer recommendation was completely machine-controlled. The experiment included pumps and stepper motors. Digital image processing technique was used to analyse the Phosphorus, Nitrogen, Potassium and pH content level present in Philippine farmlands. Artificial Neural Network offered fast and best results. The system database contained 356 captured images. Results showed that the model was 96.67 accurate.

3 METHODOLOGY

India is very much rich in cultivation. Soil plays an important role in plant growth. Soil testing is required to obtain proper and equal amount of nutrients from the soil. The model is developed to help the rural farmers to analyse the soil before planting. The system consists of five different sections i.e., Image Capturing, Image Pre-Processing, Image Segmentation, Training System, and Result.

3.1 Image capturing

A mobile phone is a good device to capture the soil image. In most of the agriculture study, a digital camera is used to capture the sample images but mobile phone is also used to capture the agriculture images. Here, mobile phone is used to capture all the soil samples and it is stored as soil dataset. The image capturing is done under visible light. The soil images are collected manually from Udupi Soil Health Lab about 4000 soil images were collected.

3.2 Image pre-processing

In this phase, the system pre-processess the input image. It checks the image quality, size, removes the unwanted noise from the image, enhances the brightness of the image and also resize the input image into fixed dimensions.

3.3 Image segmentation

The features of the image are extracted using HSV (Hue Saturation Value) techique. Hue consists of primary colors (Red, Blue, Yellow) and secondary colors (Orange, Green, violet) that appears in the color circle or color wheel. Color Saturation is the purity and strength of a color. Higher the saturation of a color it is more realistic and intense. If the color saturation is low, it is closer to pure grey on the grey scale. Color Value indicates relative darkness or lightness of a color. The color value is assumed based on the amount of light reflected off of a surface and absorbed by the human eye. Hence, we can find out HSV value for an input image.

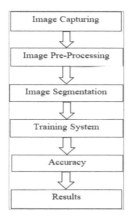
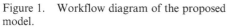

Figure 1. Workflow diagram of the proposed model.

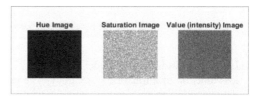

Figure 2. Hue Saturation Value (HSV) of the input image.

3.4 *Training system*

A Neural Network is a network or circuit of neurons. using the neural network, we can compare the HSV value of input image with the pre-trained datasets and the desired result is predicted. In this training section, the training model gets the input from the segmentation process. The soil pH and the soil micronutrient are based on the standards given by Soil health lab, Udupi and these images and values collected from the lab serves as a database for the system. Convolutional Neural Network is used to improve the accuracy of the result. For soil micronutrient and soil pH level identification, the dataset contains about 4000 captured soil image samples. From the image sample, 80% of the images were used for training, 20% for the images for testing the model. The CNN model consists of various layers such as Convolutional layer, ReLu layer, Pooling layer, Fully connected layer.

a) Convolutional layer- The first step is Convolution layer, the relation between the pixels is preserved by learning the image features using small squares of the input data. It is mathematical operation, that takes image matrix and filter it as an input. convolution of images with various filters will perform operation like blur, edge detection and sharper by applying filters. The convolution layer is called using cnnAddConvLayer().

b) ReLu layer- ReLu layer states Rectified Linear Unit for the non-linear operation. The basic need of is to introduce non-linearity to convent. ReLu provides better performance compared to other non linear functions such as tanh or sigmoids.

c) Pooling layer- In the Pooling layer when images are large the number of parameters are reduced. subsampling or downsampling are also called as spatial pooling, which decreases the dimensionality of every single map but takes only the required information. Max pooling works better and takes largest element from rectified feature map.

d) Fully-Connected Layer- The final layer is the Fully-Connected Layer also called as dense layer. This layer extracts only the important information from the image. with the help of FC layer, all the features are combined together to create a model. Here, there is one input layer, seven hidden layers and one output layers.

   To train the model with training dataset, we use the function: cnn=traincnn(cnn, train_x,train_y,no_of_epochs,batch_size);

   where, train_x is training data, train_y is corresponding images, batch_size should be in integer multiple of total no.of training data and no_of_epochs should be > 1 for better accuracy but it takes lot of time to train.

   To test the model, we use the function:

   testcnn(cnn, test_x, test_y);

   where, test_x is test data, test_y is corresponding values. The process of training can be repeated, till the training number is reached.

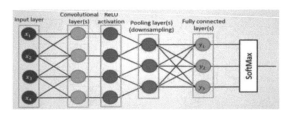

Figure 3.   Convolutional neural network architecture.

e) *Accuracy:*

   Using the loss function, model's accuracy is calculated. The machine learning algorithm uses loss function to improve the accuracy. The performance of Convolutional Neural Network is determined using an accuracy matric in a interpretable way. The model's accuracy is determined after the parameters are calculated in the form of percentage. It measures how well our model predicts by comparing true values with model prediction i.e., Predicted result with the Test

32

f) *Result:*

Finally, Using the Image Processing and Convolutional Neural Network model, it is easy to find soil micronutrient and pH value of an input image. This model is not time consuming and helps the farmers to provide the right amount fertilizers to the soil.

4 RESULTS AND DISCUSSION

The digital image based system predicts the micronutrients of the soil with the help of Convolutional Neural Network technique. Using Smartphone or digital camera's the soil images are captured and this image is provided as an input. The image is Pre- Processed to resize the image into fixed dimensions, remove the unwanted noise and adjust the brightness. Once the Pre-Processing is complete, then the Images are Segmented, the features of the image are extracted using HSV(Hue Saturation Value) techique. The HSV values are given as an input to the model and then using the Convolutional Neural Network and the result is predicted and model's accuracy is displayed. Finally, the pH level of soil and soil micronutrient value of an input image is displayed.

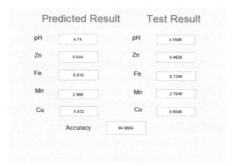

Figure 4.   Output shows the pH level of soil and soil micronutrients.

A model is pre-trained using Image Processing and Convolutional Neural Network. The data for the analysis were gathered from soil health lab, Udupi. The datasets are divided and grouped into training and testing image. Dataset consists of about 4000 images in which 80% of images are used for training and rest of the 20% of the images were used for testing the model. The predictedvalue of soil micronutrients and pH are given for every image in the dataset. The CNN model used in this system has 9 layers with the first layer has input layer and the rest 7 layers in between are the hidden layers and the last layer has output layer. Thetest soil sample image is tested against the model and the results are displayed. Apart from the test images, for any random input image given has an input the result achieved was appropriate and had better results. The test values for the different soil micronutrients are generated and loss function is used to calculate the accuracy of the result. The predicted values are the values given by the soil health lab and the test values are the values that has been tested by the CNN model. The performance of CNN model was fast and accurate. The differences in predicted value and test value are used to find the accuracy.

The Figure 5, shows Sensitivity versus Epoch graph. Here, only one epoch is considered. Sensitivity is used to assess model's performance because it shows the positive instances were able to correctly identify.

The Figure 6, shows Cross-Entropy versus Epoch graph. Here, 12 epoch are considered. Cross-Entropy is a standard measure of difference between two probability distributions for a given random variable. The main goal is to minimize the loss i.e., smaller the loss, better the model's performance.

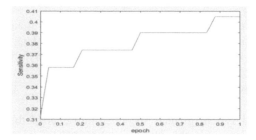

Figure 5. Sensitivity versus epoch graph.

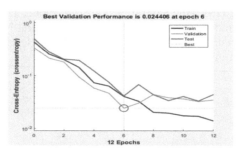

Figure 6. Cross-Entropy versus epoch graph.

The Figure 7, shows the Confusion Matrix of soil micronutrients. For summarizing the performance of classification and prediction algorithm. The matrix is compared between the actual target values with those values predicted by machine learning model.

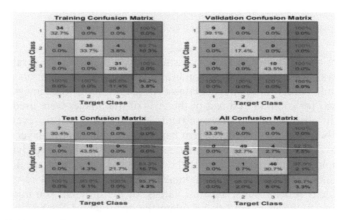

Figure 7. Confusion matrix.

5 CONCLUSION

Sometimes, farmers used soil testing laboratory or soil pH digital meter to calculate the soil pH and soil micronutrients but these methods are very time and cost consuming. The digital Smartphone based low-cost soil prediction model which will detect the soil pH and soil micronutrients easily. Here, Convolutional Neural Network technique and Image Processing is used to detect the soil micronutrients and soil pH. The soil image is captured using

Smartphone or digital camera's and then the image is fed into image preprocessing to reduce the unwanted noise, adjusting the brightness and reseizing the image. Then the images segmented, where features of the image are extracted using HSV(Hue Saturation Value) techique. Once HSV value for an input image is obtained, then it is fed into the training model with the help of CNN technique, it compares the HSV value of input image with the datasets and the desired result is predicted. This model is helpful for rural farmers. The digital image based system which accurately predicts the micronutrients of the soil with the help of Conventional Neural Network. To further improve the project, we have to Collect more soil samples and increase the size of datasets that leads to accurate results and also fertilizer recommendation for a given input soil image that will be based on the result.

ACKNOWLEDGEMENT

A special thanks to Soil Health Lab, Udupi for the constant help in providing the datasets throughout the project. We are grateful to Manipal Institute of Technology, M.A.H.E., Manipal for providing the facilities to carry out the research.

REFERENCES

[1] M.S. Suchithra and M. L. Pai, "Improving the Prediction Accuracy of Soil Nutrient Classification by Optimizing Extreme Learning Machine Parameters," *Inf. Process. Agric.*, vol. 7, no. 1, pp. 72–82, 2020.

[2] T. Wani, N. Dhas, S. Sasane, K. Nikam, and D. Abin, "Soil pH Prediction using Machine Learning Classifiers and Color Spaces," in *Machine Learning for Predictive Analysis*, Singapore: Springer Singapore, 2021, pp. 95–105.

[3] N. L. Kalyani and K. B. Prakash, "Soil Color as a Measurement for Estimation of Fertility Using Deep Learning Techniques," *Int. J. Adv. Comput. Sci. Appl.*, vol. 13, no. 5, 2022.

[4] J. C. V. Puno, R. A. R. Bedruz, A. K. M. Brillantes, R. R. P. Vicerra, A. A. Bandala, and E. P. Dadios, "Soil Nutrient Detection using Genetic Algorithm," in *2019 IEEE 11th International Conference on Humanoid, Nanotechnology, Information Technology, Communication and Control, Environment, and Management* (HNICEM), 2019.

[5] M. G. Lanjewar and O. L. Gurav, "Convolutional Neural Networks Based Classifications of Soil Images," *Multimed. Tools Appl.*, vol. 81, no. 7, pp. 10313–10336, 2022.

[6] Lili Ayu Wulandhari, Alexander Agung Santoso Gunawan, Arie Qurania, Prihastuti Harsani, Triastinurmiatiningsih, Ferdy Tarawan and Riska Fauzia Hermawan, "*Plant Nutrient Deficiency Detection Using Deep Convolutional Neural Network*", 2019.

[7] U. Barman, "Prediction of Soil pH using Smartphone Based Digital Image Processing and Prediction Algorithm," *J. Mech. Ccontin. Math. Sci.*, vol. 14, no. 2, 2019.

[8] N. Arago, J. W. Orillo, J. Haban, Jomer Juan, J. C. Puno, Jay Fel C. Quijano, Gian Matthew Tuazon, "*SoilMATe: Soil Macronutrient and pH level Assessment for Rice Plant through Digital Image Processing using Artifical Neural Network*", 2017.

[9] Sujatha and Jaidhar, "Classification of Soil Fertility Using Machine Learning-based Classifier," in *2021 2nd International Conference on Secure Cyber Computing and Communications (ICSCCC)*, 2021.

[10] Utpal Barman Ridip Dey Choudhury Ikbal Uddin, "*Prediction of soil pH using k means Segmentation and HSV Color Image Processing*", 2019.

[11] M. Chandraprabha and R. K. Dhanaraj, "Soil Based Prediction for Crop Yield using Predictive Analytics," in *2021 3rd International Conference on Advances in Computing, Communication Control and Networking (ICAC3N)*, 2021.

[12] Sandeep Kumar Sunori *et al.* "*Soil pH Prediction using Artificial Intelligence*", 2021.

[13] R. Alasco *et al.* "SoilMATTic: Arduino-based Automated Soil Nutrient and pH level Analyzer using Digital Image Processing and Artificial Neural Network," in *2018 IEEE 10th International Conference on Humanoid, Nanotechnology, Information Technology, Communication and Control, Environment and Management (HNICEM)*, 2018.

Recent Trends in Computational Sciences – Gururaj, Pooja & Flammini (Eds)
© 2024 The Author(s), ISBN 978-1-032-42685-3

A review of tracking concept drift detection in machine learning

Nail Adeeb Ali Abdu

Information Technology Department, Faculty of Engineering & Computing, University of Science and Technology, Aden, Yemen

Khaled Omer Basulaim

Information Technology Engineering Department, Faculty of Engineering, University of Aden, Jemen
ORCID ID: 0000-0002-4519-1133

ABSTRACT: The availability of time series streaming data has increased dramatically in recent years. Since the last decade, there has been a growing interest in learning from real-time data. While extracting significant information from data streams, online learning faces changes in data distribution. Concept drift refers to hidden data contexts that learning systems are unaware of it. Using previous training instances from the data stream, the classifier categorizes new instances. Hence in this regard, it is obvious to have considerable methods to discover the similarity or diversity status of the present streaming data that compared to the buffered data. This issue is defined and explored in contemporary literature as concept drift. This paper was initiated to identify the constraints of the contemporary methods devised to handle the concept drift in data streams in relation to supervised learning, we also review and categories the concept drift detectors with their key points. Eventually, the article presented observations can provide decision boundary detection at improved levels and discussing open research challenges and possible new research direction.

Keywords: concept drift, concept drift defensing, detection methods

1 INTRODUCTION

Data mining approaches are being used across multiple dynamic and real-time applications such as financial transactions, sensor networks, mobile communications, etc. However, the scalability of implementing these models in real-time applications which produce huge voluminous data in short periods of time is repeatedly challenged. As the information generated in such applications is dynamic and forms data-streams with voluminous instances, storing information prior to processing is not feasible. Further, specific applications require instant and real-time solutions. One of the most prominent issues faced by researchers during the mining process is that the data-streams in such instances can be non-stationary and witness concept drifts along the infinite length [1,2]. Due to the infinite distance and voluminous characteristics of these data-streams, storing information and utilizing the historical patterns for model training is not possible in the practical sense. Accordingly, new mechanisms that can process the information through a single scan and employ limited resources such as storage space and time are required. The several features of classical data mining over data stream mining are displayed in Table 1. In a stream with significant learning, the order of observation determines how the data are organized. Therefore, it is necessary to identify any changes in data distribution. One of the biggest problems with data stream mining is this detection, as the quality of prediction is affected if the changes is not noticed. [3]

DOI: 10.1201/9781003363781-6

Table 1. Classical data mining vs data stream mining.

	Requirement of memory	Count of concepts	Count of passes	Requirement of time	Obtained of result
data stream mining	Limited	More than one	Single	Limited	Approximate
classical data mining	Unlimited	One	Multiple	Unlimited	Accurate

In addition, concept drifts are also observed is such real-time data streams where the underlying concept changes over time. The concept drift is observed whenever the underlying concept regarding the gathered information alters over some duration of stability. In the absence of the drift it is hard to maintain the balance between the most recent concept and the previous concept, which leads to form the stability-plasticity dilemma [4]. These shifts in the inflow data stream significantly impact the precision rates of the classification model because the classifier learns by historical stable data instances. Some of the areas where the concept drift phenomenon is observed include credit card fraud identification, spam recognition, and weather forecast. Accordingly, the concept drift identification process is considered as a vital phase [5]. The abstract view of model construction in the data stream is depicted in Figure 1. This figure consists of three models. The first module is the input module, which accepts instances or examples of training data as input from the algorithms. Because data streams instances are constantly arriving, the incoming data instances are stored in memory and processed later. The model predicts the labels of newly incoming data instances allowing us to maintain a high predictive performance over time. The aforementioned procedure will be repeated until the data are accessible. [6]

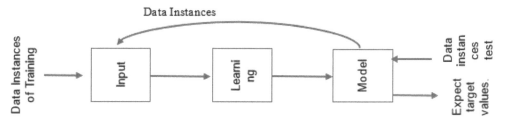

Figure 1. Construction a model in a data stream as an abstract view.

Upon detecting the concept drift in the data stream, the fixed classification models can yield inaccurate results [7]. For instance, consumer buying preferences can vary because of weather shifts or economic fluctuations. Accordingly, business intelligence entities and users of statistical prediction methods must identify such concept drifts within data-streams in a timely manner. The learning process for the classifier must be a continuous process as the model classifies the future instances by concepts learned. This continuous learning process ensures that any concept drift occurring in the streams can be tracked. The classifiers used for data-streams must be able to learn the patterns through a single scan and should be capable of performing classification at any time. Further, the model must be able to track the concept drifts over time and also preserve historical concepts. Typical classification of underlying concept drift should be by "speed of change". The change is considered as abrupt if the shift from one concept to the next concept occurs suddenly and in case the shift happens over few instances, it is considered as gradual. Alternatively, if the target concept is constant but data distribution shifts over time, it is considered as virtual [8]. However, in practical applications, both real and virtual concepts appear together [9]. The classification of drift as abrupt or gradual depends on its type upon observation. Sudden shifts over data-streams represent sudden-drifts, and on the other hand, slow transformation denoted gradual drift. An optimal classification model must be able to identify the concept drifts occurring in the data-streams to minimize accuracy loss identified during the concept-drift. Numerous methods are put forward to determine concept drift, and some of the prominent ones include online-algorithms, ensembles, and drift-detection models. Classification of error rates is the foundational concept for most of the drift detection methods [10]. Whenever any concept drift occurs then the trained classifier that belongs to the older data chunks becomes incompatible with the upcoming data chunks, and due to this the classified new chunk's error rate increases. Hence, error rate became the crucial parameter when detecting the concept drift occurrence.

2 CONCEPT DRIFT

The concept drift definition is flexible enough which can be defined informally depending on the use of the concept. For example, "the statistical properties of target concept changes over time in a situation" [11]. Finding the exact points of change can be very challenging, as the change between two distributions can be gradual and has to be differentiated from transient noise that can affect the stream. Consequently, change detection algorithms have to be robust (resistant to noise) and at the same time sensitive to true changes in the dataset, forming a classic stability-plasticity trade-off [12]. It is also essential for the algorithms to be able to reliably detect the drift with minimal delays, while also efficiently executing the detection in a shorter time than the next instance arrives and in a constant space complexity [13]. A learner should know the sequential positioning of examples in the stream [14], particularly in cases like the arrival of the samples with latency or in an out-of-order. The differentiation between the local and global drifts [15] is also identified based on the drift influence on the problem space or parts of the problem space. Over the years, the research works have proposed various methods and techniques to find concept drifts in the area of data streams [16]. The proposed techniques are classified into (1) windowing technique, (2) ensemble classifiers, and (3) drift detectors. Among these techniques, the windowing technique presents a manageable supporting mechanism for saving the latest data as well as updating the classifier of the system for maintaining accuracy. However, the size of the window hampers the drift finding process. Ensemble classifier gives the best chance to modify any of both methods, their classifier structure and aggregation methods based on the degradation of the main classifier performance because of concept drifts. A sliding window which is one of the most common windowing techniques saves the instances of the recent data in a limited number only. This Sliding window technique plus traditional learning algorithms can construct classifiers for data streams. The efficacy of drift detection is often downgraded, if fixed size sliding window technique is used. The small size window finds sudden changes quickly, however, drops stability accuracy (in periods) because the large size window gives away to see the swift changes occurring in the data streams [17]. In addition to drift detectors and windowing technique, and, for learning from data streams, the ensemble is also the most common technique. In ensembles, two types, Online ensemble, and Block-based ensemble are preferred to handle data changes. To update the classifier, online ensemble learns incrementally after every instance of data and a Block-based ensemble first processes the data which is blocked and then the classifiers are upgraded. Abrupt drifts are found by the online ensemble effectively, however, they are unable to detect the gradual drift, which appears during the data streamed over a considerable period of time. However, the Block-based ensemble finds the gradual drift efficiently but fails to detect sudden drift. The research work [17] presents Accuracy Updated Ensemble (AUE2). The concept drift of weighted base classifiers in the ensembles is adapted by the AUE2. Moreover, the AUE2 forecast class label, and records of small-sized block instances accurately. These records support for detecting the gradual drift demanding an exhaustive class labeling of these ensemble models. However, on the availability of the small amount labeled instances to build classifiers, the clustering technique is preferred. The work [18] presents ensemble classifiers as well as clusters for categorizing the test data with a perfect load as they communicate the class labels along with the class information on training data. In the work [19], using kernel density estimation for detecting the Twitter hotspot cluster activities as well as finding the hidden patterns and drifts in spatial trajectories that are chosen from spatio-temporal where the data is embedded with texts of twitter, Senaratne *et al.*, presented a framework. The majority of drift detection algorithms, which were proposed, failed in maintaining the high level of generalization between the two concepts, i.e., old and new ideas. Hence to address the issue, in the work [20] presented a system to save various diversity ensembles of the different levels. The high diversity ensembles and low diversity are joined together whenever the fresh ideas of the concepts are generalized with the existing plans of the idea. While dealing with the data streams, the detection of novelty is one of the challenges. In case of introducing either a new class or new feature to an old set of data streams, innovation is found. The general framework to handle the concept drift is explained in the Figure 2 dealing with the detection of knowledge from data collected from distributed systems and online data, which operates in a complex, dynamic and unstable environment. It explains the types of learners, their strategies and performance parameters.

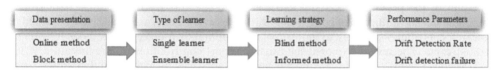

Figure 2. General framework to handle the concept drifts.

3 ENSEMBLE METHODS

The popular methods Bagging and Boosting, which are presented in the work [21] is used to improve the weak learners' algorithms accuracy. However, using different strategies, these methods utilize trained set classifiers on the training data and aggregate every classifier response; then it combines in an ensemble to give better predictions. Mainly, for training, bagging method uses various randomly produced bootstrap samples, through the repeated sampling of the training set. On the contrary, boosting method trained multiple classifiers, utilizing multiple training data distributions, provided the diversity of the classifiers and varies according to the earlier forecasting. However, attention is needed that several boosting algorithms give theoretical guarantees for their results. Likewise, the method Edwin-OZABOOST that included in MOA framework is preferred by ADWIN to detect the concept drift. In the work [22], the Leveraging Bagging named as LevBag is a newly featured version of Oza and Russell's Online Bagging. As a hard-coded concept drift detector, this Levbag is joined with ADWIN. Besides, this method presents two changes where the first change is to raise the diversity ($\lambda = 6$) value in the Poisson distribution then drives to a growth in the experts trained probabilities on the same instance. To change the instances prediction process used by the experts predict to rise in diversity and decrease the correlation in the second change. Accordingly, the LevBag with all these modifications provides better accuracies when compared to Edwin OzaBag method. Another method DDD presented in the works [20]. This is a default in the EDDM method. The researcher proclaims that the e-Detector makes better rates of the recall and false negative with minimal rise in the frequency of the false positive detection. Furthermore, it possesses a robust generalization capacity when compared with the detectors.

4 OBSERVATIONS

It is observed that many of the drift detection methods presume the complete accessibility of labeled streams to analyze data distribution as well as classifier performance. The work [23] offers an entirely unsupervised and semi-supervised approach to develop the applicability of the existing drift detection techniques supposing that streams are hardly labeled. The two methods emphasize on detecting imbalanced class distributions and statistical changes with the support of an adapted Page-Hinkley test. The work [24] depicts the dependency of their concept drift detector on data distribution change detector, instead of actual class labels. The contemporary studies emphasize the development of drift detection methods and techniques to address the real drift. Ample research hasn't been done in this area impacting the classification requirement as well as the performance. Notwithstanding the impact of drifts above on true decision boundaries, the research can provide decision boundary detection at improved levels.

5 CONCLUSION

The aim of the paper is to classify and confirm contemporary literature on data stream mining under the influence of conceptual drift. The review notes are that the current field of research contributes seriously to innovative and optimal drift detection methods for mining data streams. The review explored the existing report contributions to the literature of the last decade. The majority of these models target factors such as high-speed data flows, noisy data flows, mining accuracy, and time complexity of the mining strategy. Factors such as the context of the drift concept, the drift concept, the time validity, and the drift concept were ignored with recursive concepts. Going forward, it is clear to conclude that there is a lot of room for research in order to find new models to deal with understandable drift in mining data streams in the context of factors such as the concept of time drift, validity, context, and the impact of the iterative concept. Our contributions to future re-research will be the new or expanded models in the review competition.

REFERENCES

[1] Khamassi, M. Sayed-Mouchaweh, Hammami M., and Ghédira K., "Discussion and Review on Evolving Data Streams and Concept Drift Adapting," *Evol. Syst.*, vol. 9, no. 1, 2018.

[2] Demšar J. and Bosnić Z., "Detecting Concept Drift in Data Streams Using Model Explanation," *Expert Syst. Appl.*, vol. 92, pp. 546–559, 2018.

[3] Kifer, Daniel, Shai Ben-David and Johannes Gehrke. "Detecting Change in Data Streams." In *VLDB*, vol. 4, pp. 180–191. 2004.

[4] Rossi A.L.D., De Souza B.F., Soares C. and De Carvalho A. C. P. D. L. F., "A Guidance of Data Stream Characterization for Meta-learning," *Intell. Data Anal.*, vol. 21, no. 4, pp. 1015–1035, 2017.

[5] de L. Cabral D. R. and de Barros R. S. M., "Concept Drift Detection Based on Fisher's Exact Test," *Inf. Sci. (Ny).*, vol. 442–443, pp. 220–234, 2018.

[6] Gama, Joao. *Knowledge Discovery From Data Streams*. CRC Press, 2010.

[7] Jadhav A. and Deshpande L., "An Efficient Approach to Detect Concept Drifts in Data Streams – Semantic Scholar," 2017.

[8] Delany S.J. and Cunningham P., "A Case-Based Technique for Tracking Concept Drift in Spam Filtering A Case-Based Technique for Tracking Concept Drift in Spam Filtering," *Twenty-fourth SGAI Int. Conf. Innov. Tech. Appl. Artif. Intell.*, pp. 187–195, 2005.

[9]   Tsymbal A., Pechenizkiy M., Cunningham P. and Puuronen S., "Dynamic Integration of Classifiers for Handling Concept Drift," *Inf. Fusion*, vol. 9, no. 1, pp. 56–68, 2008.

[10] Rutkowski L., Jaworski M., Pietruczuk L. and Duda P., "A New Method for Data Stream Mining Based on the Misclassification Error," *IEEE Trans Neural Netw Learn Syst*, vol. 26, no. 5, pp. 1–12, 2014.

[11] Wang S., Schlobach S. and Klein M., "Concept Drift and How to Identify It," *J. Web Semant.*, vol. 9, no. 3, pp. 247–265, 2011.

[12] Hoens T.R., Polikar R. and Chawla N. V., "Learning From Streaming Data with Concept Drift and Imbalance: An Overview," *Prog. Artif. Intell.*, vol. 1, no. 1, pp. 89–101, 2012.

[13] Ho S. and Wechsler H., "On The Detection of Concept Changes in Time-Varying Data Stream by Testing Exchangeability," *arXiv Prepr.*, 2012.

[14] Marrs G. R., Black M. M. and Hickey R. J., "The Use of Time Stamps in Handling Latency and Concept Drift in Online Learning," *Evol. Syst.*, vol. 3, no. 4, pp. 203–220, 2012.

[15] Shaker A. and Lughofer E., "Self-adaptive and Local Strategies for a Smooth Treatment of Drifts in Data Streams," *Evol. Syst.*, vol. 5, no. 4, pp. 239–257, 2014.

[16] Krawczyk B., Minku L.L., Gama J., Stefanowski J. and Woźniak M., "Ensemble Learning for Data Stream Analysis: A Survey," *Inf. Fusion*, vol. 37, pp. 132–156, 2017.

[17] Brzezinski D. and Stefanowski J., "Reacting to Different Types of Concept Drift: The Accuracy Updated Ensemble Algorithm," *IEEE Trans. Neural Networks Learn. Syst.*, vol. 25, no. 1, pp. 81–94, 2014.

[18] Zhang P., Zhu X., Tan J. and Guo L., "Classifier and Cluster Ensembles for Mining Concept Drifting Data Streams," *Proc. – IEEE Int. Conf. Data Mining, ICDM*, pp. 1175–1180, 2010.

[19] Senaratne H., Bröring A., Schreck T. and Lehle D., "Moving on Twitter: Using Episodic Hotspot and Drift Analysis to Detect and Characterise Spatial Trajectories," Proc. *7th ACM SIGSPATIAL Int. Work. Locat. Soc. Networks*, pp. 23–30, 2014.

[20] Minku L.L. and Yao X., "DDD: A New Ensemble Approach for Dealing with Concept Drift," *IEEE Trans. Knowl. Data Eng.*, vol. 24, no. 4, pp. 619–633, 2012.

[21] Breiman L., "Bagging Predictors," *Mach. Learn.*, vol. 24, no. 2, pp. 123–140, 1996.

[22] Bifet A., Holmes G. and Pfahringer B., "Leveraging Bagging for Evolving Data Streams," *Lect. Notes Comput. Sci. (including Subser. Lect. Notes Artif. Intell. Lect. Notes Bioinformatics)*, vol. 6321 LNAI, no. PART 1, pp. 135–150, 2010.

[23] Lughofer E., Weigl E., Heidl W., Eitzinger C. and Radauer T., "Recognizing Input Space and Target Concept Drifts in Data Streams with Scarcely Labeled and Unlabelled Instances," *Inf. Sci. (Ny).*, vol. 355–356, pp. 127–151, 2016.

[24] Lindstrom P., Mac Namee B. and Delany S.J., "Drift Detection using Uncertainty Distribution Divergence," *Evol. Syst.*, vol. 4, no. 1, pp. 13–25, 2013.

Recent Trends in Computational Sciences – Gururaj, Pooja & Flammini (Eds)

Wearable electrogastrogram perspective for healthcare applications

C.P. Vijay, A. Amrutha, D.C. Vinutha, V. Rohith & N. Sriraam
VVCE
ORCID ID: 0000-0002-0525-2368, 0000-0002-5206-858X, 0000-0003-3096-2967, 0009-0002-4219-0503

ABSTRACT: The digestive system, which includes the esophagus, duodenum, small intestine, large intestine, and rectum, is responsible for the body's main chemical reactions, with the stomach playing a crucial role. Unfortunately, many people experience gastric dysrhythmia, which is characterized by abnormal myoelectrical rhythms in the stomach and can result in symptoms such as nausea, vomiting, early satiety, abdominal bloating, and abdominal pain. To better understand this condition, Electrogastrography (EGG) is recommended. EGG is a non-invasive method that records gastric myoelectrical contractions. Although EGG has been the dominant method used to date and has consistently demonstrated associations between arrhythmias and gastroparesis, it has limitations such as low signal quality and incomplete sensitivity and specificity. This article provides a comprehensive review of the methods used to analyze gastric abnormalities using EGG, the correlation between wearable electrogastrogram (EGG) and electromyogram (EMG), the anatomy and physiology of EGG, and its clinical significance. It also covers current practices in India and other Asian countries and the importance of EGG in long-term monitoring.

Keywords: Electrogastrogram (EGG), electromyogram (EMG), gastrointestinal activity, wearable device

1 INTRODUCTION

Electrogastrography is a non-invasive method used to record gastric myoelectrical contractions. In the late 1800s, Santiago Ramón y Cajal discovered interstitial cells of Cajal (ICC), which he recognized as the pacemakers of the gastrointestinal system and the regulators of gastrointestinal homeostasis. However, the challenge was to measure the activity of these cells in a meaningful way. In 1922, Dr. Walter C Alvarez conducted research on the electrical activity of the GI tract at Berkeley. Electrogastrography (EGG) has become an important tool for objectively assessing gastric motility disorders associated with various diseases. EGG is capable of measuring two types of gastric electrical activity: electrical response activity, which corresponds to the electrical activity of smooth muscle contractions, and electrical control activity, which refers to rhythmic electrophysiological events in the gastrointestinal (GI) tract. By analysing these electrical signals, healthcare providers can gain insights into the underlying physiological mechanisms of gastric motility disorders and develop targeted treatment plans to alleviate patient symptoms. The normal frequency range of gastric electrical activity is around 3 cycles per minute (CPM), and using EGG, one can observe abnormalities in electrical motility. The EGG technique involves the use of cutaneous electrodes placed on the skin of the abdomen over the stomach to record myoelectrical activity.

This technique is non-invasive and painless, making it a valuable tool for diagnosing and monitoring gastric motility disorders. By detecting and analysing the electrical signals generated by the smooth muscle cells of the stomach, EGG can provide insights into the frequency and amplitude of gastric contractions, as well as the coordination of peristalsis. The EGG technique has the potential to improve the accuracy of gastric motility disorder

DOI: 10.1201/9781003363781-7

diagnosis and inform personalized treatment plans for patients. EGG recordings can detect conditions such as Bradygastria (slow stomach contractions), Tachygastria (fast stomach contractions), dyspepsia (indigestion), ulcers, chronic nausea, and other gastric motility abnormalities, which can cause sensations of failure, vomiting, abdominal pain, stomach ulcers, and gastroesophageal reflux disorders [2].

The maximum frequency and propagation of gastric contractions are determined by the gastric slow wave. Gastric dysrhythmias include Bradygastria, Tachygastria and Arrhythmia. The average gastric slow waves in human are about 2–4 cpm. Whereas, gastric slow waves in Bradygastria ranges from 0.5–2.0 cpm and Tachygastria ranges from 4–9 cpm [1]. To find the number of cpm in a particular patient, a wearable device which calculates the gastric slow wave is used, which senses the gastric movement from the sensors placed on the abdominal region on the stomach. The observed waves are in the form of potential. In the past, measuring the electrogastrogram and intestinal electrogram for an extended period without the interference of respiration was a challenging task. However, a new invention has been introduced that involves using an apparatus and method for measuring the electro-gastrogram and intestinal electrogastrogram. This method involves using a low-pass filter that eliminates high-frequency signals that are not included in the potential of the stomach and bowel. The filter is inserted between an amplifier for biological signals and a recorder that records the amplified signals. The method also ensures a linear relationship between the input signal's frequency of the low-pass filter and the output signal's.

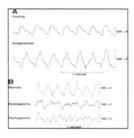

Figure 1.   A) Gastric slow wave at fasting and at postprandial. B) Normal gastric slow wave, Bradygastria, and Tachygastria.

2   IMPORTANCE OF ELECTROGASTROGRAM (EGG) AND ELECTROMYOGRAM (EMG)

Electromyography (EMG) is a diagnostic method which evaluates the health condition of muscles and the nerve cells (motor neurons) that control them. Motor neurons transmit electric signals that cause muscles to contract and relax. These signals are translated into graph form using EMG to diagnose. The results of EMG help one to diagnose disorders of muscle, nerve and disorders affecting the neuro-muscular connection [4]. Some symptoms that require EMG include: tingling, numbness, muscle weakness, muscle cramp or pain, paralysis, involuntary muscle twitching. The results of EMG help doctor to determine the underlying cause for the symptoms which possibly include: muscular dystrophy (group of conditions that weakens or damage the muscle over time), myasthenia gravis (weakness in the voluntary muscles), radi-culopathies (pinched spin nerve), carpal tunnel syndrome, amyotrophic lateral sclerosis.

EMG electrodes are utilized to detect the bioelectrical activity within a person's muscle. There are two primary types of EMG electrodes, namely surface electrodes (also known as skin electrodes) and inserted electrodes. The latter can be further classified into two types: needle electrodes and fine wire electrodes.

The objective of this EMG system is to assist individuals who are weak or elderly to assess their muscle strength level and obtain beneficial muscle signals for rehabilitation purposes [3].

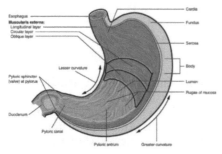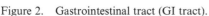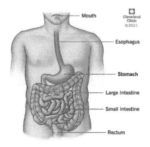

Figure 2. Gastrointestinal tract (GI tract).   Figure 3. Parts and layers in stomach.

As said earlier, EGG is a non-invasive method used for recording the gastric myoelectrical contractions. EGG has a wide range of applications in electrophysiological studies and clinical practice. It is commonly used to assess the efficacy of interventions and therapies aimed at improving gastric motility in patients with digestive disorders. By measuring the electrical signals generated by the smooth muscle cells of the stomach, EGG can provide objective data on the effectiveness of these interventions. In addition, EGG is a valuable tool for detecting slow wave abnormalities in patients with suspected or confirmed gastric motility disorders. By analysing the frequency and amplitude of electrical signals, healthcare providers can identify patterns associated with gastric dysmotility and develop personalized treatment plans for patients. Overall, EGG has the potential to improve the accuracy of gastric motility disorder diagnosis and inform effective treatment strategies. [1].

2.1 *Physiology of stomach*

The stomach is as a J-shaped muscular bag curved in the frontal plane. The process of gastric emptying, which refers to the movement of food from the stomach into the small intestine, is regulated by a complex interplay of neural, hormonal, and mechanical factors. The rate of gastric emptying can affect nutrient absorption and blood glucose levels, and can also play a role in conditions such as gastroparesis and dumping syndrome. The study of gastric motility and gastric emptying is important in understanding the physiology of the digestive system and in diagnosing and treating related disorders. The volume of stomach is determined by the amount of stored digesta and normally ranges between 25 mL in the fasted and 1500 mL in the fed state. After a typical meal of~1000 mL, the stomach measures around 30 cm along its greater curvature and exhibits a maximal width of around 10 cm. Peristaltic muscular contraction waves, that is, ring-like contraction that constrict the stomach circumferentially, running from the corpus through the antrum to the pylorus, are mixing and grinding the food in the distal part of the stomach. Temporal coordination of these so-called antral contraction waves (ACWs) with the aperture of the pylorus controls releases the digesta into the duodenum [5].

The function of stomach include:

- Acts as a reservoir for ingested food
- Mixes food with gastric secretions
- Empties gastric contents into the duodenum [6].

3   CLINICAL IMPORTANCE OF EGG

Endoscopy involves the insertion of a long, flexible tube with a camera and light source through the mouth or rectum to view the inside of the digestive system. It can help diagnose various gastrointestinal disorders such as ulcers, inflammation, tumors, and polyps.

However, endoscopy may cause discomfort, and in rare cases, complications such as bleeding or perforation of the gastrointestinal tract may occur. Imaging techniques such as X-rays, CT scans, and MRI scans are also used to evaluate the digestive system. These techniques provide detailed images of the digestive system, but they are expensive and may involve exposure to ionizing radiation. In comparison, EGG is a non-invasive and cost-effective technique that can provide valuable information about gastric motility disorders [6].

4   CURRENT PRACTICE OF RECORDING GASTRIC SIGNALS

4.1   *Spatial sampling*

In addition to electrode layout and size, other factors that can affect the accuracy of EGG recordings include patient position, electrode preparation, and electrode impedance [4]. Patient position can affect the amplitude and frequency of EGG signals, and it is recommended that patients be in a seated position during recording to minimize motion artifacts. Proper electrode preparation and placement are also important for obtaining accurate recordings, and impedance should be monitored and kept low to minimize noise.

Overall, EGG is a non-invasive and cost-effective method for evaluating gastric motility and identifying gastrointestinal disorders. While there are limitations to single channel EGG analysis, HR-EGG shows promise in overcoming these limitations and providing more accurate and detailed information about gastric function. Further research is needed to establish biomarkers for specific gastrointestinal disorders and to optimize electrode layout and recording parameters for maximum accuracy.

The concept of sampling continuous time-series data in a discrete manner has been thoroughly described [7]. According to the Nyquist criterion, in order to avoid losing data during digitization, the sampling rate must be at least twice the maximum frequency of the signal (fs > 2fmax, where fs represents the sampling rate and fmax represents the highest frequency in the signal). If a time series is aliased, it becomes impossible to retrieve the lost information using any signal processing technique.

The same principle underlying the Nyquist criterion for temporal sampling is also relevant for spatial sampling. In spatial sampling, the highest detectable spatial frequency without experiencing aliasing is determined by the density and measurement area of the electrodes. The electrode acts as an analog filter by removing spatial frequencies that have wavelengths shorter than its actual measurement diameter. This is because the electrode averages the potentials within the area in contact with the gel or measurement area. For instance, if an array of electrodes has uniform centre-to-centre spacing of d and electrode diameter of D [7], applying the Nyquist criterion to the edge-to-edge distance between neighbouring electrodes yields the following limitation:

$$d - D < \lambda_{min}/2$$

where λ_{min} is the shortest spatial wavelength of the signal.

4.2   *Fast Fourier transforms*

The normal frequency range of gastric electrical activity, or normogastria, is typically between 2.25–3.75 cpm. Frequencies that are lower or higher than this range are referred to as arrhythmias, with abnormally low frequencies known as Bradygastria (1.00–2.25 cpm) and abnormally high frequencies known as Tachygastria (3.75–9.00 cpm). Slow waves with a frequency of approximately 3 cpm are typically observed in healthy volunteers, indicating normal gastric electrical activity.

Despite its potential utility in clinical practice, EGG has some limitations due to the use of cutaneous electrodes. The EGG signals recorded using these electrodes can be affected by various types of artifacts. For instance, electrical activity originating in the small intestine and respiratory muscles can produce frequencies in the range of 9–13 cpm and 12–24 cpm, respectively. Cardiac action potentials can also produce frequencies around 60 cpm. These artifacts can potentially affect the accuracy and reliability of EGG analysis, making it more challenging to differentiate gastric slow waves from other sources of electrical activity. To minimize these artifacts, researchers and clinicians should carefully design and perform EGG studies, and develop appropriate methods for signal processing and analysis. Digital filters Are used to remove these artifacts.

(Ch4) 4 cm below Ch3 on the midline. Additional electrodes can be placed as needed for more detailed recordings. Once the electrodes are in place, the patient should be in a comfortable and relaxed position, and any potential sources of interference, such as electrical equipment or metal objects, should be kept away from the recording area. The recording typically lasts between 30 minutes to several hours, depending on the purpose of the examination. During the recording, the patient may be asked to take deep breaths or swallow water to stimulate gastric activity. After the recording is complete, the data is analysed and interpreted by a healthcare professional to diagnose any abnormalities in the ga electrical activity [8].

5 LITERATURE REVIEW

Sl. No.	Author-Year	Sample/Source	Evaluation algorithms	Results
[1]	Yin *et al.* 2013	Review paper	–	These findings suggest that there is a strong association between abnormal gastric electrical activity and abnormal antral motility in patients with functional dyspepsia. The excessive dysrhythmias seen in the EGG recordings of these patients may contribute to their symptoms of nausea, bloating, and early satiety. Gastroduodenal manometry can provide additional information about the motor function of the stomach and duodenum, and may be useful in identifying specific abnormalities that can guide targeted therapy. Overall, simultaneous recordings of EGG and gastroduodenal manometry can provide valuable information about the pathophysiology of functional dyspepsia and guide treatment decisions.
[2]	Gandhi *et al.* 2021	Data was provided on request from the author	Logistic Regression, SVM, and KNN Genetic Algorithms in Feature Selection Classification Using Adaptive Resonance Classifier Network	The performance of the SVM algorithm used in the study was remarkable, achieving a maximum F1 score, precision, and recall value of 100 percent and 92.11 percent, respectively, for the two and three classes of EGG signals. These results indicate the high accuracy and effectiveness of the SVM algorithm in classifying EGG signals into different categories, suggesting its potential utility in clinical practice. However, it is important to note that the performance of the SVM algorithm may vary depending on the size and quality of the dataset, the selection of features, and the parameters used for training the model. Therefore, it is important to evaluate the performance of the SVM algorithm on a larger and more diverse dataset and to compare its performance with other classification algorithms before applying it in a clinical setting.

(continued)

Continued

Sl. No.	Author-Year	Sample/Source	Evaluation algorithms	Results
[3]	Saad, *et al.* 2015	–	–	It's good to hear that the experiment was conducted on 10 people with different genders to ensure the performance of the algorithm. Comparing the simulated and experimental results is also a crucial step in validating the performance of the algorithm. It's great that the results showed a similar pattern for each element, which further validates the accuracy and reliability of the algorithm.
[4]	Strzecha *et al.* 2021	EMG signal Samples were collected from abdominal muscles during the research.	–	–
[5]	Brandstaeter *et al.* 2019	–	–	–
[6]	Krishnasamy *et al.* 2018	–	–	–
[7]	Gharibans *et al.* 2017	The study included a total of eight healthy subjects, consisting of five males and three females, with an average age of 26 ± 4 years and an average BMI of 22 ± 3. None of the participants reported experiencing gastrointestinal symptoms or discomfort at the time of the study.	Laplacian method, forward electrophysiology model, non-parametric bootstrapping method.	The phase gradient directionality (PGD) is a measure of the spatial alignment of the phase gradients in the signal. A PGD value of 1 indicates that the phase gradients are perfectly aligned, while a value of 0 indicates no alignment. A PGD value greater than 0.9 for all time points indicates that the phase gradients are almost perfectly aligned throughout the signal. This suggests that the estimated direction and speed of 187 degrees and 5.3 mm/s, respectively, are reliable, as they are based on a signal with well-aligned phase gradients
[8]	Murakami *et al.* 2013		Fast Fourier transformation,	
[9]	Alagumariappan *et al.* 2020	30 patients with diabetes. 20 – 50 years Global Hospitals & Health City, Chennai. (HR/2017/MS/002)	EMD algorithm, Fast Fourier transform, Genetic algorithm,	Mobility was highly correlated with the spectral entropies of both the normal EGG signals (R = 0.96741) and the diabetes EGG signals (R = 0.90993). Healthy EGGs had a strong association (R = 0.91737) between mobility and HFD (Hausdorff box-counting fractal dimension) values. Diabetes-related EGGs' HFD levels and mobility had a strong connection (R = 0.77178). Healthy EGGs have a strong association (R = 0.88976) between mobility and their MFD (Maragos fractal dimensions) values. Mobility and MFD levels of diabetic EGG signals have a strong connection (R = 0.8077).

6 CORRELATION OF ELECTROGASTROGRAM AND ELECTROMYOGRAM

Based on the information provided, it seems that the study found significant correlations between the amplitude of EGG, electromyogram, and mechanical activity of the gastric antrum in both digestive and interdigestive states. In addition, the EGG reflected both electrical control activity (ECA) and electrical response activity (ERA). However, it appears that there were some discrepancies between the EGG waves and myoelectrical waves, suggesting that the EGG may not always be an accurate reflection of the underlying gastric activity.

ACKNOWLEDGEMENT

The authors would like to thank Dr. N Sriraam, Head-R&D, Ramaiah Institute of Technology, Bengaluru, India for his generous support in guiding us.

REFERENCES

[1] Yin, J., & Z Chen, J. D. "Electrogastrography: Methodology, Validation and Applications," *Journal of Neurogastroenterology and Motility*, *19*(1), 5–17, 2013.

[2] Gandhi C, Ahmad SS, *et al.* "Biosensor-Assisted Method for Abdominal Syndrome Classification Using Machine Learning Algorithm," *Comput Intell Neurosci.* 2022:4454226. doi: 10.1155/2022/4454226. PMID: 35126492; PMCID: PMC8816582, Jan 28, 2022

[3] Saad, Ismail *et al.* "Electromyogram (EMG) Signal Processing Analysis for Clinical Rehabilitation Application," 105–110., doi: 10.1109/AIMS.2015.76, 2015.

[4] Strzecha K. *et al.* "Processing of EMG Signals with High Impact of Power Line and Cardiac Interferences," *Applied Sciences*, vol. 11, no. 10, p. 4625, doi: 10.3390/app11104625, May 2021

[5] Brandstaeter, S, Fuchs, et al. "Mechanics of the Stomach: A Review of an Emerging Field of Biomechanics," *GAMM – Mitteilungen*; 42: e201900001, 2019.

[6] Krishnasamy S, Abell TL, *et al.* "Diabetic Gastroparesis: Principles and Current Trends in Management," *Diabetes Ther.*, doi: 10.1007/s13300-018-0454-9. PMID: 29934758; PMCID: PMC6028327, 22 July 2018.

[7] Gharibans AA, Kim S, *et al.* "High-Resolution Electrogastrogram: A Novel, Noninvasive Method for Determining Gastric Slow-Wave Direction and Speed," *IEEE Trans Biomed Eng.* doi: 10.1109/TBME.2016.2579310. PMID: 27305668; PMCID: PMC5474202, April 2017.

[8] Murakami H, *et al.* "Current Status of Multichannel Electrogastrography and Examples of Its Use," *J Smooth Muscle Res.*; 49:78–88. doi: 10.1540/jsmr.49.78. PMID: 24662473; PMCID: PMC5137273, 2013.

[9] Alagumariappan P, *et al.* "Diagnosis of Type 2 Diabetes Using Electrogastrograms: Extraction and Genetic Algorithm–Based Selection of Informative Features", *JMIR Biomed Eng*, doi: 10.2196/20932, 2020.

Big data processing and applications

Recent Trends in Computational Sciences – Gururaj, Pooja & Flammini (Eds)
© 2024 The Author(s), ISBN 978-1-032-42685-3

Computer vision based home automation

M. Srinivasan
Department of ECE KSIT, Bengaluru, India

R. Srividya
Department of ISE CMR Institute of Technology, Bengaluru, India
ORCID ID: 0000-0001-9673-0785

N. Rekha
Department of ECE KSIT, Bengaluru, India
ORCID ID: 0000-0002-7870-277X

ABSTRACT: Computer vision gives us the ability to detect and analyze human actions from images and videos. Artificial Intelligence combined with computer vision can be used to perform certain tasks based on the information detected from a video or an image. This paper introduces a home automation framework using image processing and computer vision. The work done will derive meaningful gestures of a person from a video and use it to control the home appliances. Based on the hand gesture detected, the corresponding appliance can be controlled or automated. The proposed works focuses on developing a framework in which the appliances can be automated based on desired parameters by the user.

Keywords: Microcontroller, Gesture, Arduino, Computer Vision, Python

1   INTRODUCTION

Automation is the key to progress in any industry. The ability to control and automate the tasks will reduce the time and effort. One such task is the home automation. The controlling of home appliances can be automated according to human needs. To automate the switching of appliances we have Bluetooth, Internet of Things (IoT), smartphone applications. Although all these methods are able to automate home appliances, this paper focus on developing a more effective trend efficient alternate solution for the same. The idea of proposed work is developing an automation system in such a way that anybody can control the home appliances hassle free by using their hand gestures.

An automation framework is developed here, on which any feature can be added. Through computer vision we will be able to detect the hand gestures, upon detecting the hand gestures we need to process it in order to get desired and meaningful information. The corresponding appliance is controlled based on the gesture detected. There is an interface between computer vision and the microcontroller that controls the appliances. This method provides an eco-system inside a house such that anything can be automated as desired by the user.

2   LITERATURE SURVEY

This section discusses the existing literature and the shortcomings and problems in existing work. The paper uses segmented image processing to automate a home, a http server is used to transfer the data between devices and control it [1]. This paper incorporates Artificial intelligence in the process of imaging through computer vision. The calculation methods and

DOI: 10.1201/9781003363781-8

approaches that are basically used to derive information are discussed [2]. Using the internet of things protocol the automation of home appliances is developed. The different scenarios are considered and analyzed. Based on the analyzed data the approaches for the automation of home are modified. The security aspect of IoT were taken care [3]. The paper proposes using methods to vary brightness of lights in home. Potentiometer is used to control the intensity of light. Based on the analog value received from the potentiometer the values are mapped to a bulb. The paper discussed on the different ways and approaches that can be used to reduce the electricity wastage [4]. The paper throws light on a system that detects activity of human and can control the bulbs accordingly. This system uses Internet of things to solve this problem. The mobile application developed can be used to control the auto-mation of the appliances [5]. The paper discusses hand gesture detection based on the parameters that corresponds to the shape of the hand. The location of each finger in a hand and orientation are tracked. Implementation of this algorithm on different lighting condi-tions was also explored [6]. The paper discusses smart home design incorporating features of Computer Vision. Presence of intruders are detected using Computer vision techniques [7]. The paper focuses mainly on gesture biometric. The approach is about 3-Dimensional hand gesture recognition [8]. The paper uses gesture biometric and an event based hardware is used to implement gesture recognition system. Event based cameras provide a new hand-gesture dataset [9]. The paper uses concept of Convolution Neural Networks (CNNs). Many Gesture Recognition tasks are based on CNN [11]. The paper is based on heterogeneous networks and investigates a different architecture. 3D ConvLSTMs and ConvNets are used in building these networks and later combined using late fusion [10]. This paper discusses the usage of Arduino board for home automation. Raspberry Pi is comparatively better for real-time and advanced projects [16]. The paper focuses on human computer interaction through Vision based Gesture Recognition (VGR). Visual interface is useful in establishing interface for non-verbal interaction between human and computer. The key component in such interface is the gesture recognition [15]. The paper elaborates on how IoT has been used for effective data exchange and storage. The paper putforth implements IoT network for home appliances control. The appliances status will be reflected on the cloud platform [14]. Human face recognition methods belongs to any of the two following classes, the global feature-based method and the local feature-based method. Recent innovations in face recognition system uses multiple-face views in recognition process. This multi-view face recognition method has been proposed for detecting each side of face [12]. IoT is basic idea behind Home automation system. Nodes in IoT communicate with little or no human intervention or interaction. A smart automation system, monitors and controls units in and around living area. IoT is a vision for a smart inter-network objects and communication between them [13]

3 METHODOLOGY

3.1 *Proposed model*

The concept proposed here incorporates Computer Vision and Arduino. The user will be able to show few hand gestures towards the camera and control the appliances. In order to recognize and detect the gesture, Python OpenCV library and mediapipe library are made use of. The commands are then sent by python to Arduino using USB serial communication.

The Arduino microcontroller, controls the corresponding appliance based on the com-mand received through serial communication. The state of the appliance can be controlled and any analog parameter can also be varied. In case of a bulb the brightness of a bulb can be varied. The speed of the fan, temperature can be set in air conditioners. This is clearly depicted using the Figure 1 below.

The system can be divided into software and hardware. Using python OpenCV library and mediapipe library we have developed a hand gesture detection module.

Smartphone camera is used instead of webcam for better video quality. The video cap-tured by the smartphone is processed by the hand gesture detection module. The processing

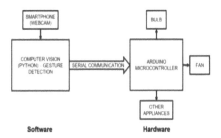

Figure 1. Home automation using CV.

of the video is based on the hand node points. The node points are the points marked on the vital motionary parts in a hand. Each finger and within the finger each movable part is marked as a node point. For the brightness level of the bulb, the distance between index finger and thumb finger is calculated. That distance is mapped to the analog value of the appliance and sent to Arduino. The commands and parameters are sent to Arduino through USB serial communication. The python program developed will detect the hand gesture and send the commands through serial communication to Arduino.

Based on the command received by the Arduino microcontroller bulb, fan and other home appliance will be controlled. Arduino microcontroller uses the PWM pins to vary the brightness level. DC appliances can be directly controlled by the Arduino. AC appliances can be controlled by a driver or a triac.

Hardware components used in developing the proposed model are, Arduino uno micro-controller board, L298N motor driver, LEDs, Electrical Bulbs, DC motor, 4 channel Relay, Smartphone as webcam. The Software tools include, Python language, OpenCv library, MediaPipe library, Pycharm IDE, Arduino IDE.

3.2 Flow charts

The working of the proposed model is represented using Figure 2. Initially, system detects hand as input. If it is not detected then the system waits until hand is recognized. Later hand gesture is detected. The detected gesture value has to be mapped to the parameter, that in turn maps to a home appliance that can be controlled.

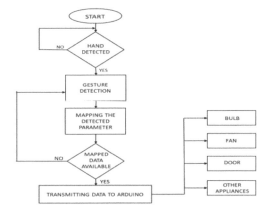

Figure 2. Working principle of the proposed model.

Once the value is mapped then it has to be sent to Arduino through USB serial communication. Arduino receives the command and the parameter that has to be varied. Upon receiving the values, Arduino will control or automate the corresponding appliance. The

Arduino microcontroller uses pulse width modulation (PWM) to control the intensity of light or vary the speed of a fan.

4 RESULTS AND DISCUSSION

The implementation results obtained from the proposed model are detailed in this section.

4.1 *CV module*

The computer vision module developed in the project detects the presence of hand. It draws a bounding box around the hand as a indication that within the box area the gesture has to be detected. It also marks the nodal points within the hand, that is on the fingers and important motion joints in our hand, as depicted in Figure 3.

Figure 3.   Hand detected.

4.2 *Gesture identification*

The python module then tracks the node points and identify the gesture shown by the user. In Figure 4. gesture the distance between index finger and thumb finger is calculated. This distance can be used to indicate the brightness of the home lights or can be used to vary the speed of the fan. In Figure 5 gesture we can observe that distance between the index finger and thumb finger is reduced, indicating to reduce the brightness or the intensity of the bulb.

Figure 4.   Gesture detected.

Figure 5.   Gesture varied.

4.3 *Gesture based light control*

The python module will then send the commands through serial communication to Arduino. As we can see in the image, the gesture is same as Figure 5, indicating to turn off the bulb. The leds are not glowing in Figure 6.

When we increase the distance between the index finger and thumb finger, this value gets mapped to corresponding brightness level range (0 to 255) and this value is pulse width modulated to vary the brightness. There by we can see the intensity of the led being varied in accordance with distance between index and thumb finger, in Figure 7.

Figure 6.   Lights off based on gesture input.

Figure 7.   Light intensity varied based on gesture input.

5   CONCLUSION AND FUTURE SCOPE

The framework developed will be able to automate and control all the home appliances. This system will detect all gestures as desired by the user and can be used to control or automate the appliance based on given parameters. The efficiency of detecting the gesture and auto-mating the appliance can be improvised. The response time of hand gesture detection and automation can be further reduced.

REFERENCES

[1]   Agarwal Mohammad Hasnain, Rishab S, Mayank P, Swapnil G. "Smart Home Automation Using Computer Vision and Segmented Image Processing." In *2019 International Conference on Communication and Signal Processing (ICCSP)*, IEEE, 2019.

[2]   Faris Alsuhaym, Tawfik AL-Hadrami, Kenny Awuson-David. "Toward Home Automation: An IoT Based Home Automation System Control and Security." *2021 International Congress of Advanced Technology and Engineering(ICOTEN)*, IEEE.

[3]   Xin Li, Yilang Shi. "Computer Vision Imaging Based on Artificial Intelligence." *2018 International Conference on Virtual Reality and Intelligent Systems (ICVRIS)*.

[4]   G.V. Nagesh Kumar, A. Bhavya, J. Balaji, S. Dhanunjay, P. Vikasitha, V. Rafi. "Smart Home Light Intensity Control Using Potentiometer Method for Energy Conservation," *2021 International Conference on Recent Trends on Electronics, Information, Communication & Technology (RTEICT)*

[5]   Nursyazwani Adnan, Noorfazila Kamal, Kalaivani Chellappan. "An IoT Based Smart Lighting System Based on Human Activity." *2019 IEEE 14th Malaysia International Conference on Communication (MICC)*,

[6]   Meenakshi, Panwar. "Hand Gesture Recognition Based on Shape Parameters." *2012 International Conference on Computing, Communication and Applications.*

[7] Syed Kashan Ali Shah, Waqas Mahmood. "Smart Home Automation Using IOT and Its Low Cost Implementation." In *International Journal of Engineering and Manufacturing-Octiber 2020*, doi: 10.5815/ijem.2020.05.03.

[8] Quentin De Smedt, Hazem Wannous, Jean-Philippe. "Skeleton-based Dynamic Hand Gesture Recognition.", *IEEE Computer Society Conference on Computer Vision and Pattern Recognition Workshops (CVPRW)*, 2016.

[9] Arnon Amir, Brian Taba, David Berg, Timothy Melano, Jeffrey McKinstry, Carmelo Di Nolfo, Tapan Nayak, Alexander Andreopoulos, Guillaume Garreau. "*A Low Power, Fully Event-Based Gesture Recognition System.*" published in IEEE Xplore, 2017.

[10] Qiguang Miao, Yunan Li, Wanli Ouyang, Zhenxin Ma, Xin Xu, Weikang Shi, Xiaochun Cao. "*Multimodal Gesture Recognition Based on the ResC3D Network.*"published in IEEE Xplore, January 2018.

[11] Huogen Wang, Pichao Wang, Zhanjie Song, Wanqing Li. "*Large-scale Multi-modal Gesture Recognition Using Heterogeneous Networks.*", published in IEEE Xplore, January 2018.

[12] Cristina Stolojescu-Crisan, Calin Crisan Bogdan-Petru Butunoi. "An IoT-Based Smart Home Automation System." In *Sensors, Communication Department, Politehnica University of Timisoara, 300223 Timis,oara, Romania, 30 May 2021*, doi: 10.3390/s21113784

[13] Biplab Ketan Chakraborty, Debajit Sarma, M.K. Bhuyan, Karl F MacDorman. "Review of Constraints on Vision-based Gesture Recognition for Human–computer Interaction." In 14th November 2017, *The Institution of Engineering and Technology*, doi: 10.1049/iet-cvi.2017.0052

[14] Ahmad Sinali Abdulraheem, Azar Abid Salih, Abdulrahman Ihsan Abdulla, Mohammed A. M. Sadeeq, Nareen O. M. Salim, Hilmi Abdullah, Farhad M. Khalifa Rebin Abdullah Saeed. "*Home Automation System based on IoT.*" In Technology Reports of Kansai University, June 2020, ISSN: 04532198, Volume 62, Issue 05

[15] Siddharth Wadhwani, Uday Sing2, Prakarsh Singh, Shraddha Dwivedi. "Smart Home Automation and Security System using Arduino and IOT." *In International Research Journal of Engineering and Technology (IRJET)*, Feb-2018, Volume: 05 Issue: 02, ISSN: 2395-0056.

[16] Manisha M. Kasa, Debnath Bhattacharyya, Tai-hoon Kim. "Face Recognition Using Neural Network: A Review." *In International Journal of Security and Its Applications* Vol. 10, No.3, March 2016, doi: 10.14257/ijsia.2016.10.3.08

Recent Trends in Computational Sciences – Gururaj, Pooja & Flammini (Eds)
© 2024 The Author(s), ISBN 978-1-032-42685-3

Early autism detection using ML on behavioural pattern

Renuka Patwari, Suyog Havare, Vibodh Bhosure, Yash Jagdale & Divya Surve
Vidyalankar Institute of Technology, Mumbai, India
ORCID ID: 0009-0004-6408-4618, 0009-0005-9513-3803, 0009-0007-4777-1223, 0009-0008-3350-2206

ABSTRACT: Autistic Spectrum Disorder (ASD) is one of the main neurological disorders. The use of verbal, expressive, intellectual, and social skills is restricted by ASD, a mental condition. Unfortunately, there are substantial wait times for an autistic spectrum disorder diagnosis because the current diagnostic processes are time-consuming and inefficient. We can now anticipate autism early on because of advances in artificial intelligence (AI) and machine learning (ML). This research attempts to investigate the potential of Decision Tree, K-Nearest Neighbour, and Logistic Regression for the analysis and prediction of ASD issues in toddlers. To create a web application that can accurately identify an individual's autistic features, the study proposes an autism prediction model leveraging ML approaches. On a publicly accessible, non-clinical dataset, the proposed strategies are evaluated. The dataset is related to ASD screening in toddlers which comprises 1054 instances and 18 attributes (including class variable). Results from the application of various ML techniques point to the better performance of a logistic regression-based model on the dataset, with 100% accuracy under both scenarios including and excluding the QChat-10-Score attribute.

Keywords: Autism Spectrum Disorder (ASD), Decision Tree, K-Nearest Neighbour (KNN), Logistic Regression (LR), Q-Chat, Machine Learning (ML), Childhood Screening

1 INTRODUCTION

Autism spectrum disorder (ASD) is a developmental impairment that can lead to serious social, communicative, and behavioural difficulties. Research suggests that both genes and environment play important roles, and early identification of this neurological issue can help maintain physical and mental health. With the increased use of machine learning-based models for illness prediction, it is now possible to identify diseases early based on various physiological and health characteristics. This element encouraged us to become more interested in the identification and examination of ASD disorders to develop more effective treatment approaches.

In their research and prediction processes, the present studies rely only on the attribute "QChat-10-Score", which is an aggregate of the values allotted to the selected options to the questions. There is a need for a system that predicts autism based on the individual answers to the questions and related information of the subject. Our research shows that although the QChat-10-Score is the one factor that contributes the most, it is still essential to find additional factors by eliminating the QChat-10-Score and creating prediction models in its absence.

All the study papers that we have cited are listed in the second section, which is devoted to a review of the literature, along with any adjustments we made and the important findings from each. The third section, which is about implementation, has information about the system we used. The results part, which is the fourth component, contains all the pertinent findings relating to our application. The fifth section, which is about the future scope, lists all the further alterations and adjustments that can be made to the current implementation to boost its accuracy, performance, and scalability. The conclusion, which is the sixth section, summarises all the pertinent information we learned from our study. The references section, which is the seventh, lists every research publication that was used as a source for our analysis.

2 LITERATURE SURVEY

The emphasis of most ASD studies is on genetic factors. In a study by V. Pream Sudha and M. S. Vijaya [2], Support Vector Machine (SVM), Decision Tree Classifier, and Multilayer Perceptron (MLP) were utilized. They looked at various syndromes, monogenetic illnesses and pathogenic gene mutations. SVM accuracy was 0.96, MLP accuracy was 0.95, and Decision Tree accuracy was 0.98.

Support vector machines (SVM), K-nearest Neighbours (KNN), and Random Forest (RF) machine learning algorithms are used by A. Demirhan [3] to evaluate ASD adolescent scan data to quickly and accurately diagnose the ASD status. Binary classification with 10-fold cross-validation (CV) using SVM, KNN, and RF techniques results in respective accuracy rates of 95%, 89%, and 100%. This work demonstrated that utilizing adult screening data for ASD, classification with the RF approach may successfully identify ASD patients.

Using classification techniques, M. S. Mythili, A. R. Mohamed Shanavas [4] conducted a study on ASD. The paper's primary goals were to identify the autism problem and the degrees of autism associated with it. SVM, Fuzzy, and WEKA tools are utilized in this neural network to study the social interaction and behaviour of the students.

Using three levels of assessment, an interactive framework is presented by Khondaker *et al.* [5]. It first screens the children through a pictorial questionnaire via a mobile application. If suspected, a video is displayed, and the child's reactions are captured and sent to the cloud for remote expert review. On further suspicion, the child is directed to the nearest Autism Resource Center (ARC) for an actual assessment.

The major purpose of the research by Aishwarya *et al.* [6] is to create an autism screening tool for detecting ASD characteristics in children and young adults under the age of three. Using the AQ-10 dataset (1054 instances), the suggested model can reliably diagnose autism with 92.89%, 96.20%, 100.00%, and 79.14% for the Decision tree, Random Forest, Adaboost, and SVM algorithms, respectively. The results showed that the Random Forest and Adaboost algorithms excelled.

3 IMPLEMENTATION

The screening system is built on a web application with a flowchart shown in Figure 1. The user interface has 10 questions for the user to answer, along with other parameters such as age, sex, ethnicity, medical history of the child, and family history of Autism. The user can select three algorithm options – Logistic Regression, K-Nearest Classifier, and Decision Tree. The front-end is built using react.js and deployed on Vercel, while the back end is built using Flask and hosted on Render. The response is processed using the scikit-learn library and sent back to the user as a prediction. Two databases are maintained – one is a CSV file containing data used for prediction, and the second is a NoSQL database that stores all the requests received by the backend.

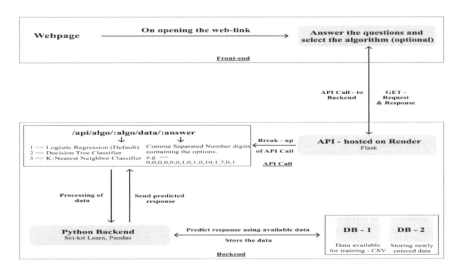

Figure 1.   Flowchart of the screening system.

Figure 2.   Steps in the suggested ASD detection solute.

Figure 2 depicts the steps in the suggested workflow, which includes the pre-processing of data, training and testing with specified models, evaluation of results and prediction of ASD. This work is performed in Python 3.

3.1   *Data collection*

The dataset used in this study was collected from the Kaggle repository [8]. The actual dataset was compiled by Dr.Fadi Fayez Thabtah who, has multiple papers in the relevant field of Machine Learning and Autism to his name. This data was collected by Tabtah F. utilizing a mobile application for diagnosing ASD [7]. The dataset includes information on ten behavioural traits as well as additional individual traits that have proved to be effective in detecting the ASD cases from controls in behaviour science. The dataset had 1054 instances and 18 attributes (including the class variable).

3.2 Data pre-processing

Pre-processing entails putting raw data in an understandable and meaningful structure. Real-world data is frequently inconsistent and lacking because of errors and null values. Well pre-processed data always results in positive results. The dataset contained no null or missing values. However, some of the dataset's features like 'Sex', 'Ethnicity', 'Jaundice', 'Family_mem_with_ASD', 'ClassASD_Traits' were of the object data type. These categorical features were converted to numeric using the LabelEncoder() class from the sklearn python library for better analysis and machine learning purposes.

3.3 Load pre-processed data

The processed data that was cleaned and converted to a suitable format, the dataset was then divided into 80% training purposes and 20% testing purposes. The attribute "ClassASD_Traits" was set as the Class Variable.

3.4 Apply machine learning algorithms

The machine learning algorithms such as the Decision Tree, K-Nearest Neighbour, and Logistic Regression were applied. The models were trained and tested in this phase. The goal of using a Decision Tree is to create a training model that can be used to predict the class or value of the target variable. KNN stores all the available data and classifies a new data point based on the similarity. Based on a given dataset of independent factors, logistic regression calculates the likelihood that a specific event will occur. The dependent variable's range is 0 to 1 since the outcome is a probability. The Confusion Matrix and Classification report were obtained to understand the performance of the models better.

3.5 Evaluation of ASD and non-ASD classes

The trained and tested models predicted the input as Autistic or Not Autistic with different accuracies.

n the first iteration, we realized that the model considered the "QChat-10-Score" attribute, which is an aggregate of the values allotted to the options of the questions. So, in the second iteration, we decided to drop the "QChat-10-Score" attribute and then train and test the models. The results now displayed; considered and classified the input based on answers to the questions and the related information of the subject. The second iteration also allowed us to test real-time data input and classify it as Autistic or Non-Autistic.

4 RESULTS

The model's performance was calculated using the variables from the confusion matrix. The following figures display the classification reports for the Decision Tree, K-Nearest Neighbour, and Logistic Regression models without the "QChat-10-Score" attribute.

The Table 1 shown below describes the performance evaluation metrics of the different models that were employed. In models, with the "QChat-10-Score" Attribute, the Decision Tree and Logistic Regression models show an Accuracy of 100% along with Sensitivity, i.e., True Positive Rate and Specificity, i.e., True Negative Rate as 1. In models, without the "QChat-10-Score" attribute, only the Logistic Regression model shows an Accuracy of 100% along with Sensitivity and Specificity as 1.

Without QChat-10-Score

Table 1. Performance measures.

Model	Accuracy	Report				
Decision Tree	86.73%		precision	recall	f1-score	support
		0	1.00	0.57	0.73	65
		1	0.84	1.00	0.91	146
		accuracy			0.87	211
		macro avg	0.92	0.78	0.82	211
		weighted avg	0.89	0.87	0.85	211
K-Nearest Neighbour	93.83%		precision	recall	f1-score	support
		0	0.93	0.86	0.90	65
		1	0.94	0.97	0.96	146
		accuracy			0.94	211
		macro avg	0.94	0.92	0.93	211
		weighted avg	0.94	0.94	0.94	211
Logistic Regression	100%		precision	recall	f1-score	support
		0	1.00	1.00	1.00	65
		1	1.00	1.00	1.00	146
		accuracy			1.00	211
		macro avg	1.00	1.00	1.00	211
		weighted avg	1.00	1.00	1.00	211

5 FUTURE SCOPE

Currently, we have implemented three algorithms. As a future scope of development, we can implement multiple algorithms into a single system, which is publicly accessible. A method for user authentication can be implemented, through which we can store the medical reports of the child whose ASD is being predicted. The prediction can thus be verified with the genetically tested reports, thereby increasing the algorithm's accuracy. Modifications can be made in a way in which we are accepting the answers from the user, and it can be implemented in such a way that we assign individual unique scores to each of the options available to the user and then normalize it to predict the outcome. This way of implementation would consider the prominence of the individual options and thus thereby increasing the accuracy of the algorithm. The API we have implemented is publicly accessible, and this means that anyone can make use of our API to integrate the prediction system into their system. This public access can be restricted to a specific limit and, therefore, will help scale it to more platforms and save more computational power.

6 CONCLUSIONS

In this study, different machine-learning approaches were used to identify and detect autism spectrum disorder. The accuracy of the model was used to analyse the performance of the models implemented for ASD detection on a non-clinical dataset from the age group of Toddlers. When first performing the analysis with the presence of QChat-10-Score attribute, it was identified as the most contributing of all attributes with the threshold value of 3.5 and was by itself sufficient to categorize the values as Autistic or Non-autistic. The Decision Tree and the Logistic Regression models showed the accuracy of 100% while the accuracy of the K-Nearest Neighbour model was 98.57%. In the absence of the QChat-10-Score attribute, the attribute A9 was discovered to be the most contributing, followed by A7 and A5. The Table 2 describes the comparison of the proposed models with the existing models.

Table 2. Comparison of Accuracy of models with and without QChat-10.

Model	With QChat-10-Score	Without QChat-10-Score
Decision Tree	100%	86.73%
K-Nearest Neighbor	98.10%	93.83%
Logistic Regression	100%	100%

Model	With QChat-10-Score*	With QChat-10-Score**
Support vector machine	95%	79.14%
Random Forest	100%	96.20%
Adaboost	–	100%

* Performance of Machine Learning Methods in Determining the Autism Spectrum Disorder Cases.
** Engagement Detection with Autism Spectrum Disorder using Machine Learning.

REFERENCES

[1] Fadi Thabtah., "Autism Spectrum Disorder Screening: Machine Learning Adaptation and DSM-5 Fulfillment," In *Proceedings of the 1st International Conference on Medical and Health Informatics*, pp. 1–6. ACM, 2017.

[2] Sudha, V. Pream, and Vijaya M. S. "Machine Learning-Based Model for Identification of Syndromic Autism Spectrum Disorder." *Integrated Intelligent Computing, Communication and Security*. Springer, Singapore, 2019.

[3] Demirhan A., "Performance of Machine Learning Methods in Determining the Autism Spectrum Disorder Cases," *Mugla Journal of Science and Technology*, No. 6, June 2018.

[4] Mythili M. S., and Mohamed Shanavas AR., "A Study on Autism Spectrum Disorders Using Classification Techniques," *International Journal of Soft Computing and Engineering (IJSCE)*, 4: 88–91, 2014.

[5] Khondaker A., Sharmistha B., Md. Anwar U., Evdokia A., Jessica B., Shaheen A. and Mohammod G., "Smart Autism – A mobile, interactive and integrated framework for screening and confirmation of autism," IEEE, (2016).

[6] Aishwarya J, Akshatha N, Anusha H, Shishira J, Deepa Mahadev, "Engagement Detection with Autism Spectrum Disorder Using Machine Learning," *International Research Journal of Engineering and Technology (IRJET)*. 2020.

[7] Fadi Thabtah., 2017, *"ASD Tests. A Mobile App for ASD Screening." [Online]*. Available: www.asdtests.com

[8] Fadi Thabtah., 2018, *"Autism Screening Data for Toddlers,"* Kaggle. [Online]. Available: https://www.kaggle.com/datasets/fabdelja/autism-screening-for-toddlers.

Recent Trends in Computational Sciences – Gururaj, Pooja & Flammini (Eds)
© 2024 The Author(s), ISBN 978-1-032-42685-3

Implementation of application prototypes for human-to-computer interactions

N. Soujanya, G. Sourabha, M.R. Pooja, K.R. Swathi Meghana & C. Jeevitha
Vidyavardhaka college of Engineering, Mysuru, India

ABSTRACT: In this increasing age of reduced human-to-human communication, human-to-computer interaction has a high range in the contactless edible refreshment industry. The work presented in the paper aims at developing a desktop application prototype of a vending machine that employs fundamental human traits like a person's age and gender in order to maximize computer-to-human interaction. In order to accomplish this, we explore technologies including computer vision and convolution Neural Network to recognize the vending machine user. We aim to provide a welcoming customer experience to the vending machine's customers, giving them a little more opportunity for better mental health. The proposed system will also cater to specially challenged customers who are willing to embrace the introduction-less edible refreshment industry.

Keywords: human-computer interaction, deep learning, convolutional neural network

1 INTRODUCTION

The barrier between customers and food vending companies is broken down via human-computer interaction [1], improving the user experience. A better user experience would boost both the company's and the sector's economies. Due of the sharp decline in interpersonal communication, this will improve mental health by giving users a comfortable environment. In this study, we address python, deep learning [7], and convolution neural networks. The philosophies on which this work will be based focus on the emphasis placed on the mental health of the customers who are a part of this industry and the need to make the items that are a part of it more easily understandable for customers who are mentally challenged. The difficulty in the nature of the data that is required to train is the primary cause of the disparity. Age and gender classification is an inherently difficult subject. The photos used here are for training datasets with labels for age and gender [5]. We are approaching the CNN technique [4] since they require the personal information of the subject in the photos for such photographs. This system's objective is to categorize the faces in an image according to their age and gender.

2 RELATED WORK

2.1 *Purpose*

The main goal of this application is to find an image pose prediction and recognition of his or her images using convolutional neural network algorithms. It enhances performance and high face detector for improving the speed of the model significantly better in performance and performing many more tasks.

The automatic age and prediction of gender extracted from human face images. The task of estimating age and gender is carried out using the Convolutional Neural Network method. When predicting age and gender (his/her) for usage in online social networking sites

DOI: 10.1201/9781003363781-10

and social networks, this model makes use of a variety of hardware and software tools. Convolutional neural networks and a classification algorithm are both used in this.

2.2 *Methodology*

The suggested system is concerned with identifying a person's fundamental traits, such as age and gender. Utilizing computer vision, this system uses facial recognition to identify individuals. The present version of the proposed system is a desktop application that uses the user's face image as input. The system then processes this image to produce speech through the speakers in the proposed system. Vending machines use it to welcome their customers. The team developing the suggested method will teach the computer to comprehend the face photos so that it can recognize and distinguish between different people. We use data to train the system to understand the social differences that an individual gives.

Convolutional neural networks [2,8], are utilized in the proposed system. The conversion of an image into a form achieved by minimization is a technology choice that is simple to compute without losing the properties of the image, which produces a good indexIs important for, and then the image itself. CNN is part of this process, a laborious procedure. This technology is significant since it can be applied to many different datasets and is acceptable even while learning indicators or features.

We employ arbitrary groups of people with various ages, facial features, and genders. It offers a great amount of information that is typically in the form of a picture and may be converted into a form that is easily calculable without losing important image characteristics that help to categorize a person's age and gender.

We make sure that identification occurs as correctly as possible and, if anything goes wrong, we agree that the system will function as intended provided the person's age and gender can be determined by comparing the criteria used to train our system. Since the output obtained and the expected output differ significantly, we train the system using larger amounts of data. Once we are satisfied with our system's performance, we prototype it on the desktop and test it using real-time models.

The output obtained is in the form of a speech. Identify the expected customer from our system and appreciate them accordingly.

Additionally, our group considers privacy. The proposed approach does not keep track of the photos used to gauge customer privacy trust. Since we no longer greet one another in person for basic communication, we also aim to promote human-computer contact in today's environment. In the ever-changing world, this proposed technology attempts to transmit greetings from person to person.

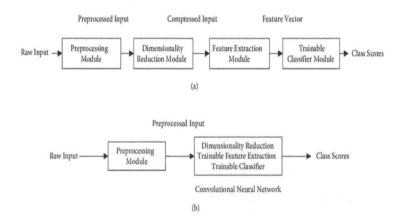

Figure 1. Pattern recognition approaches: (a) conventional, (b) CNN-based.

It is meant for real-time gender categorization based on facial photos in Figure 1 Gender classification using CNN [9]. For real-time gender classification, a backpropagation algorithm based on face images was utilized [10,11]. The CNN methodology was also proposed, and they mix convolutional & subsampling layers in the sample identification as compared to previous CNN methods. The network is trained using neural networks, which requires less computational labor. Pattern recognition is the process of finding patterns with a machine-learning system. The process of sample identification involves identifying data patterns and sequences using computer algorithms. The input types for this type of identification can include biometric recognition, image recognition, and facial recognition. It has been used in a variety of industries, including seismic analysis, picture analysis, and computer vision. It is very accurate at spotting patterns and unidentified items.

The sample identification system can find patterns that are partially hidden. In order to identify samples, the idea of learning is being used. Model identification systems can be modified for training and improved accuracy with sample recognition. Digital image analysis uses sample recognition to analyze photos and extract essential data from them.

The healthcare industry will employ the identification model prototype to enhance health services, and data will be saved and used by medical specialists for study in the future. Model recognition could evolve into a more intelligent procedure that supports different digital technologies. The main benefit of CNN over conventional pattern detection techniques is that it concurrently collects attributes, shrinks the quantity of the data, and organizes them into a network structure.

2.3 CNN

A neural network is a collection of artificial neurons that are interconnected in order to communicate with one another. Connections incorporate numerical weights that emerge during training, assisting in accurate identification when an image or model is presented to a trained network. CNN Speech recognition [3], natural language processing, video, image, and pattern recognition, and used in many different domains, such as analytics, and feature extractors created by hand, and why convolutional neural networks are vital in conventional models for model identification. Applications with local correlation perform better when using CNN input (e.g. image & speech). The best identification rates in prototype and image recognition applications are attained using CNN [12].

Figure 2.   Data flow model.

In Convolutional Neural Network (CNN) algorithm:

2.3.1 Reading the images.
2.3.2 Decode JPEG content into an RGB grid of pixels.
2.3.3 Converting into Floating points tensors taking input to the neural network.
2.3.4 Rescale the pixel values between 0 to 255. Trained the neural network efficiently or effectively.

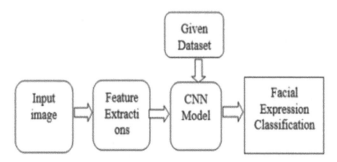

Figure 3.   Working model.

The CNN algorithm is used for model classification for automatic live gender recognition using a support vector machine. Execution of work results on fEI, live images, and detection accuracy 97% of FEI datasets, 95% of own datasets This proposed method is comparable to the previous method for better estimation and helps in a real-time system.

For the classification dataset for age and gender, our method improves performance in both age and his or her assessment and classification of gender, significantly exceeding the performance of models. For future improvement of work support. We then use the Convulsive Neural Network Algorithm [6] for age assessment and gender assessment. Technology has been growing in the past in relation to security-related issues in our daily life. In this paper, we will discuss biometric features in the gender assessment model for his or her identity and use it to minimize search space tests.

2.4 *Adience dataset*

The Adience Dataset, which was developed to support the research of age and gender identification [11], may also be used as a benchmark dataset for face pictures. The dataset is as accurate a representation of the face imaging environment as possible.

The dataset includes 26,580 photographs with 8 age groups (labels) shot by 2,284 subjects using an iPhone 5 (or later) and images set for Flickr albums. The dataset contains a wide range of faces with a variety of positions, appearances, noise, lighting, and other characteristics.

There are way too many pictures of children under four. The model would not adequately match the other ages since it would fit these ages far too well.

3 IMPLEMENTATION

The dataset utilized in this project is taken from the Kaggle website.

This dataset contains a complete 26,580 images of 2,284 subjects in eight age ranges as follows:

- 0 – 2
- 3 – 6

- 8 – 12
- 15 – 20
- 25 – 32
- 38 – 43
- 48 – 53
- 60 – 100

The dataset has a total of 8 columns. The CNN architecture has 3 convolutional layers [13,14] Implementation is divided into 2 modules:

- Training and testing – Here data is split into training and testing datasets. after the split data cleaning, data processing, and data analysis are performed on the training dataset
- Deployment – After the Caffe model is trained, it is then deployed to the Age and gender prediction website where the output is displayed.

The following are the steps to implement this project:

- Firstly, detect the faces
- Classify the image into Male or Female
- Classify the image into one among the 8 mentioned age ranges
- Put the results on the image and display the image The implementation of this project is split into following steps:

Data Preprocessing There are 19370 entries within the dataset. By using the .info() method we are able to find the information about the dataset and also the null values are removed by using dropna().

```
RangeIndex: 19370 entries, 0 to 19369
Data columns (total 12 columns):
 #   Column             Non-Null Count   Dtype
---  ------             --------------   -----
 0   user_id            19370 non-null   object
 1   original_image     19370 non-null   object
 2   face_id            19370 non-null   int64
 3   age                19370 non-null   object
 4   gender             18591 non-null   object
 5   x                  19370 non-null   int64
 6   y                  19370 non-null   int64
 7   dx                 19370 non-null   int64
 8   dy                 19370 non-null   int64
 9   tilt_ang           19370 non-null   int64
 10  fiducial_yaw_angle 19370 non-null   int64
 11  fiducial_score     19370 non-null   int64
dtypes: int64(8), object(4)
```

Figure 4.   Age and gender dataset information.

Building and Training Model The training and testing of the dataset begins by firstly dividing the dataset into xtrain, ytrain and xtest, ytest. From sklearn. preprocessing we import OneHotEncoder. Building of the model is done with the help of model.fit (xtrain, ytrain). After running 28 epochs the accuracy obtained was 0.8839.

Prediction After the model is totally built using above process, we've done the prediction part using model.predict(xtest). The accuracy is calculated using accuracy score (xtest, ytest)

Gender

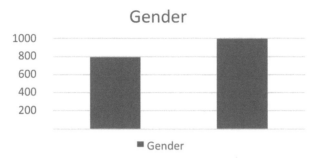

Figure 5. Bar chart of gender.

Visualization Data Visualization is done using the matplotlib library from sklearn. Analysis of the Age and gender dataset is completed by plotting various graphs.

Gender

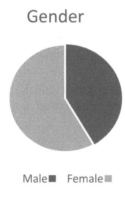

Male■   Female■

Figure 6. Pie chart for gender.

Age

■ (0 , 2)■ (4 , 8) ■ (8 , 12)■ (12 , 25)■ (25 , 32)■ (33 , 48)■ (48 , 53)■ (60 , 100)

Figure 7. Pie chart for age.

4 RESULT

The main aim of our system is to recognize the age and gender of our customers which is a key feature that allows us to customize the way we greet them. We choose this method of greeting individuals because of the situation in today's world where talking and greeting anyone face to face has decreased drastically, which has a huge effect on health and businesses in the industry. The figure below shows us the recognition that our system has done which gives us the opportunity to greet our customers depending on their age and gender. By doing this we can see a boost in the happiness of our customers and hence give a boost to the economy of the industry.

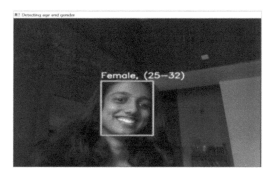

Figure 8.   Output obtained for age and gender prediction.

5   CONCLUSION

Our goal is to use the proposed expertise to develop a new system for the consumers of the edible beveragesector. According to the articles we have read and considered, the notion of human-computer interaction starts with the client, and it is this that we hope to further improve. As welcoming the client is the greatest method to begin any interaction in the industry, our suggested solution is designed forthe computer to do so. Another crucial component of the suggested method is that it would enable clientswho are specially challenged, to experience the same as regular customers, thereby improving the atmosphere for both the clientele and the business.

REFERENCES

[1]   De Bogotá, Cámara de Comercio. *"Observatorio de Innovación–Estrategias y Tácticas de Marketing."* (2020).
[2]   Ishaq, Muhammad, Guiyoung Son, and Soonil Kwon. *"Utterance-Level Speech Emotion Recognition Using Parallel Convolutional Neural Network with Self-Attention Module."*
[3]   SATO, Kunio, *et al.* "Voice Control of a Tractor Using Vowel Pitch Characteristics." *Journal of the Japanese Society of Agricultural Machinery* 63.1 (2001): 35–40
[4]   Alkhawaldeh, Rami S. "DGR: Gender Recognition of Human Speech Using One-dimensional Conventional Neural Networks." *Scientific Programming* 2019 (2019).
[5]   Zhu, Zijiang, *et al.* "Age Estimation Algorithm of Facial Images Based on Multi-label Sorting." *EURASIP Journal on Image and Video Processing* 2018. 1 (2018): 1–10.
[6]   Mamyrbayev, Orken, *et al.* "Neural architectures for Gender Detection and Speaker Identification." *Cogent Engineering* 7. 1 (2020): 1727168.
[7]   Dhomne, Amit, Ranjit Kumar, and Vijay Bhan. "Gender Recognition Through Face Using Deep Learning." *Procedia Computer Science* 132 (2018): 2 10
[8]   Sumi, Tahmina Akter, *et al.* "Human Gender Detection from Facial Images Using Convolution Neural Networks." *International Conference on Applied Intelligence and Informatics.* Springer, Cham, 2021.
[9]   Kabil, Selen Hande, Hannah Muckenhirn, and Mathew Magimai-Doss. "On Learning to Identify Genders from Raw Speech Signal Using CNNs." *Interspeech.* 2018.
[10]   Liew, Shan Sung, *et al.* "Gender Classification: A Convolutional Neural Network Approach." *Turkish Journal of Electrical Engineering & Computer Sciences* 24. 3 (2016): 1248–1264.
[11]   Djemili, Rafik, Hocine Bourouba, and Mohamed Cherif Amara Korba. "A Speech Signal Based gender Identification System Using Four Classifiers." *2012 International Conference on Multimedia Computing and Systems.* IEEE, 2012.
[12]   Kuppusamy, Karthika, and Chandra Eswaran. *"Speaker Recognition System Based on Age-related Features Using Convolutional and Deep Neural Networks."* (2020).
[13]   Ekmekji, Ari. *"Convolutional Neural Networks for Age and Gender Classification."* Stanford University (2016).
[14]   Soujanya, N., *et al.* "Application Prototypes for Human to Computer Interactions." *Sentiment Analysis and Deep Learning: Proceedings of ICSADL 2022.* Singapore: Springer Nature Singapore, 2023. 27–36.

Recent Trends in Computational Sciences – Gururaj, Pooja & Flammini (Eds)
© 2024 The Author(s), ISBN 978-1-032-42685-3

Recommendation system for anime using machine learning algorithms

Harsh Choudhary, S. Raghavendra & Ramyashree
Department of Information and Communication Technology, Manipal Institute of Technology, Manipal Academy of Higher Education, Manipal, India
ORCID ID: 0009-0008-9678-997X, 0000-0003-2733-3916, 0000-0002-0237-2444

ABSTRACT: These days, many people watch anime, especially the younger generations. With so many different sorts of programming available, this specialist area of the entertainment industry is drawing an increasing number of people. The word "anime," which was derived from the word "animation," has a growing fan following all over the world. The anime industry has been growing rapidly in recent years, bringing in billions of dollars annually. Significant streaming providers like Netflix and Amazon Prime are paying attention to this industry. Researchers are working hard to apply machine learning algorithms to offer the viewer with the appropriate anime because it is currently a popular trend among the younger generation. The purpose of the research presented in the research article that follows is to further current field research. From Kaggle, two datasets have been utilized. The first dataset for an anime comprises 7 columns and 12293 rows. The second rating dataset comprises 3 columns and 7813735 rows. Anime may be rated by users, who can then add it to their finished list. The top 15 matching suggestions and forecasted ratings were then obtained using machine learning approaches (content-based filtering, collaborative filtering, and popularity-based filtering).

Keywords: Recommendation system, Machine learning, KNN, SVD

1 INTRODUCTION

Recommender structures are a type of data filtering technology that tries to infer the "rating" or "interest" of a customer for a product. They adjust the information presented to a customer based on his preferences, the material's relevance, and other factors [1]. In addition to other things, recommender systems are frequently used to promote books, movies, restaurants, travel destinations, and purchases. These systems eliminate a large amount of data by concentrating on the most common data based on the input of the user and other factors such as the user's preferences and interests. It evaluates the similarity between consumers and items as well as the compatibility between the person and the thing to make suggestions. On the other hand, the younger generation is now very interested in anime. In the days before the internet, anime fans used to discover new titles through word of mouth. Individualized advices were therefore not required [2]. Additionally, there weren't enough titles available to provide a data-driven approach to customized suggestions. Due to the growth of OTT services and other factors, people may now watch as much anime as they want. how many new anime titles have been released in recent years. Major contributions of this work is listed below:

1. To engross the latest generation of anime fans, a more and enhanced personalized recommendation system is required [3].
2. Thus, in this research, we attempt to recognize a person's preferences and anticipate which goods are most likely to be highly rated by that user.

DOI: 10.1201/9781003363781-11

3. The recommendation engines utilize a variety of filtering methods, including content-based, collaborative filtering, and popularity-based filtering [4–6].
4. The latter is a combination of many different filtering algorithms [21].

We used content provided by users over time to recommend similar anime that are almost the same genre and type; similarly, the collaborative advice gadget has models: memory and model based. We additionally employed the K-Nearest Neighbor method, which shows 5 anime for every anime.

2 LITERATURE SURVEY

Chate *et al.* [1], primarily focused on filtering algorithms based on the neighborhood of users or objects and based on content. The description of these algorithms includes similarities, disadvantages, and advantages, measures for evaluating the algorithm, and calculation of the sample value of the evaluation prediction. Advantage of this paper is: Provides a basis for the implementation of more complex algorithms. Disadvantages of this paper is: Accuracy is low from what is desired.

Rajashree Shedge *et al.* [2], Done the collaborative filtering algorithm that underlies online personalization is the major topic of this research. Web site personalization is the process of tailoring a website's content and layout to the unique demands of each user while taking advantage of that user's navigational habits. Here the Algorithms used are: Collaborative Filtering, Hybrid Collaborative Filtering.

Nancy Victor *et al.* [3], There have been more than 250 research paper recommender systems published, and the number of papers published each day is rapidly rising. So that researchers can save time and effort, it is necessary to have an effective search and filtering process to select the best research publications. The three main building blocks for the recommender system that is here proposed are datasets, prediction ratings based on users, and cosine similarity. Advantage of this paper is: Saves researchers time and effort by providing best research publications. Disadvantage is: No accurate measurements are shown.

Erich Elsen *et al.* [4], provides three methods for limiting the loss of crucial information. First off, it advises keeping a single-precision copy of the weights that collects the gradients following each optimizer step. To retain gradient values with tiny magnitudes, it also suggests loss-scaling. Thirdly, it does half-precision calculations that result in single-precision outputs that are then transformed to half-precision before being stored in memory.

Hornung *et al.* [5] Recommending music tracks is a difficult study endeavor since musical taste is quite subjective. This study shows a weighted hybrid recommender method, the live prototype system, which combines three different recommender algorithms into one thorough score. Additionally, their technology sprinkles recommendations based on a straightforward serendipity heuristic across the resulting result list. Users obtain some exploratory variety while receiving recommendations that are in line with their present musical preferences in this way.

Gupta VK *et al.* [6], they promote MOVREC, a movie recommendation system. It is founded on a collaborative filtering methodology that uses the data supplied by users, analyses them, and then suggests the movies that are most appropriate for the user at that moment. Here the Algorithms used are: Collaborative filtering, Content-based filtering, k-means. Advantage of this paper is: The interface is user-friendly. Disadvantage is: The system is not very diversifiable.

Hande, R. *et al.* [7], they have presented Movie Mender, a mechanism for movie recommendation. The goal of Movie Mender is to give consumers reliable movie suggestions. Here the Algorithms used are: CBF-Content-based filtering, CF-Collaborative filtering, Hybrid systems. Advantage of this paper is: This approach overcomes drawbacks of each individual algorithm and improves the performance of the system. Disadvantage is: Clustering could've been used to improve performance.

Hoshino, J.I. *et al.* [8], The learning curve to efficiently use the present solution is too steep, according to the study in this paper, which looked carefully at the current approach to fixing this problem. They have created a software to guarantee simpler solutions to the problem. They have also conducted several iterations of various experiments to examine how the accuracy changed when other conditions were applied. Here the Algorithms used are: SVM. Advantage of this paper is: Can read data from user's profile on MyAnimeList. Disadvantage is: Accuracy is very low at 57%.

3 METHODOLOGY AND IMPLEMENTATION DETAILS

This section briefs about the dataset details, techniques used and the applied workflow. The workflow is given in Figure 1 as below. First step includes data collection.

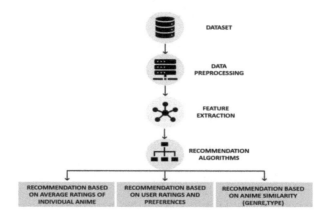

Figure 1. Architecture diagram.

3.1 *Problem formulation*

Anime being an emerging form of entertainment, there are many types of Anime movies and shows to choose from. There isn't enough information about people's preferences and viewing habits as anime has only recently gained popularity round the globe. It's a difficult task to create a recommendation engine for this relatively unknown form of entertainment. Table 1 is provided to understand the basic structure of the subject. It contains three users (X, Y and Z) and five Animes (A, B, C, D, and E). User X assigns a score of 10, 10, 3, and 5 to the Animes A, B, C, and D respectively. User X likes Animes A and B the most, assigning them a rating of 10. However, he/she dislikes the lowest-rating Animes D and E.

Table 1. Anime rating by users.

Animes	User X	User Y	User Z
A	10	10	2
B	10	?	2
C	?	9	?
D	3	1	9
E	4	1	?

3.2 Proposed solution

Users can waste a lot of time scrolling through hundreds, if not thousands, of anime all without discovering something they enjoy. Our goal is to fill the gap in the existing research and construct a recommendation system that proposes anime to viewers based on their preferences and needs, resulting in a better streaming experience that increases income and time spent on a website. In this effort, The system can be implemented in multiple ways. We intend to address Popularity Based Recommendation System, Collaborative Filtering using KNN.

3.3 Dataset description

We worked with two datasets. The first is Anime, while the second is Rating. These two datasets were collected from the website, Kaggle. This data collection contains customer choices for 12,293 anime from a total of 73,516 users. The ratings in this data set are a collection of them. The first anime dataset has 12293 rows and 7 columns. The second rating dataset has 7813736 rows and 3 columns. Each user may rate anime and add it to their finished list.Anime.csv and Rating.

Next step is Data Preprocessing which is a technique used to clean the raw data and make it suitable for classification models. This will make the data to be easily parsed by the machine and additionally improve the accuracy of the results. Real-world datasets are highly prone to inconsistent, noisy, or missing data. If the data mining algorithms are applied on such a dataset, they will not give quality results as they will not be able to identify patterns effectively. Hence, data preprocessing is a crucial step to improve the data quality before applying any classification algorithm to it.

3.4 Algorithms

3.4.1 Popularity-based

It's a type of recommendation system that works on the basis of popularity or anything that's currently popular. These algorithms search for items or movies that are currently hot or popular among customers and propose them right away [6]. For Example If a product is regularly purchased by the majority of people, for example, the system will learn that it is the most popular and will promote it to every new user who has just joined up, increasing the possibility that the new user would purchase it as well.

3.4.2 Collaborative-based filtering

There are probably lots of people who rate a product the way the user intended. This comparable trend of their ratings with the user serves as the basis for collaborative filtering [5]. The foundation of collaborative filtering is the idea of making product recommendations based on the tastes of others with similar interests.

3.4.3 K-nearest neighbour for collaborative filtering

KNN is a type of algorithm that may be used to solve issues in classification and regression [14]. Based on a provided distance parameter such as cosine, Euclidean, Jaccard similarity, minkowsky, or custom distance measures, K-nearest neighbor determines the k most similar objects to a single instance.

In our model, we employed the cosine similarity measure to discover the k-nearest neighbors [19].

4 RESULTS AND DISCUSSION

This section summarizes the outcomes of using the strategies to obtain similar anime and ratings.

4.1 Popularity based algorithm

The Figure 2 shows the top 15 most popular shows based on their given by the users. In the graph given below, the Y-axis represents different Animes and on the X-axis, the corresponding popularity score is given from 0-10.

Based on our recommendation system, we have obtained the top 15 most popular anime, according to people's ratings. The Table 2 is drawn, with columns as Name, Average Ratings and the Score based on the 15 most popular Animes.

Table 2. Fifteen most popular shows.

Name	Average Ratings	Score
Fullmetal Alchemist Brotherhood	9.26	9.12926197
SteinsGate	9.17	9.022239918
Clannad After Story	9.06	8.890068519
Code Geass Hangyaku no Lelouch R2	8.98	8.843125868
Hunter x Hunter 2011	9.13	8.826738505
Sen to Chihiro no Kamikakushi	8.93	8.76503438
Code Geass Hangyaku no Lelouch	8.83	8.714967667
Shigatsu wa Kimi no Uso	8.92	8.663781243
Death Note	8.71	8.636635884
One Punch Man	8.82	8.634723759

4.1.1 KNN

We obtained the following result using the KNN recommender. The Table 3 represents the top 5 recommended anime (similar genre/ratings), based on a user given Anime i.e., Anime 0. This is an effective way to club Anime based on specific taste.

Table 3. Top 5 recommended anime based on Anime 0.

Anime0	Anime1	Anime2	Anime3	Anime4	Anime5
Choisuji	Princess Tutu Recaps	Paizuri Cheerlead er vs Sakunyuu Ouendan	PeroPero Teacher	Please Me	Osaru no Sankichi Boukuuse n
Gekigange r 3 The Movie	Ookiku Furika- butt e Special	009 ReCyborg	Gundam Build Fighters Try Island Wars	Avengers Confidenti al Black Widow to Punisher	Xmen
91	3 × 3 Eyes	3 × 3 Eyes	MazeBa kunetsu Jikuu TV	Hana yori Dango	Koihime Musou OVA Omake
Candy Candy	Omae Umasou da na	Candy Candy Haru no Yobigoe	Okusama wa Joshikousei	Candy Candy Movie	Ashita no Joe Pilots
07Ghost	Vampire Knight Guilty	Pandora Hearts	Nabari no Ou	Vampire Knight	Dantalian no Shoka
11eyes	IS Infinite Stratos	Dragon Crisis	Mayo Chiki	IS Infinite Stratos 2	11eyes Mo- moiro Genmu- tan
11eyes Momoiro Genmutan	11eyes	Boku wa Tomodach i ga Sukunai Episode 0	IS Infinite Stratos	Asobi ni Iku yo	IS Infinite Stra- tos 2

4.2 Collaborative based filtering

The above bar graph depicts the top 10 animes, with the same genres as "Kimi No Na Wa" and their respective percentage of likeliness to the source Anime. The similarity of animes decreases as we go downwards which is shown in Figure 2.

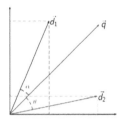

Figure 2.   Collaborative based filtering graph. q = Anime 0 vector; d1, d2 = Recommended Animes vectors α, θ = Deviations of d1, d2 from q.

4.3 *KNN*

KNN is a type of algorithm that may be used to solve issues in classification and regression [14].

In classification issues, we use the distance metric to find the k closest examples to the supplied one, then use the majority vote method or weighted majority voting to anticipate the labels. Based on a provided distance parameter such as cosine, Euclidean, Jaccard similarity, Minkowski, or custom distance measures, K-nearest neighbor determines the k most similar objects to a single instance. In our model, we employed the cosine similarity measure to discover the k-nearest neighbors [19].

5   CONCLUSION AND FUTURE WORK

Anime is a vast class of entertainment that consists of a huge variety of genres and topics to attract a huge variety of viewers. The animation enterprise employs 430 firms. As a result, developing private anime recommendations turns into a hard task. To cope with this problem, we evolved an anime recommendation algorithm after a good deal of thought. The results have been created through our algorithms based on the customers' various hobbies. Collaborative filtering techniques based on recognition, content material, collaboration, and reminiscence have all been used. We focused on the subsequent strategy: the consumer is given items that previous customers with comparable pursuits and possibilities liked. Online companies have to now assist clients deal with facts overload via means of providing tailor-made ideas, content material, and services. We developed the top 15 most popular shows based on average user ratings in the popularity-based model; similarly, in the content-based model, we used content provided by users over time to recommend similar anime that are almost the same genre and type; similarly, the collaborative advice gadget has models: memory and model based. We additionally employed the K-Nearest Neighbor method, which shows 5 anime for every anime. As indicated through this research, all those algorithms have proven promising results. A more potent anime advice device arises while the user's looking records are considered. Hybrid models may be applied to offer extra unique results. An excessive degree of accuracy is needed for real-time superior advice systems. A sort of machine learning algorithm can assist those real-time systems. We wish to examine the other algorithms to look if they'll assist to enhance the precision even further.

REFERENCES

[1]   Tewari, Anand Shanker, Abhay Kumar, and Asim Gopal Barman. "Book Recommendation System Based on Combine Features of Content Based Filtering, Collaborative Filtering and Association Rule Mining." In *2014 IEEE International Advance Computing Conference (IACC)*, pp. 500–503. IEEE, 2014.

[2]   Saadati, Mojdeh, Syed Shihab, and Mohammed Shaiqur Rahman. "Movie Recommender Systems: Implementation and Performance Evaluation." *arXiv preprint arXiv*:1909.12749 (2019).

[3]   Chate, Parinita J. "The Use of Machine Learning Algorithms in Recommender Systems: A Systematic Review." *IJRAR-International Journal of Research and Analytical Reviews (IJRAR) 6*, no. 2 (2019): 671–681.

[4]   Shinde, Urmila, and Rajashree Shedge. "Comparative Analysis of Collaborative Filtering Technique." *IOSR Journal of Computer Engineering* 10 (2013): 77–82.

[5]   Murali, M. Viswa, T. G. Vishnu, and Nancy Victor. "*A Collaborative Filtering based Recommender System for Suggesting New Trends in Any Domain of Research.*" In 2019 5th International Conference on Advanced Computing & Communication Systems (ICACCS), pp. 550–553. IEEE, 2019.

[6]   Wang X, Xu Z, Xia X, Mao C 2017 Computing User Similarity by Combining Simrank++ and Cosine Similarities to Improve Collaborative Filtering *IEEE 14th Web Information Systems and Applications Conference (WISA)* pp 205–210.

[7]   Kuchaiev, & Ginsburg. (2017). Micikevicius, Paulius, *et al.* "Mixed precision training." (2017). *arXiv* preprint arXiv:1710.03740

[8]   Hornung, Thomas, *et al.* "Evaluating hybrid music recommender systems." *2013 IEEE/WIC/ACM International Joint Conferences on Web Intelligence (WI) and Intelligent Agent Technologies (IAT).* Vol. 1. IEEE, 2013.

[9]   Mishra, N., Chaturvedi, S., Mishra, V., Srivastava, R., Bargah, P. (2017). Solving Sparsity Problem in Rating-Based Movie Recommendation System. In: Behera, H., Mohapatra, D. (eds) *Computational Intelligence in Data Mining. Advances in Intelligent Systems and Computing*, vol 556. Springer, Singapore. https://doi.org/10.1007/978-981-10-3874-7_11

[10]   Kumar M, Yadav DK, Singh A, Gupta VK 2015 A Movie Recommender System: Movrec International Journal of Computer Applications Jan 1 124(3)

[11]   Hande, R., Gutti, A., Shah, K., Gandhi, J., & Kamtikar, V. (2016). MOVIEMENDER-A movie Recommender System. *International Journal of Engineering Sciences & Research Technology* ISSN, 2277-9655.

[12]   Ota, S., Kawata, H., Muta, M., Masuko, S., & Hoshino, J. I. (2017, September). AniReco: Japanese Anime Recommendation System. In *International Conference on Entertainment Computing* (pp. 400–403). Springer, Cham.

[13]   Geetha, G., Safa, M., Fancy, C., & Saranya, D. (2018, April). A Hybrid Approach using Collaborative Filtering and Content based Filtering for Recommender System. In *Journal of Physics: Conference Series* (Vol. 1000, No. 1, p. 012101). IOP Publishing.

[14]   Cintia Ganesha Putri, D., Leu, J. S., & Seda, P. (2020). Design of an Unsupervised Machine Learning-based Movie Recommender System. *Symmetry*, 12(2), 185.

[15]   Na, S., Xumin, L., & Yong, G. (2010, April). Research on k-means Clustering Algorithm: An Improved k-means Clustering Algorithm. In *2010 Third International Symposium on Intelligent Information Technology and Security Informatics* (pp. 63–67). Ieee.

[16]   Shahjalal MA, Ahmad Z, Arefin MS, Hossain MR 2017 A user rating-based collaborative filtering approach to predict movie preferences *IEEE 3rd International Conference on Electrical Information and Communication Technology (EICT)* pp 1–5.

[17]   Qaiser, S., & Ali, R. (2018). Text Mining: Use of TF-IDF to Examine The Relevance of Words to Documents. *International Journal of Computer Applications*, 181(1), 25–29.

[18]   Bafna, P., Pramod, D., & Vaidya, A. (2016, March). Document Clustering: TF-IDF Approach. In *2016 International Conference on Electrical, Electronics, and Optimization Techniques (ICEEOT)* (pp. 61–66). IEEE.

[19]   Pandya, S., Shah, J., Joshi, N., Ghayvat, H., Mukhopadhyay, S. C., & Yap, M. H. (2016, November). A Novel Hybrid Based Recommendation System Based on Clustering and Association Mining. In *2016 10th International Conference on Sensing Technology (ICST)* (pp. 1–6). IEEE

[20]   Zarzour, H., Al-Sharif, Z., Al-Ayyoub, M., & Jararweh, Y. (2018, April). A New Collaborative Filtering Recommendation Algorithm Based on Dimensionality Reduction and Clustering Techniques. In *2018 9th International Conference on Information and Communication Systems (ICICS)* (pp. 102–106). IEEE.

[21]   N. S. Patankar, Y. S. Deshmukh, R. D. Chintamani, K. Vengatesan and N. L. Shelake, "An Efficient Machine Learning System using Sentiment Analysis for Movie Recommendations," *2022 3rd International Conference on Smart Electronics and Communication (ICOSEC)*, Trichy, India, 2022, pp. 1284–1291, doi: 10.1109/ICOSEC54921.2022.9951906.

Predicting bitcoin price fluctuation by Twitter sentiment analysis

Hardik Choudhary, Mrityunjay Shukla, S. Raghavendra & Ramyashree
Department of Information and Communication Technology, Manipal Institute of Technology, Manipal Academy of Higher Education, Manipal, India
ORCID ID: 0009-0007-3537-4810, 0009-0007-2282-638X, 0000-0003-2733-3916, 0000-0002-0237-2444

ABSTRACT: The development of the fintech industry has transformed cryptocurrencies into intangible assets and opened many opportunities in the fields of financial research and quantitative markets. Cryptocurrencies are a type of electronic currency used to conduct transactions in the financial system. In addition to the technical analysis that a trader typically does, it has been established over time that market mood is extremely important in determining market conditions. This document provides a method for estimating cryptocurrency prices based on historical data and user sentiment. To achieve this, a long short-term memory (LSTM) model and sentiment analysis of tweets were used. Furthermore, it was supported by the outcomes, as the LSTM model demonstrated a precision of 69.32%, which is respectable when it comes to the forecasting of financially risky assets like bitcoin. The final accuracy attained was 70%, indicating that the model will accurately recommend buying or selling in about 3 out of every 4 scenarios that it is presented with. Traders can achieve a high alpha with a risk reward ration of 1:2 to benefit from this research finding and can combine the findings with technical indicators to produce better trades. This research has a very large application in the field of quant trading. Findings in this research can be used to build multiple models with multiple attributes which will improve the overall accuracy and precision of trades.

Keywords: *Bitcoin Price*, cryptocurrencies, LSTM, *Data sets, Accuracy, Precision*

1 INTRODUCTION

The development of the fintech industry has transformed cryptocurrencies into intangible assets and opened up many opportunities in the fields of financial research and quantitative markets. Cryptocurrencies are a type of electronic currency used to conduct transactions in the financial system [1] Money. All these cryptocurrencies are encrypted and cannot be duplicated. It can be distinguished from traditional currency as it is not issued by a government. It is a decentralized cryptocurrency with active crypto conversion functionality that allows users to convert currencies to other formats using a decentralized exchange. This system uses blockchain technology. Due to the high volatility of cryptocurrencies, it is very difficult to predict price movements. Many cryptocurrencies are used every day to invest in and purchase products and services around the world. Bitcoin volatility makes it difficult to predict. We now find that we can build a better model by combining the Tweet analysis model and the LSTM model [2]. This article combines LSTM machine learning algorithms with sentiment analysis to help retail investors and hedge funds decide whether to buy cryptocurrencies and improve investment predictability.

This document provides a method for estimating cryptocurrency prices based on historical data and user sentiment. To achieve this, a long short-term memory (LSTM) model and sentiment analysis of tweets were used. The introduction and literature review of this article will help the reader to have a basic understanding of the topic. After researching, the next step is the programming part. Additionally, the results section shows how the machine learning

DOI: 10.1201/9781003363781-12

and sentiment analysis models combine to provide the final suggestions. Due to the high volatility, the price movements of cryptocurrencies are very difficult to predict [3]. Hundreds of cryptocurrencies are used every day to invest in and buy products and services around the world. Bitcoin price volatility and dynamics make it difficult to predict. All research to date concludes that machine learning and sentiment analysis must be combined for the best results. Therefore, the study combines LSTM machine learning algorithms with sentiment analysis to help investors decide whether to buy cryptocurrencies and better predict their Bitcoin investments. This document provides a method for estimating cryptocurrency prices based on historical data and user sentiment. To achieve this, a long short-term memory (LSTM) model and sentiment analysis of tweets were used. The introduction and literature review of this post will help readers gain a basic understanding of the topic. For the datasets, performed the data pre-processing, and the models are used to obtain the results [4]. After that, the next step is the programming part. Additionally, the results section shows how the machine learning and sentiment analysis models combine to provide the final suggestions.

2 LITERATURE SURVEY

Twitter is getting better at predicting public opinion, especially some things related to financial markets. How well Twitter can predict Bitcoin returns is investigated by Sattarov *et al.* [4]. Found that high price predictability can be achieved by using sentiment analysers for tweets and financials. This information confirms the existence of the link. The accuracy of predictions made using historical prices and tweet sentiment was 66.51%. A study in [5] attempts to predict price movements by analyzing Twitter sentiments and calculating possible correlations between them. News accounts are tweeting positivelyscores or negatively about Bitcoin. Combine the historical price with the percentage of good and bad tweets to predict the price for the next period. We found an accuracy of 81.39% for sentiment classification of tweets and an overall accuracy of 77.62% for price prediction using RNN [1]. LSTM, Bi-LSTM and GRU models provide examples of how to predict the price of BTC, ETH and LTC cryptocurrencies. Comparing GRU with other algorithms shows higher accuracy on the MAPE scale. The performance of investment portfolios using gradient-based decision trees and his LSTM model was also tested in this study [3], proving that both approaches are beneficial even when transaction costs are considered. Over 1,600 different currencies go through this process. [6] The results of this study show that the combined autoencoder and retraining method performs very well on the same task (F-measure 78%). [7] uses Twitter and Google Trends data to try to predict short-term cryptocurrency prices. We also look for correlations between tweet volume and cryptocurrency value. We found a strong inverse relationship between price and volume of tweets. Finally, [8] provided an overview of previous research and an overview of Bitcoin prediction using machine learning models. This allowed me to understand the idea and develop it further. The conclusions of [7] on the same topic are confirmed by this study [9]. [5] proposes an active trading approach based on machine learning. He tested this article and the methods [1], [3] on his five cryptocurrencies. According to this article [10], the Levenberg-Marquardt algorithm is the best predictor of high-frequency Bitcoin compared to conjugate gradients and robust algorithms. Similarly the modified binary autoregressive tree (BART) model [11], which he uses ARIMA-ARFIMA models for short-term forecasting. Our analysis showed better performance than the ARIMA ARFIMA model.

3 METHODOLOGY

The development of the fintech industry has transformed cryptocurrencies into intangible assets and opened up many opportunities in the fields of financial research and quantitative markets. Cryptocurrencies are a type of electronic currency used to conduct transactions in the financial system. Money. All of these cryptocurrencies are encrypted and cannot be duplicated.

3.1 *Data collection*

We obtained tweets in JSON format using the GET command, transformed the data into strings, and added the results to a csv file for later use in NLP. Developers can access public Tweet data for requested available Tweets via the GET /tweets endpoint. JSON-formatted Tweet objects are included in the response. Among other public Tweet metadata, a Tweet object includes the following: id, text, created at, lang, source, and public metrics. HTTP requests can be organised, tested, and debugged using REST services like Postman. A command-line utility for delivering or receiving files using the URL syntax is called cURL [5].

3.2 *Data pre-processing*

Since disposition of the data required in this scenario for sentiment analysis is text data, pre-processing the data becomes a vital step. The nature of text data is packed with noise, subsequently causing it to be difficult to intelligently decontaminate. Furthermore, pre-processing immensely reduces the size of information inputted. The process occurs in the following stages:

3.2.1 *Data cleaning*
The first and most important phase of sentiment analysis is data preprocessing for customers to express their opinions and thoughts using natural language syntax. Add letters, alphabets, words, and special characters to express your emotions. These excesses should be removed from the data before classification. It is done step by step. Each check begins with a pre-processing phase where ambiguous data such as stop words, numbers and special characters are removed. Both the Porter Stemmer and NLTK libraries were imported to remove stop words, numbers, and special characters from it [6].

3.2.2 *Using Regular Expression (RE)*
A set of characters called a regular expression is commonly used to match and replace patterns in strings. Written words are often represented as strings when stored in a computer. Each word is made up of strings. Some strings contain characters specific to spaces and newlines. It is critical to utilize special methods to manipulate strings. For example a text can use a wildcard to allow for ease of searching in characters, furthermore it can be surrounded by notations to be used as a link to another webpage. To conduct such feats, regular expressions need to be used. In processing and analysing the large amount of random review documents, regular expressions play a vital role.

3.2.3 *Elimination of stop-words*
Stop words are convenience words that are frequently used in textual languages (for example, English "a", "the", "an", and "of*) and are not useful for classification. We use the Natural Language Toolkit package to reduce the use of stop words. You don't want these expressions to take up valuable database processing time or disk space [7]. All you have to do is maintain a list of terms that you consider to be stop words. The list of stop words for 16 different languages is maintained in Python by the NLTK package. From which we'll use English language stop-words.

3.2.4 *Stemming*
Stemming is a technique used in information retrieval that reduces inflected words to their stems (or roots) so that words with similar meanings map to the same stem. By reducing the number of linked terms on each page naturally, this tactic shrinks the feature space. The tests use an implementation of the Porter stemming technique. For example, the English "generalizations" would be formed from the stems "generalization" "generalize", "general", and "general".

3.3 Sentiment analysis

A vocabulary and rule-based sentiment analysis tool called VADER (Valence Aware Dictionary and Sentiment Reasoner) is customized precisely to the sentiments expressed in social media. Negations and contractions are taken into account by VADER (bad, wasn't good). CAPS, emotes, emojis, intensifications (very, kind of), and acronyms (lol) are all acceptable forms of punctuation. The relevance of each tweet is represented by a score that is calculated using the sentiment analysis of the tweets [8].

3.4 Polarity and subjectivity

Polarity is a type of floating-point number that falls inside the range $[-1,1]$, where 1 denotes a positive assertion and -1, a negative one. Objective statements pertain to factual information, but subjective sentences typically allude to personal opinion, emotion, or judgement. Additionally, subjectivity is a float with a [0, 1] range.

3.5 Tokenization

Tokenization is the process of dividing a written document into tokens, which are very little pieces of text. Words, word fragments, or simple characters like punctuation can all be considered tokens. These tokens are later used to train and predict the model [9].

3.6 Model building flowchart

The Figure 1 is a graphical representation of adding different layers to the model and building the model which is to be used to train in order to predict prices of bitcoin based on the results of the sentimental analysis obtained earlier.

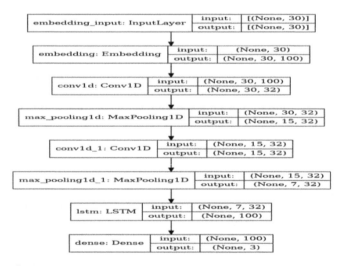

Figure 1. Flow diagram.

3.6.1 Maxpooling1D
Maximum pooling operations for temporary 1D data.

3.6.2 Conv1D
To produce the output tensor, this layer produces a convolution kernel that is convolved with the layer's input in a single spatial (or temporal) dimension. If Use Bias is set to True, a

bias vector will be created and added to the output. Last but not least, it also applies to the output if the activation is not None.

3.6.3   *Input layers*

Input neurons in the input layer communicate with the embedding layer. Data is sent from the hidden layer to the output layer. Each neuron has an output, an activation function (which determines the output given the input), and weighted inputs (synapses).

Long Short-Term Memory (LSTM) Recurrent Neural Networks (RNNs) have emerged as efficient and scalable methods for a variety of sequential input learning problems which is shown in Figure 2. They are essential for recognizing long-term temporary dependencies. LSTM is an RNN-inspired model [10] with control gates for information flow between cells.

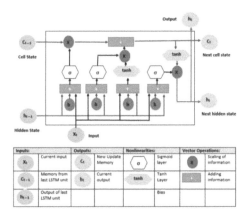

Figure 2.   Long Short-Term Memory (LSTM) Recurrent Neural Networks (RNNs).

3.7   *Hyper parameter tuning*

During hyper parameter tuning, learning rate and momentum were tweaked and checked up to 300 times the default value to obtain the optimum parameters for the model to predict bitcoin price [11].

3.8   *Default parameter*

Vocab_size=5000,Embedding_size=32,epochs=50,LearningRate=0.01,Decay_rate=learning_rate/epochs,Momentum=0.8 are considered as Default Parameters.

3.9   *Optimum hyper parameter*

Vocab_size=5000,Embedding_size=32,epochs=50,LearningRate=0.3,Decay_rate=learning_rate/epochs,Momentum=0.9 are considered as Optimum Hyper Parameters.

4   RESULTS AND DISCUSSION

The model was tested using a range of various sorts of data for testing purposes. It was the one with four LSTM layers of 64,32 units each, then a dense layer on top of which we forecasted near using time series of the close column with a time step of 110 seconds.

We suggested eight layers and sixty units for the LSTM. Neurons are concealed within each unit. Every layer of the regressor has been connected. As an optimizer, adaptive moment optimization has been employed.

To examine the fluctuation in bitcoin prices caused by the polarity of tweets, each sentiment, whether positive or negative, has been examined in a dataset. The actual and anticipated values for each timeframe about the feelings and polarity of each tweet are displayed in Figure 3. Figure 3 displays the actual sentimental analysis value as a blue line and the anticipated value as an orange line.

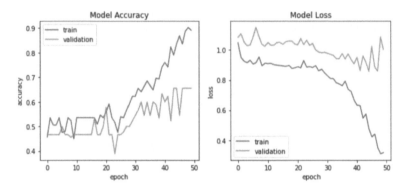

Figure 3.   Actual and Anticipated values for each timeframe.

We can achieve the following after multiple parameter tuning shown in Table 1:

Table 1.   Crime rates according to different places.

Accuracy	0.700
Precision	0.6932
Recall	0.6778
F1-Score	0.6854

We used confusion matrix to depict number of predicted label and actual label which is shown in Figure 4.

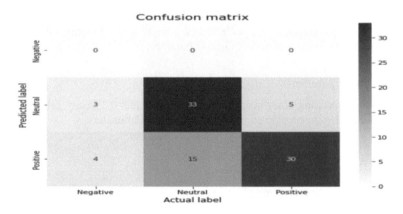

Figure 4.   Confusion matrix.

82

5 CONCLUSION AND FUTURE WORK

In addition to the technical analysis that a trader typically does, it has been established over time that market mood is extremely important in determining market conditions. Furthermore, it was supported by the outcomes, as the LSTM model demonstrated a precision of 69.32%, which is respectable when it comes to the forecasting of financially risky assets like bitcoin. The final accuracy attained was 70%, indicating that the model will accurately recommend buying or selling in about 3 out of every 4 scenarios that it is presented with. Traders can achieve a high alpha with a risk reward ration of 1:2 to benefit from this research finding and can combine the findings with technical indicators to produce better trades.

This research has a very large application in the field of quant trading. Findings in this research can be used to build multiple models with multiple attributes which will improve the overall accuracy and precision of trades as these models will provide a ground sentiment of the public. Flock trading can be achieved through this model in combination with various other high frequency trade systems.

REFERENCES

[1]   Xinwen Zhang *The Evolution Of Natural Language Processing And Its Impact On AI Forbes*, Nov 2018
[2]   Muxi Xu *NLP for Stock Market Prediction with Reddit Data*, 2021
[3]   Toni Pano and Rasha Kashef *A Complete VADER-Based Sentiment Analysis of Bitcoin (BTC)*, 9 Nov 2020
[4]   O. Sattarov, H. S. Jeon, R. Oh and J. D. Lee, "Forecasting Bitcoin Price Fluctuation by Twitter Sentiment Analysis", *2020 International Conference on Information Science and Communications Technologies ICISCT* 2020, pp. 1–4, November 2020.
[5]   D. R. Pant, P. Neupane, A. Poudel, A. K. Pokhrel and B. K. Lama, "Recurrent Neural Network Based Bitcoin Price Prediction by Twitter Sentiment Analysis", *2018 IEEE 3rd International Conference on Computing*, 2018.
[6]   Hamayel M. J. and Owda A. Y., "A Novel Cryptocurrency Price Prediction Model Using gru lstm and bi-lstm Machine Learning Algorithms", AI, vol. 2, no. 4, pp. 477–496, 2021, [online] Available: https://www.mdpi.com/2673-2688/2/4/30.
[7]   Jaquart P., Dann D. and Weinhardt C., "Short-term Bitcoin Market Prediction Via Machine Learning", *The Journal of Finance and Data Science*, vol. 7, pp. 45–66, 2021.
[8]   Alessandretti L., ElBahrawy A., Aiello L. M. and Baronchelli A., "Anticipating Cryptocurrency Prices Using Machine Learning", *Complexity*, vol. 2018, 2018.
[9]   Lengare K., "*Introduction to Cryptocurrencies*", vol. 12, pp. 01–08, 03 2022.
[10] Koker T. E. and Koutmos D., "Cryptocurrency Trading Using Machine Learning", *Journal of Risk and Financial Management*, vol. 13, no. 8, pp. 178, 2020.

Recent Trends in Computational Sciences – Gururaj, Pooja & Flammini (Eds)
© 2024 The Author(s), ISBN 978-1-032-42685-3

Microarchitecture design and verification of co-processor for floating point operation

K. Harshitha, A.S. Hithen, T.A. Jagath Madappa, B.G. Karthik & Narasimhamoorty S. Bhat
Department of Electronics and Communication Engineering, Vidyavardhaka College of Engineering, Mysuru, India
ORCID ID: 0000-0002-8094-1913

ABSTRACT: In the microprocessor's chip, additional circuits can be introduced for unique purposes to execute particular tasks or to carry out numerical operations to offload work from the main processor. Then the CPU can work faster. To make the most of mainframe CPU time, I/O responsibilities are assigned to other systems called Input/Output channels. The central computer doesn't require any Input / Output processing, but simply sets the constraint for the input or output operation and then signals the channel processor to perform the entire operation. By dedicating relatively simple sub-processors to time-consuming formatting and I/O processing, the overall system performance has been improved. We can use the treadmill for extra work while the engine is running. Therefore, the engine is used more efficiently. Likewise, an additional processor i.e., co-processor handles the math part of the job when we run complex applications. Co-processors for the floating point arithmetic operation first appeared in the desktop computers in the 1970s and it became popular whole of the 1980s and early 1990s. Earlier generations used 16-bit and 8-bit processor software to perform the FPU arithmetic. When the co-processor supports, floating point calculations that can be perform much faster. The math co-processors are common adoption for computer-aided designing (CAD) users and the scientific and the engineering computer software. Some of the Floating-Point Units, like AMD 9511, the Intel 8231/8232 and Waitek FPU, are considered peripherals, while others such as Intel 8087, Motorola 68881 are tightly integrated. than the processor.

Keywords: microprocessor chip, unique purposes, time consuming, Channel I/O, floating-point arithmetic

1 INTRODUCTION

Co-processors are often called math processors. The coprocessor performs routine math tasks, freeing the core processor from this computation and saving its time. Coprocessors offload specialized processing tasks from the core CPU, allowing them to run faster by offloading the main microprocessor. A co-processor is nothing but a computer processor which is used to assist the functionality of the main processor (CPU). The operations which are performed by the co-processor are fed into a floating point-based Smart Water Quality Monitoring (SWQM) system that serves for non-stop measurement of the quality of the arithmetic, graphics, signal processing, string processing, cryptography, or I/O interfaces. can. Peripheral equipment. Co-processors can speed up system performance by reducing the processor-risky tasks from primary processor. Coprocessors allow you to customize a variety of computers so customers who don't need extra power don't have to pay for it. The quickness of the floating-point calculation(operations), are commonly measured in FLOPS, which is the main characteristic of a system, mainly in the apps involving intense math based computations. The FPU (Floating-point unit, usually known as the math based coprocessor) is a part of a device specifically integrated to perform floating-point number operations.

DOI: 10.1201/9781003363781-13

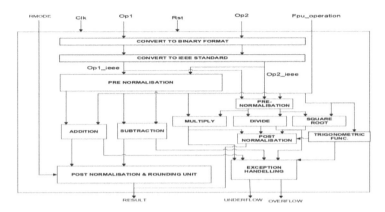

Figure 1.   Basic system architecture.

2   CO-PROCESSOR FOR FLOATING POINT OPERATION

Coprocessors offload specialized processing tasks from the core CPU, allowing them to run faster by offloading the main microprocessor. Coprocessors can perform specialized tasks such as complex mathematical calculations and graphics display processing. These tasks run faster than the core CPU. Microprocessors developed in such a way that it need to work fast in real-time, such that the desired output is seen without any delay. If we add floating point operations to the instruction path it leads to slow down the process of microprocessor. In order to offload the processor-rigorous errands from the primary processor, the coprocessor is being developed which can accelerate the system performance, save the time as well. Coprocessor executes the computationally more intensive instructions. We use the cadence tool to write the RTL code as per the requirement of our project, performing all the necessary arithmetic operations. The flow of operation takes place as shown in the below system architecture. After successful execution of the RTL code, we will verify the output of all arithmetic operations for the given testbench values.

3   LITERATURE REVIEW

Metin Mete Ozbilen & Mustafa Gok [1] proposed design named "A Multi-Precision Floating-Point Adder".

This document presents a multi-precision floating-point adder that can perform a single high-precision floating-point addition or the multiple low-precision floating-point additions in parallel. The proposed design eliminates time-consuming format conversion operations when operating in low-precision mode. The delay of the proposed multi-precision floating-point adder is almost the same as the standard double-precision floating-point adder.

Park *et al.* [2] published a paper named "Design, Implementation and On-Chip High-Speed Test of SFQ Half-Precision Floating-Point Multiplier" In this paper, they explained that, to overcome delay, a Large-Scale Reconfigurable Data Path (LSRDP) has been proposed as a suitable processor architecture for single flux quantum (SFQ) circuits. SFQ-LSRDP consists of thousands of SFQ floating point units connected by reconfigurable SFQ network switches to achieve high performance with low power consumption. In this study, we designed and implemented one of the key components of SFQ LSRDP, the SFQ floating point multiplier (FPM).

Hani Saleh and Earl E. Swartz lander [3] developed "A Floating-Point Fused Add-Subtract Unit" from which A fused floating-point addition/subtraction unit as mentioned in this section, that simultaneously performs the operation of floating-point addition and subtraction on a common pair of single-precision data in approximately the same time it would take a conventional floating-point adder to perform a single addition. The fused adder/subtractor unit performed both ADD and SUB operations at nearly the same speed as the floating point adder with less than 40% area overhead.

Vinayak Patil, Aneesh Raveendran, Sobha, David Selvakumar and Vivian [4] proposed "Out of Order Floating Point Coprocessor for RISC V ISA" This reviewed paper details the architecture of floating-point coprocessors that can use RISC-V floating-point instructions. A floating-point coprocessor is integrated into the integer pipeline. A proposed architecture for afloating-point coprocessor with out-of-order execution, in-order commit, and completion/retire was synthesized, tested, and verified on a Xilinx Virtex 6xc6vlx550t-2ff1759 FPGA. FPGAs have seen system frequencies of 240 MHz for single-precision floating-point arithmetic.

Cheol-Ho Jeong, Woo-Chan, Sang-Woo Kim and Moon-Key Lee [5] developed a method to fix "In-Order Issue Out-of-Order Execution Floating-Point Coprocessor for CalmRISC32." The CalmRISC32 Floating Point Unit is a Reduced Instruction Set Computer type coprocessor which is basically an open source. Configured to work with CalmRISC32 microcontrollers. IEEE-754 based single precision floating point addition/subtraction and multiplication are supported coprocessor design. Statements can be executed out-of-order and completed if some restrictions such as the data dependencies, resource contention exception prediction are fixed. It has a simple coprocessor interface and the exception predictors to support precise exceptions. Standard cell-based design techniques are used to reduce design time and cost.

Remya Jose & Dhanesh [6] developed a design "Single Precision Floating Point Co-Processor for Statistical Analysis." This published paper describes the design and implementation of a single-precision floating-point coprocessor that can be used for statistical analysis. This paper presents a comparative analysis of different arithmetic modules which is used to implement the efficient coprocessors and also different floating-point adders and multipliers based on area and delay. Latency is one of the key factors in processor implementation. Based on comparative analysis, the Koggestone adder and the modified cabin multiplier were selected. A powerful coprocessor was implemented using these computational modules.

Aditi Sharma, Sukanya Singh and Abhay Sharma [7] developed a method "Implementation of Single Precision Conventional and Fused Floating Point Add-Subunit Using Verilog." This deals with Floating-point adder and floating-point subtractor architecture for the single precision format has been compared to its fused counterpart known as the fused add-subtract unit. Floating point addition and subtraction were implemented using Verilog HDL. Fused add-sub FPUs have been observed to yield better latency results compared to conventional FPUs. Shrink area, increase speed, and reduce power consumption.

Aneesh Raveendran, Vinayak Baramu Patil, David Selvakumar and Vivian Desalphine [8] proposed "A RISC-V Instruction Set Processor-Microarchitecture Design and Analysis." This paper has described the micro-architectural design and the analysis of the RISC-V instruction set processors. RISC-V is one of the open-source ISA. This paper details the micro-architecture design and analysis of a 5-stage pipelined RISC-VISA compatible processor. The proposed processor with FPU unit has a 171 K logic cells and has area 557 mm2.

Taek-Jun Kwon, Jeff Sondeen and Jeff Draper [9] developed "A 0.18 μm implementation of a floating-point unit for a processing-in-memory system" This was Rudolf Usselmann's open source project. His FPU described a single-precision floating-point unit capable of addition, subtraction, multiplication, and division. It consists of the two pre-normalization units that can normalize fractions. The main drawback of this model is that most of the code is written in MATLAB, so it cannot be synthesized.

M. Sangwan and A. A. Angeline [10] proposed a "Design and implementation of single precision pipelined floating point co-processor." This paper presents a comparison of the various computational modules and an optimized floating-point ALU implementation. Here, we use a pipelined architecture to improve performance and achieve a design that increases operating frequency by a factor of 1.62. Ultimately, these individual blocks are combined to create his ALU, which is pipelined and floating-point based, to increase operating frequency while minimizing power consumption. At the same time, the number of gates increases and so does the area, due to the number of iterations used in the algorithm.

S. Palekar and N. Narkhede [11] proposed "High-speed, area-saving single-precision floating-point arithmetic unit". The 32-bit floating point unit is designed for a wide variety of applications that require outstanding speed. The performance of the proposed design is compared to research papers related to critical path delays, with up to 59% floating point multiplier critical path delay optimization and 50% floating point adder/subtractor optimization is assumed. The results show that the proposed design is optimized for both speed and area efficiency.

K.V. Gowreesrinivas and P. Samundiswary [12] proposed "Comparative study on performance of single precision floating point multiplier using vedic multiplier and different types of adders." Single-precision floating-point performance is the main and also it is primarily based on the footprint and multiplier delay. Reducing the complexity and the connections, can improve overall performance of the system. In this paper, a new type of single-precision floating-point multiplier is developed by integrating Vedic multipliers with various types of adders.

4 RESULTS

We've completed till the half way of our project, that is the RTL and testbench codes for addition and subtraction operations have been written, tested and verified. The simulated output along with the manual calculation is shown below

✓ **Addition**
EX 1: A = 48.265 B = −48.362
Binary representation of the given inputs
A = 01000010010000010000111101011100
B = 11000010010000010111001010110000
Adding the mantissa part,
A_mantissa = 10000010000111101011100
B_mantissa = 10000010111001010110000
After addition ,The final result,
A + B = 48.265 + (−48.362) = −0.097
A+B= 10111101110001101010100000000000
Hexadecimal equivalent of the final result A+B= **bdc6a800 (-0.097)**
==================

EX 2 : A = 3.256 B = −4.321
Binary representation of the given inputs
A = 01000000010100000100010010011110
B = 11000000100010100100010101000010
After adding the mantissa part, The final result,
A + B = 3.256 + (−4.321) = −1.065
A + B = 10111111100010000101000111101100
Hexadecimal equivalent of the final result, A + B = **bf8851ec (–1.065)**

Note: To revert to EPWave opening in a new browser window, set that option on your user page.

✓ **Subtraction**

EX 1: A = 4.5 B = 2.32
Binary representation of the given inputs
A = 01000000100100000000000000000000
B = 01000000000101000111101011100001
Subtracting the mantissa part, The final result is, A-B = 4.5 – 2.32 = 2.18
A-B = 01000000000010111000010100011111
Hexadecimal equivalent of the final result, A-B = **400b851f (2.18)**

EX 2: A = 4.5 B = 6.32
Binary representation of the given inputs
A = 01000000100100000000000000000000
B = 01000000110010100011110101110001
Subtracting the mantissa part, The final result is, A-B = 4.5 – 6.32 = −1.82
A-B = 01000000000010111000010100011111
Hexadecimal equivalent of the final result, A-B = **bfe8f5c3 (–1.82)**

Note: To revert to EPWave opening in a new browser window, set that option on your user page.

6 CONCLUSION

We are designing a Coprocessor specifically for floating point calculations. A coprocessor offloads the main microprocessor, the greater run speed has been achieved. Special tasks like complex mathematical calculations are performed with good speed. The faster performance is achieved than core CPU. Coprocessor offloads the Processor-intensive tasks from the main processor, This speeds up system performance. and save the time. As the area, power, timing are the three challenging factors in the processor, we are mainly focused on these three areas. The accuracy of the results plays major role. In this design, as we designed Coprocessor specifically for floating point calculations only, we can see the greater accuracy.

REFERENCES

[1] Ozbilen M.M. and Gok M., "A Multi-precision Floating-point Adder," *2008 Ph.D. Research in Microelectronics and Electronics*, 2008, pp. 117–120, doi: 10.1109/RME.2008.4595739

[2]   Park H. *et al.* "Design and Implementation and On-Chip High-Speed Test of SFQ Half-Precision Floating-Point Adders," in *IEEE Transactions on Applied Superconductivity*, vol. 19, no. 3, pp. 634–639, June 2009, doi: 10.1109/TASC.2009.2019070

[3]   Saleh H. and Swartzlander E.E., "A Floating-point Fused Add-subtract Unit," *2008 51st Midwest Symposium on Circuits and Systems*, 2008, pp. 519–522, doi: 10.1109/MWSCAS.2008.4616850

[4]   Patil V., Raveendran A., Sobha P.M., David Selvakumar A. and Vivian D., "Out of Order Floating Point Coprocessor for RISC V ISA," *2015 19th International Symposium on VLSI Design and Test*, 2015, pp. 1–7, doi: 10.1109/ISVDAT.2015.7208116.

[5]   Cheol-Ho Jeong, Woo-Chan Park, Tack-Don Han, Sang-Woo Kim and Moon-Key Lee, "In Order Issue Out-of-order Execution Floating-point Coprocessor for CalmRISC32," *Proceedings 15th IEEE Symposium on Computer Arithmetic. ARITH-15 2001*, 2001, pp. 195–200, doi: 10.1109/ARITH. 2001.930119

[6]   Remya Jose, Dhanesh M.S., 2015, Single Precision Floating Point Co-Processor for Statistical Analysis, *International Journal Of Engineering Research & Technology (IJERT) NCETET – 2015* (Volume 3–Issue 05)

[7]   Sharma A., Singh S. and Sharma A., "Implementation of Single Precision Conventional and Fused Floating Point Add-sub Unit Using Verilog," *2017 International Conference on Wireless Communications, Signal Processing and Networking (WiSPNET)*, 2017, pp. 169–171, doi: 10.1109/ WiSPNET.2017.8299741

[8]   Raveendran A., Patil V.B., Selvakumar D. and Desalphine V., "A RISC-V Instruction Set Processor-micro-architecture Design and Analysis," *2016 International Conference on VLSI Systems, Architectures, Technology and Applications (VLSI-SATA)*, 2016, pp. 1–7, doi: 10.1109/VLSI-SATA.2016.7593047

[9]   Taek-Jun Kwon, Jeff Sondeen, Jeff Draper, "Design Trade-Offs in Floating-Point Unit, Implementation for Embedded and Processing-In-Memory Systems", *USC Information Sciences Institute*, 4676 Admiralty Way Marina del Rey, CA 90292 U.S.A.

[10]   Sangwan M. and Angeline A.A., "Design and Implementation of Single Precision Pipelined Floating Point co-processor," *2013 International Conference on Advanced Electronic Systems (ICAES)*, 2013, pp. 79–82, doi: 10.1109/ICAE

[11]   Palekar S. and Narkhede N., "High Speed and Area Efficient Single Precision Floating Point Arithmetic Unit," *2016 IEEE International Conference on Recent Trends in Electronics, Information & Communication Technology (RTEICT)*, 2016, pp. 1950–1954, doi: 10.1109/RTEICT.2016.7808177

[12]   Gowreesrinivas K.V. and Samundiswary P., "Comparative Study on Performance of Single Precision Floating Point Multiplier Using Vedic Multiplier and Different Types of Adders," *2016 International Conference on Control, Instrumentation, Communication and Computational Technologies (ICCICCT)*, 2016, pp. 466–471, doi: 10.1109/ICCICCT.2016.7987995S.2013.6659365

Blockchain technology and application

Blockchain based higher education ecosystem

V. Abhishek, N. Pavan, K. Soundarya, J. Shreya & V. Sarasvathi
PES University
ORCID ID: 0009-0003-5741-4933, 0009-0009-5013-1424, 0009-0001-9737-3254, 0000-0002-0494-9825

ABSTRACT: A blockchain-based higher education ecosystem can create a decentralized, transparent, and accessible learning environment. It includes a web portal for connecting students and instructors, digital certificates stored on the Interplanetary File System (IPFS), and digital credits in the form of a cryptographic token. Instructors transfer tokens to students' blockchain wallets via the Ethereum network upon successful completion of a course. This system stores academic records on the blockchain, allowing recruiters to verify educational achievements and offer relevant jobs. Smart contracts and the immutability of blockchain technology enhance the effectiveness and credibility of the higher education sector.

Keywords: blockchain, ipfs, ethereum, smart contracts

1 INTRODUCTION

Blockchain technology has the potential to revolutionize higher education by creating a secure and transparent system for managing academic records, verifying credentials, and facilitating communication among students, educators, and institutions. This paper examines the implementation of a decentralized, blockchain-based higher education ecosystem for direct interactions among students, instructors, and recruiters. Blockchain is a distributed database of linked and secured blocks containing a timestamp and a reference to the previous block. This technology has been used in various fields to provide a secure and transparent way of storing and exchanging data and value, including finance, supply chain management, and the Internet of Things (IoT). In contrast to traditional university education, which relies on centralized systems and institutions for managing academic records, a decentralized, blockchain-based higher education ecosystem offers several advantages, such as:

- Decentralization: Blockchain technology can create a decentralized and democratic education ecosystem where students and educators have control over their educational data, and can connect directly without centralized institutions.
- Security and immutability: Blockchain technology provides secure features like hashing and consensus mechanisms, making it difficult for unauthorized entities to alter academic records.

Overall, a decentralized blockchain-based higher education ecosystem can create a secure, efficient, and transparent system for managing academic records and facilitating learning and collaboration. In the following sections of this research paper, we will explore associated technical, economic, and social challenges and opportunities.

2 LITERATURE SURVEY

Blockchain's adoption in education necessitates addressing gaps and ethical issues. Recent works examine its advantages and disadvantages in higher education. A pragmatic yet simple system is needed to store student credentials, certificates, and course credits.

Chen, G., Xu, B., Lu, M. *et al.* explored blockchain's potential uses in education, discussing benefits, challenges, and opportunities in their paper [1]. Srivastava, P. Bhattacharya *et al.* in [2] proposed a distributed credit transfer educational framework based on blockchain, discussing limitations of traditional education and blockchain's potential benefits for academic record management and credit transfer. M. Turkanovic *et al.* presented EduCTX, a blockchain-based higher education credit platform with potential benefits for students, educators, and institutions, discussing its features and architecture in [3]. Afrianto *et al.* presented a work training certificate authentication system based on a public blockchain platform, discussing the advantages of using blockchain technology for secure and transparent verification in [4]. In paper [5], Bahrami *et al.* proposed a blockchain-based solution for managing academic certificates, discussing the advantages of blockchain for certificate security and transparency and presenting a prototype implementation. I. Zhou *et al.* proposed a blockchain-based file-sharing system for academic paper review, introducing Acadcoin and using smart contracts for monetary transactions between publishers and reviewers in [6]. Jain R. Sarasvathi, V. *et al.* in [7] proposed blockchain technology for profile verification, using digital signatures and hash functions and employing the Proof of Work (PoW) consensus algorithm. N. Yadav and S. V. introduced the concept of smart contract-based crowdfunding for improved transaction transparency and security in blockchain technology in [8]. M. Nithin, S. Shraddha, N. Vaddem, and V. Sarasvathi in [9] proposed the HyperIoT framework for securing IoT transactions with private permissioned blockchain technology, evaluating its performance and comparing it with existing blockchain-based IoT security solutions. Haugsbakken, Halvdan, and Inger Langseth critically analysed blockchain's potential in higher education, dis cussing challenges such as interoperability, standardization, and governance in [10].

3 PROPOSED METHODOLOGY

The proposed decentralized system aims to create a user-friendly environment where students and instructors can interact efficiently without involving a central authority. The system uses public blockchain and IPFS to store and manage digital assets, such as course certificates and credits. The students' academic records consisting of courses completed, skills, and certifications are logged into the blockchain, allowing authorized individuals to access them easily and securely. This system is designed to prevent tampering and fraud, thanks to the immutability of the blockchain.

3.1 *Proposed architecture*

Blockchain can transform higher education by creating a decentralized system for managing academic records, credential verification, and information exchange. The high-level architecture of the proposed system is presented in Figure 1. which depicts the stakeholder's interaction, without the need for any central institutions like university, government, etc. Different stakeholders of the proposed system involve

- Student: Students can sign in, view credentials, register for courses of choice by paying the enrolment fees
- Instructor: Admit interested students for the course, and award digital credits and certificates upon successful completion.
- Employer: Register on blockchain university network, access student credentials, and boffer jobs based on their skills and certifications.

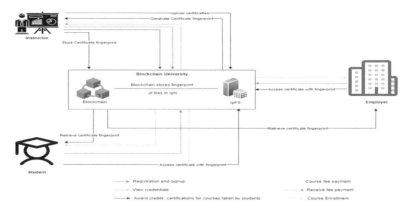

Figure 1. System architecture.

3.2 *System requirements*

The requirements for building the proposed system include:

- A web portal for students and course instructors: Built with React and Node.js at the back-end, the web portal provides an interactive interface for students and course instructors to connect and manage academic records.
- Smart contracts for creating digital certificates and credit transfer: Solidity is used to write and deploy contracts on the Ethereum blockchain network. These smart contracts enable secure and automated execution of agreements between multiple parties, such as the issuance of digital certificates and the transfer of academic credits.
- A local blockchain network for testing and development: This is set up using Ganache, which allows developers to test and debug smart contracts in a controlled environment before deploying them on a public blockchain network.
- A decentralized storage solution for digital certificates: The IPFS is a decentralized protocol for storing and sharing digital certificates.
- A digital token in ERC-20 standards: Ethereum blockchain enables the creation of ERC-20 tokens. In this system, these tokens are used as a digital credit that course instructors award to students upon successful completion of a course.
- A backend database for storing certificate information: This is setup using MongoDB, which stores digital certificate data.

3.3 *IPFS interaction with the system architecture*

IPFS is a distributed file system that uses content-identification (CID) and a Distributed Hash Table to uniquely identify and access files in a shared global namespace.

- When a file is uploaded to IPFS, it is divided to smaller blocks and a unique CID associated with each block is being generated, CID of the root block is returned from the IPFS. CID utilized here constitutes as a location and integrity verifier.
- The root CID is used to find the location of the contents while retrieving the file from IPFS. Hashing mechanism is used to verify the received blocks against the CID hash.

3.4 *Storing certificate credentials in MongoDB*

Using MongoDB in the backend system can help to improve the efficiency and security of the system by providing a central repository for digital certificate information.

Procedure for storing and verifying digital certificates on MongoDB:

- Define the schema for digital certificate which include fields for the student's name, social security number, course name, organization name, date of certificate issuance, expiration date, and certificate ID.
- Certificate ID is added to the corresponding document in the MongoDB database and blockchain as a unique identifier.
- To verify the authenticity of a digital certificate, use the certificate ID to fetch the document from MongoDB and compare it with the blockchain data.

3.5 *Creation and transfer of digital credits*

Procedure for creating digital credits:

- Import ERC20 solidity file from Openzepplin: This involves importing the ERC20 solidity file from the OpenZeppelin library. This file contains the required methods and data members to create an ERC-20 token contract.
- Declare contract and create constructor: A new contract called 'Credit' is declared and its constructor is created by providing the name and symbol of token.
- Set the initial token supply: The initial supply of tokens that will be minted for the contract is defined. This initial supply will be transferred to the course instructors who will then transfer the tokens to the students on successful course completion.
- Deploy contract: Compile and deploy the contract on the blockchain network to make it available for use. Then, course instructors can transfer tokens to students upon successful course completion using the deployed contract.

In order to transfer digital credits to students, the instructor first logs into the web portal using their MetaMask account, then specifies the student's public address and the number of credits to be transferred. The web portal then creates a transaction which includes the details of the credit transfer, such as the recipient's address and the number of credits. The transaction is signed using the course instructor's private key stored in his/her MetaMask account. The instructor needs to pay a transaction fee in Ether, the native cryptocurrency of the Ethereum network, before the transaction is recorded on the blockchain network and the student's credit balance is updated to reflect the transfer of credits.

4 RESULTS AND DISCUSSION

In this paper, we have implemented a decentralized platform for higher education that assists students in skill development, facilitates permanent storage of academic records and digital credit transfer. The web portal in this system provides a platform for students and course instructors to interact and participate in the decentralized learning environment. The users can choose to login as either a student or a course instructor.

- Login as student – Users can browse, register, and pay for courses on the dash board, accessing course materials upon registration.
- *Login* as course instructor – Course instructors can post courses, set fees, and make learning materials available to interested students. They can track student progress and award digital certificates and credits upon completion.

A digital certificate is generated followed by the course instructor providing the student's credentials to acknowledge their active participation in the course. The portal created also allows users to validate the certificates using the unique CID provided by the IPFS.

The proposed system does have a few drawbacks.

- Difficult to assess this sort of learning activities using pre-programmed smart contract without human mediation.
- Immutability: This can be both a benefit and a drawback, as it prevents fraudulent changes to educational records but also makes legitimate changes impossible.
- Opposition from central authorities: There is a lack of clear regulatory guidance regarding its use in the higher education sector.

In order to address these challenges, a potential solution is to develop standardized learning objectives and assessment methods for consistent assessment of student learning. It is also important to engage with central authorities to gain their support and educate them about blockchain's potential benefits.

5 CONCLUSION AND FUTURE INSIGHTS

The proposed system that uses blockchain to connect students with the course instructors in a decentralized manner without any intermediaries is a promising development in the field of higher education. The use of smart contracts for creating and storing digital certificates on IPFS and for the creation and transfer of digital credits to students for successful course completion has the potential to improve the accessibility, affordability, and transparency of higher education.

The future scope the system entails to include further development and integration of blockchain technology in research collaboration, administration of scholarships and financial aid.

REFERENCES

[1]   Chen, G., Xu, B., Lu, M. *et al.*, "Exploring Blockchain Technology and Its Potential Applications for Education," *Smart Learn. Environ.*, pp. 1–5, 2018.
[2]   Srivastava A., Bhattacharya P., Singh A., Mathur A., Prakash O. and Pradhan R., "A Distributed Credit Transfer Educational Framework Based on Blockchain," pp. 54–59, 2018 *Second International Conference on Advances in Computing, Control and Communication Technology.*
[3]   Turkanovic M., Holbl M., Kosic K., Hericko M. and Kamisalic A., "EduCTX: A Blockchain-Based Higher Education Credit Platform," in *IEEE Access*, vol. 6, pp. 5112–5127, *2018.*
[4]   Afrianto, Irawan, and Yayan Heryanto, "Design and Implementation of Work Training Certificate Verification Based On Public Blockchain Platform." *2020 Fifth International Conference on Informatics and Computing (ICIC).*, *IEEE, 2020.*
[5]   Bahrami, Alireza Movahedian, Arash Deldari, "A Comprehensive Blockchain-based Solution for Academic Certificates Management Using Smart Contracts", *10th International Conference on Computer and Knowledge Engineering, IEEE, 2020.*
[6]   Zhou, Makhdoom I., Abolhasan M., Lipman J. and Sharıatı N., "A Blockchain-based File-sharing System for Academic Paper Review," *13th International Conference on Signal Processing and Communication Systems (ICSPCS)*, pp. 1–10, 2019.
[7]   Jain R., Sarasvathi V. (2021). Profile Verification Using Blockchain. In: Thampi S.M., Lloret Mauri J., Fernando X., Boppana R., Geetha S., Sikora A. (eds) *Applied Soft Computing and Communication Networks. Lecture Notes in Networks and Systems*, vol 187. Springer, Singapore.
[8]   Yadav N. and S. V., "Venturing Crowdfunding using Smart Contracts in Blockchain," *2020 Third International Conference on Smart Systems and Inventive Technology (IC SSIT)*, 2020, pp. 192–197, doi: 10.1109/ICSSIT48917.2020.9214295.
[9]   Nithin M., Shraddha S., Vaddem N. and Sarasvathi V., "HyperIoT: Securing Transactions in IoT through Private Permissioned Blockchain," *2020 IEEE International Conference on Electronics, Computing and Communication Technologies (CONECCT)*, 2020, pp. 1–6, doi: 10.1109/ CONECCT50063.2020.9198474.
[10] Haugsbakken H. and Langseth I., "The Blockchain Challenge for Higher Education Institutions," *Eur. J. Educ.*, vol. 2, no. 3, pp. 41–46, 2019.

Recent Trends in Computational Sciences – Gururaj, Pooja & Flammini (Eds)

Document verification using blockchain

Danish Hashmi, Atul Anurag, Reyyala Chethan, Sahitya Sagiraju & V. Sarasvathi
PES University

ABSTRACT: A student produces several certificates over their entire academic career, whether they are in high school, college, or even a postgraduate program. These certificates could include results, diplomas, or transcripts. Students must provide these certificates to universities or employers to be admitted. It becomes tiresome to track these certificates and manually verify their legitimacy. The lack of a suitable anti forge system results in a situation where it is discovered that the diploma is forged. Everything needs to be digitalized with the principles of Confidentiality, Reliability, and Availability to make the data more secure and safe. Our proposed system will use the Blockchain technology along with the Inter Planetary File System (IPFS) to solve this problem. The flow of our system will, in a nutshell, include a Certificate Issuer who will upload documents along with student details, onto the IPFS, which breaks the file into multiple parts and returns a CID as an IPFS link. The uploading of the document involves a transaction performed by the issuing authority that is being recorded on the blockchain. Each document will have a distinct IPFS link that may be utilized by any official through the portal to verify the document's authenticity. Our system will have the advantage of reducing the possibility of a student losing or destroying or forging a document and making document verification simple.

1 INTRODUCTION

The issue of the veracity of digital information can potentially be solved with the help of blockchain, a recently created revolutionary technology. Currently, the blockchain which is majorly used by cryptocurrencies is thought of as a distributed ledger on which anybody can initiate a transaction without the aid of middlemen, and everything is carried out independently by the miners. Paper [8] gives detailed information about using permissioned blockchain which is used to limit the access to the blockchain so that only approved members can add a block. This helps in minimizing the security risks associated with giving access to the public.

Blockchain can also be used to store data on it, however, blockchain is a costly way of storing data, particularly huge data, due to complex mathematical calculations. To overcome this problem, IPFS can be implemented for the effective storing of vast amounts of data and content. IPFS is a framework based on blockchain technology for storing data and files with integrity and resilience that is distributed and decentralized. In its most basic form, IPFS is an open source, peer-to-peer, widely dispersed file system that can be utilized to store and share volumes of files quickly. The IPFS returns a CID (Content Identifier) which is a combine hash of all the hashes, which makes it easy to access the document and to give authenticity and traceability. Our suggested approach uses both IPFS and blockchain smart contracts. The hash of a digital document changes whenever its contents change, indicating that the original content has been changed and amended. This is collision-free, which means that any two different values will always produce two different hashes.

Certificates that are crucial to a person's skill set are frequently the target of forgeries. To identify between a phony certificate and an actual certificate will require a lot of focus and

DOI: 10.1201/9781003363781-15

time. Blockchain technology can overcome this drawback because it can't be altered in a typical situation. If the data is altered, it only takes a moment to alert the change. As a result, the system would always be authenticated and reliable. This paper aims to work on a blockchain based platform that allows a modern and hassle-free solution to manage certificates, verify them and keep its integrity.

2 LITERATURE SURVEY

The objective of paper [1] was a detailed explanation of the blockchain and its applications and the terms related to blockchain, the methodology they used was research on the top to down stack of the blockchain technology and uses. The authors in paper [2] have proposed a system that uses the blockchain's immutable characteristics to make it possible to issue digital certificates. The main problem with the system implemented in paper [3] is validating and verifying a person's birth record. In paper [4] The document is split into multiple minuscule data objects. All the hashes are rehashed into a single hash key, and it is being provided to the authority. This creates an immutable data object which can be only accessed by the user with the hash key. Block IPFS provides data integrity, and an alternate solution for a decentralized data storage. The authors in paper [5] have proposed a blockchain-based identity verification system in which verifying user details and the certificate issuing authority plays a crucial role. The system uses Proof of Work consensus algorithm to verify and upload the document onto the blockchain, which makes it very costly to implement as it will require heavy machinery for the miners to verify the transaction and add it to the block. The system architecture proposed in paper [6] contains two entities, student, and mentor. The student will be given an option to upload the image of the certificates, and which will then be put on wait for approval from the respective mentor under which the certificate was issued. On successful verification, the certificate will then be hashed using MD5 algorithm and then added on to the blockchain. The drawback with this system is that it requires 3 validators i.e., the mentors to verify the certificate each time a student uploads a certificate which makes it a very tedious approach for a variety of certificates issued by multiple institutions and uploaded by numerous students. Paper [7] provides the information about implementation of smart contracts in the blockchain technology, using which, the system can be automated to execute the functions on receiving valid inputs from the user or when certain tasks are executed on the blockchain.

3 PROPOSED METHODOLOGY

Our proposed system uses Blockchain in connection with IPFS. In this system, the document uploaded is divided into smaller chunks, which are then cryptographically hashed and stored into the peer-to-peer network of nodes by the IPFS. The combined hash of all these hashes is then stored into the Blockchain with a consensus forming via the proof of stake algorithm. The main advantages of this system are that it protects the document's integrity, data is stored based on a P2P network, hence chances of data loss and data compromising is very less, and the last important advantage is that we do not require any heavy computational machinery to add a block or put up any high stakes as well.

Figure 1 portrays the basic system architecture of the system which comprises 3 entities, university admin, student, and recruiter. The rights to upload the certificate will be reserved for admin who will upload the student details along with the certificates onto the IPFS. The IPFS link returned after successful upload will then be stored against the unique student id.

Figure 2(a) Gives a detailed visualization about the flow of the system. The users can sign up and login via the credentials which they have entered. The credentials are stored in the MongoDB database, which is a NoSQL database in which the schema can be altered as per the user's requirements. The college admin logs in using their credentials and uploads the

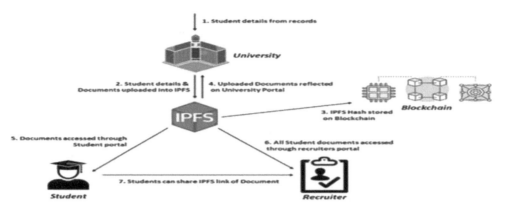

Figure 1. System design.

certificates. Students and Recruiters will be able to view and download the certificate using the IPFS link. The recruiter can filter the students with the help of student ID or based on their departments as well.

4 IMPLEMENTATION AND RESULTS

4.1 *Technologies used*

- **Blockchain** – served as the basis for the entire project, is best understood as an immutable database. It offers a reliable setting where what has been stored is concise and unchangeable.
- **IPFS** – A distributed file system can be used to store and share data utilizing the Inter Planetary File System protocol and peer-to-peer distributed network.
- **Smart Contract** – programs that are saved on a blockchain and are only activated under specific circumstances. In most cases, they are utilized to automate contract execution so that all parties can be confident in the result right away, without the requirement for an intermediary or additional time.
- **Solidity** – The object-oriented programming language Solidity is used to create smart contracts. It is applied when putting in place smart contracts on different Blockchain systems. Although it has more particular data types, it is very similar to Typescript.
- **MetaMask** – With the help of the MetaMask browser plugin, distributed programs, or Dapps, are used on the browser. To enable Dapps to read from the blockchain, the add-on injects the web3 API into the JavaScript context of every webpage.
- **Node.js** – used to create backends and oversees managing user authentication, serving frontend assets, and serving frontend pages. Additionally, it enables web3, which in turn enables to run the solidity code.
- **Mongo DB** – Users' personal information, session data, and encrypted authentication records are kept in MongoDB, which is a NoSQL database.

4.2 *IPFS*

The fundamental steps in the upload and access of files via IPFS are displayed in the Figure 2(b) below. The user uploads the file, and the file is being sent to IPFS. In IPFS it is broken down into multiple parts and the hash of each of the fragments is combined into a single CID which is being returned to the system for storage.

100

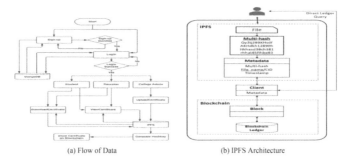

(a) Flow of Data   (b) IPFS Architecture

Figure 2. IPFS.

4.3 *Web interface*

Initially, the users will be shown a landing page from which they can redirect to the home page where a brief introduction about the system can be found. From there, if they are a new user, they can sign up, else they can just log in.

Once they log in, they will be redirected to their respective profile page i.e., based on their user type which is student, admin, or recruiter. Accordingly, a different page will be rendered for different user type.

The admin will be able to upload the certificate along with the student details as shown in the Figure 3(a) such as SRN, College name, document description, and the Document to be uploaded.

Once the transaction is sent through on the Polygon Mumbai Testnet, the document will be uploaded and will be available to view and download them. The details of the individual transaction can also be verified on the polygonscan official website by entering the trans-action id as displayed in Figure 3(b).

(a)   (b)

Figure 3. (a) Admin Uploading Documents (b) Transaction Details on PolygonScan.

The Polygon Mumbai Testnet is a layer-two (L2) scaling platform for Ethereum. The testnet replicates the Polygon mainnet. It enables developers to deploy, test, and execute their dApps in the blockchain environment risk-free and at no cost. Like Polygon, which launched in 2017, this testnet also uses the proof-of-stake (PoS) consensus mechanism to agree upon the state of the blockchain. It offers very high throughput and extremely low transaction fees while leveraging Ethereum's security.

In the student portal they can view all their details, along with certificates uploaded by the admin which are attached to their Unique id.

The recruiter can view all the students' certificates which are there in our system. They also have an option to search for a particular student using the unique id or filter them with respect to the departments. The recruiter will also have an option to verify the certificate using the IPFS link which has been shared to them by the verify option.

The code responsible to verify the certificate is described by the algorithm below-

Step 1 – The function to verify a certificate is invoked if the IPFS link of the corresponding certificate is entered in the input field.

Step 2 – The query i.e., the CID of the certificate in question is passed to a find function.
Step 3 – If the function can find a node corresponding to the CID of the certificate, it flashes a success message saying, "Document Verified."
Step 4 – If the function fails to find a node corresponding to the CID, it flashes an error saying, "Document Not Verified" which mean the certificate is not stored on the blockchain, or it has been tampered.

5 CONCLUSION

The ability to create immutable ledgers is one of the Blockchain's primary features. This behaviour assists us in creating a system where every step of the process is transparent, unalterable, and easily traceable. The amount of manual and computational labor required to do so is also very minuscule since the consensus algorithm used is Proof of Stake. Additionally, there is a comparatively low chance of for the loss of certificate by the students since the document is broken down into smaller chunks, cryptographically hashed and scattered across a peer-to-peer distributed network powered by the Inter Planetary-File-System which is a distributed file storage protocol.

Hence, the system provides the necessary tools required to upload, store, and verify the important documents all while maintaining the integrity of the document which is of the utmost significance in the scope of document verification.

The proposed model is tamper-proof and works well against all database attacks to modify the student academic details since all the documents are stored on IPFS whose CID are stored on Blockchain. So, any attempt on altering the document will result in subsequent change in the hash value which will not be allowed by the blockchain. The model is immune to the 51% attack, since the right to upload, the document is reserved to the college admin only. Hence, even if someone acquires the testnet MATIC, they won't be able to upload new documents. It is also protected against the sybil attack since the users are not permitted to create an admin account and execute transactions on the blockchain.

REFERENCES

[1]   Zibin Zheng, Shaoan Xie, Hong-Ning Dai, Xiangping Chen, "An Overview of Blockchain Technology: Architecture, Consensus, and Future Trends", *IEEE 6th International Congress on Big Data*, June 2017.

[2]   Jiin-Chiou, Narn-Yih Lee, Chien Chi, YI-Hua Chen, "Blockchain and Smart Contract for Digital Certificate", *Proceedings of IEEE International Conference on Applied System Innovation*, 2018.

[3]   Maharshi Shah, Priyanka Kumar, "Tamper Proof Birth Certificate Using Blockchain Technology", International Journal of Recent Technology and Engineering (IJRTE), Volume 7, Issue-5S3, February 2019.

[4]   Emmanuel Nyaletey, Reza M. Parizi, Qi Zhang, Kim-Kwang Raymond Choo, "Block IPFS – Blockchain-enabled Interplanetary File System for Forensic and Trusted Data Traceability", *IEEE International Conference on Blockchain*, 2019.

[5]   Gunit Malik, Kshitij Parasrampuria, Sai Prasanth Reddy, Dr. Seema Shah, "Blockchain Based Identity Verification Model", *International Conference on Vision Towards Emerging Trends in Communication and Networking (ViTECoN)*, 2019.

[6]   Jain, R., Sarasvathi, V., "Profile Verification Using Blockchain", In: Thampi, S.M., Lloret Mauri, J., Fernando, X., Boppana, R., Geetha, S., Sikora, A. (eds) *Applied Soft Computing and Communication Networks*. Lecture Notes in Networks and Systems, vol 187. Springer, Singapore, 2021.

[7]   Yadav, N., V., S. (2020). "Venturing Crowdfunding Using Smart Contracts in Blockchain", *Third International Conference on Smart Systems and Inventive Technology (ICSSIT)*, 2020.

[8]   Nithin, M., Shraddha, S., Vaddem, N., Sarasvathi, V., "HyperIoT: Securing Transactions in IoT through Private Permissioned Blockchain", *2020 IEEE International Conference on Electronics, Computing and Communication Technologies (CONECCT)*, 2020.

Recent Trends in Computational Sciences – Gururaj, Pooja & Flammini (Eds)
© 2024 The Author(s), ISBN 978-1-032-42685-3

Blockchain-based traceability system for readymade food products

G.U. Disha, H.S. Dharthi Gowda, J. Priya, H.S. Srushti & M.S. Usha
NIEIT

ABSTRACT: Customary recognizability framework have issues of incorporated adminis-tration, murky data, dishonest information and simple age of data islands. To take care of the above issues, this paper plans a recognizability framework given blockchain innovation for capacity and question of item data in-store network of instant food items. Utilizing the qualities of decentralization, sealed and discernibility of blockchain innovation, the straightforwardness and believability of detectability data expanded. The developing number of issues connected with sanitation and tainting gambles has laid out a massive requirement for viable detectability arrangement that goes about as a fundamental quality administration apparatus guaranteeing satisfactory security of items in the instant food store network. Blockchain is a troublesome innovation that can give an imaginative answer for item discernibility in food supply chains. The current food supply chains are a convoluted climate including a couple of accomplices making it blundering to endorse a couple of huge standards like the country of starting, stages in snacks improvement, conformance to quality rules, and screen yields. In this paper, we propose a system that impacts the blockchain and savvies successfully perform bargains for moment food thing following and perceptibility across the food creation organization.

Keywords: Traceability, Food Safety and Standards Authority of India (FSSAI), Blockchain, Tamperproof, Food, Supply Chain, Decentralization, Food Safety, Transparency, Security

1 INTRODUCTION

Blockchain innovation has acquired enormous ubiquity among store networks and coordi-nated operations in local areas due to straightforwardness and unchanging nature of exchanges improving trust among partaking partners. Because of its sealed, trusted, secure and recognizable nature, blockchain can be conveyed really in the food production network on the board. The reception and execution of blockchain innovation can likewise guarantee instalment security to merchants while sharing significant rules about the beginning, accreditation of natural or non-GMO (Genetically Modified Organisms).

Checking the improvement of instant food items and effective coordinated operations on the board in the food store network is basic to guarantee item security. The utilization of exact information assortment through data specialized devices, for example, scanner tags and RFID empowers information securing and better detectability in food supply chains. Individual stages in food supply chains frequently have great detectability yet the trade of data between stages ends up being troublesome and tedious.

The general design and working of the food store network are huge and complex including various partners going from producers, processors, and buyers. Consequently, there is a need to make a protected structure for following insights concerning the beginning, cultivating strategies embraced, and wellbeing of the food item all through the production network cycle without an outsider control.

Late innovation improvements through the utilization of blockchain innovation can give a significant and useful arrangement guaranteeing detectability of food production and killing the requirement for a confided in unified power. We present a blockchain-based arrangement and structure for detectability and permeability in the food store network. We talk about and feature key parts of our blockchain arrangement as far as the general framework plan and design, including principal communications among the fundamental members, with substance relations and grouping outlines. We present, and execute, SHA calculations that oversee and guarantee the appropriate connections among key partners in the food production network.

2 RELATED WORK

Detectability is portrayed as the ability to get to any or all information interfacing with that which is getting checked out, all through for as far back as it can recall cycle, through recorded distinctive bits of evidence. Perceptibility has transformed into a test for all food and food related associations, and the perceptibility system has transformed into a strong strategy for the quality organization in the food store network.

Blockchain as an emerging development that has properties of decentralization, fixed and obviousness gives the probability to handle the issues existing in the force traditional readymade food thing perceptibility system. The plan of readymade food things blockchain obviousness structure was essentially divided into limit layer, organization layer, interface layer and application layer. The blockchain system stores the mixed ciphertext of the secret information and the hash worth of the public information to resolve the issues of speedy data questions of blockchain. The help layer integrates data examination, reputation based smart understanding, key organization, key endorsement, PBFT arrangement framework, etc. The mark of communication layer is made from sagacious agreements, which basically do the data move blockchain structure and data question capacities.

The principal motivation behind the ongoing paper is to make sense of how to apply blockchain innovation to the detectability of Readymade food items. The paper utilized the strategies of Consensus Mechanism, On-Chain, and Off-chain information capacity innovation. It enjoys the benefit of three likelihood limits, hypergeometric and binomial conveyances, and burdens of adaptability and decentralization. The paper utilized the strategy of Systematic writing planning. It enjoys the benefit of Tracking/Tracing, Fraud Prevention, Inventory Management 2 Collaborative SCM, Financing and drawbacks of Information Asymmetry, Unknown Provenance, Counterfeit, IT Security, Connectivity.

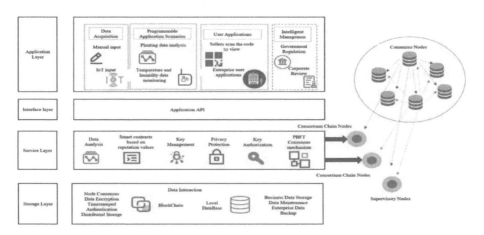

Figure 1.   Blockchain-based traceability system architecture.

3 PROPOSED METHODOLOGY

Proposed System: The proposed plan revolves around the use of blockchain to regulate and control all interchanges and trades among all of the individuals expected inside the store network climate. All trades are recorded and taken care of and subsequently giving every one of the raised levels of straightforwardness and detectability in the creative network climate in a strong, trusted and reliable way. The food creation network organic framework trade nuances has given SHA estimation, AES Rijndael computation and blockchain-based set aside in informational collection.

Methodology: The proposed framework gives deals to pressed food items following, detectability and straightforwardness across the food production network biological system in a protected, trusted, solid, and effective way.

1. *System Manager*: This is the approval that grants admittance to the items with the fixings and hence gives verification to the makers to deliver. Thus, forestalls robbery and vagueness in additional means.
2. *Manufacturing Company and FSSAI*: Manufacturing organization is the substance that gives items to a retailer on request. Alongside items, this element likewise gives different supplements that help pressed food creation.
3. *Distributor*: The Distributor's job is to buy the handled items from the maker and appropriate them to retailers. With the assistance of the proposed framework, the merchant will want to pick the loaded food items effortlessly and trust as the entire history of the items will be made open to him.
4. *Retailers*: The Retailer purchases the item from the merchant in mass and further offers it to the purchaser in more modest amounts.
5. *Customer*: The client is the end client who buys the items from the retailer. At this stage, the client will approach every one of the information about the item.

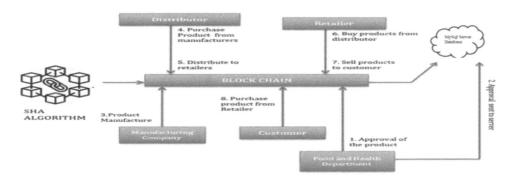

Figure 2. Block diagram for traceability of readymade food products.

4 SYSTEM DESIGN IMPLEMENTATION

The production network environment for bites can be gotten to by any programmer thus there is no security for such records. There ought to be an answer adjusted to build the security of this exchange, consequently lessening the possibilities of hacking.

Framework Manager is the approval that grants admittance to the items with the fixings and consequently gives verification to the makers to create. The Distributor's job is to buy the handled items from the producer and convey them to retailers. With the assistance of the proposed framework, the merchant will want to pick the loaded food items effortlessly and trust as the entire history of the items will be made open to him. The Retailer purchases the item from the merchant in mass and further offers it to the customer in more modest amounts. With the proposed framework, the retailer will have more choices concerning how new they maintain that the items should be as the retailer should watch out for the timeframe of realistic usability of the items. The client is the end client who buys the items from the retailer. At this stage, the client will approach every one of the information about the item. This gets straightforwardness to the client which expands the trust and dependability of every partner.

Algorithms Used: SHA 256 (Secure Hash Algorithm): SHA-256 is a licensed cryptographic hash capability that outputs a value that is 256 pieces in length. In encryption, information is changed into a protected configuration that is unintelligible except if the beneficiary has a key. The hash code of a safeguarded record can be posted uninhibitedly so clients who download the report can confirm they have a trustworthy structure without the things in the record being uncovered. Taking care of clients' passwords in a plain-text record is a fiasco in the works. That is the explanation. Exactly when a client enters a mystery word, the hash is not entirely set in stone and subsequently differentiated and the table. In case it matches one of the saved hashes, it's a genuine mystery state and the client can be permitted permission.

AES (Advanced Encryption Standard) Rijndael Algorithm: Rijndael is an iterated block figure, implying that it encodes and decodes a block of information by the emphasis or round of a particular change. It upholds encryption key sizes of 128, 192, and 256 pieces and handles information in 128-cycle blocks.

Working of Rijndael: In Rijndael, encryption occurs through a progression of framework changes or adjustments. The number of rounds are variable, contingent upon the key or block sizes utilized: 128 pieces = 9 rounds, 192 pieces = 11 rounds, 256 pieces = 13 rounds

The Rijndael calculation depends on byte-by-byte substitution, trade and XOR activities. The technique is as per the following: The calculation creates 10 128-digit keys from the 128-bit key, which are put away in 4 × 4 tables, The plaintext is separated into 4 × 4 tables, every one of 128-digit sizes. The code is created after the tenth round.

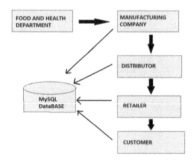

Figure 3.   Flow diagram of our model.

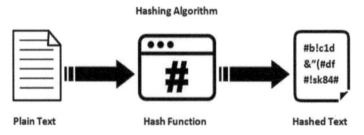

Figure 4.   SHA-256 algorithm.

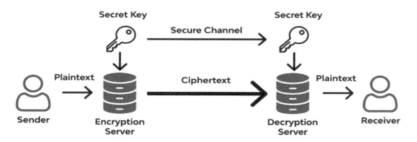

Figure 5.   AES Rijndael algorithm working.

5   APPLICATIONS

- Increased food Safety: Foodborne diseases are a preventable and underreported general medical issue. These diseases are a weight on general well-being and contribute fundamentally to the expense of medical care.
- Better quality control: Supply chain quality administration is the cycle through which organizations evaluate, screen, and proactively deal with the nature of items and cycles in the store network.
- Better inventory visibility and tracking: Empowers carriers to additionally foster client backing and cost controls through the organization of stock moving and proactive.
- Better customer service: It's tied in with giving clients what they need, when they need it, in the most ideal way.
- Better ability to meet customer needs: Can be accomplished by finding and utilizing the right information, All-around commitment, expanding personalization, comfort and straightforwardness.

6   CONCLUSION

We have planned and executed the detectability arrangement of readymade food items because of the non-altering and recognizable qualities of SHA calculation, AES Rijndael calculation and blockchain assists with meeting genuine business needs. The proposed game plan clears out the necessity for a trusted concentrated power, and goes between and gives trades records, updating efficiency and prosperity with high uprightness, enduring quality, and security. The reason for this application is to exhibit the way that our response can be applied to endlessly follow the store network for squeezed food things. Regardless, the presented perspectives and nuances are adequately nonexclusive and can be applied to give trust and perceptibility to snack things in the creation organization.

REFERENCES

[1]   Chuanheng Sun, Minting Wang, and Xinting Yang, "A Trusted Blockchain-based Traceability System forfruits and Vegetable Agricultural Products," *Phytochemistry*, vol. 173, March 2021.
[2]   P. Zhu, J. Hu, Y. Zhang, and X. Li, "A Blockchain Based Solution for Medication Anticounterfeiting and Traceability," *IEEE Access*, vol. 8, pp. 184256–184272, 2020, doi: 10.1109/ ACCESS.2020.3029196.

[3]   Abdelatif Hafid, Abdelhakim Senhaji Hafid, Mustapha Samih, "New Mathematical Model to Analyse Security of Sharding-Based Blockchain Protocols," *Food Control*, vol. 30, no. 1, pp. 341–353, Mar. 2013.

[4]   Tan Guerpinar, Gilberto Guadiana, Philipp Asterios Ioannidis, Natalia Straub, "The Current State of Blockchain Applications in Supply Chain Management," *Food Control*, vol. 39, pp. 172–184, May 2014.

[5]   K. Christidis and M. Devetsikiotis, "Blockchains and Smart Contracts for the Internet of Things," *IEEE Access*.

[6]   *Food Safety Law of the People's Republic of China, Order No. 21 of the President of the People's Republic of China C.F.R.*, Standing Committee NPC, Beijing, China, 2009.

[7]   Li X., Lv F., Xiang F., Sun Z., and Sun Z., "Research on Key Technologies of Logistics Information Traceability Model Based on Consortium Chain," *IEEE Access*, vol. 8, pp. 69754–69762, 2020.

[8]   Qian J.-P., Yang X.-T., Wu X.-M., Zhao L., Fan B.-L., and Xing B., "A traceability system incorporating 2D barcode and RFID technology for wheat flour mills," *Comput. Electron. Agricult.*, vol. 89, pp. 76–85, Nov. 2012, doi: 10.1016/j.compag.2012.08.004.

[9]   Christidis K. and Devetsikiotis M., "Blockchains and Smart Contracts for the Internet of Things," *IEEE Access*, vol. 4, pp. 2292–2303, 2016, doi: 10.1109/ACCESS.2016.2566339.

[10]   Salah K., Nizamuddin N., Jayaraman R., and Omar M., "Blockchainbased Soybean Traceability in Agricultural Supply Chain," *IEEE Access*, vol. 7, pp. 73295–73305, 2019, doi: 10.1109/ACCESS. 2019.2918000.

A cloud based interactive framework for emergency medical data sharing

Ayesha Taranum
Department of Computer Science and Engineering, Vidyavardhaka College of Engineering, Mysore, Karnataka, India
ORCID ID: 0000-0002-3171-6656

S. Manishankar
Department of Computer Science and Engineering, Sahrdaya College of Engineering and Technology, Kodakkara, Thrissur, India

K. Padmaja
Department of Computer Science and Engineering, GSSS Institue of Engineering and Technology for Women, Mysore, Karnataka, India

C. Balarengadurai
Department of Computer Science and Engineering, Vidyavardhaka College of Engineering, Mysore, Karnataka, India

ABSTRACT: Pandemic bringing a change in the medical system and medical infrastructure. This requires a complete revamping of medical data collections and storage. In such a scenario there has to be a system which enables the efficient transport of a patient to a distant hospital with minimal human support. In such a remote telehealth service, the inclusion of a system which is capable of sending and receiving medical data like live image of patients, X-Rays and multimedia images would result in better and fast pre-hospitalization. In such an emergency situation, the information has to be exchanged between the ambulance and the doctor in the least time possible. An efficient image compression algorithm will ensure the faster communication. The proposed framework is a combination of mobile application an interactive OpenTok API, cloud based platform Heroku and Firebase. The proposed system ensures a dedicated high speed data sharing environment based on cloud and mobile API to make an interactive framework for Healthcare Industry.

Keywords: Remote telehealth, Image compression, Cloud computing, LRJPEG, OpenTok SDK

1 INTRODUCTION

Cloud computing is a model for enabling convenient, on-demand network access to a shared pool of configurable computing resources (e.g., networks, servers, storage, applications, and services) that can be rapidly provisioned and released with minimal management effort or service provider interaction. A 'cloud' is basically a collection of resources (e.g., CPU, storage) which are provided as utilities that the user can request and access through the internet in an on demand fashion. 75% of the cloud comprises of data i.e., cloud mostly comprises of storage. This helps the client to access his/her data whenever he/she wishes to do so. This reduces the complexity of storing the client's data within his/her computer or system. A client might choose to store valuable data in the cloud. Therefore, the CSP (Cloud service provider) has to make sure that his/her cloud services are highly available so that the client can access it whenever he/she wants. Our target is to set up a highly available and reliable server which can be accessed by clients on multiple platforms. A highly reliable server can help the users to connect and exchange information on demand. This project can be implemented in real time to set up a connection which will have the maximum throughput and bandwidth. Also clients on multiple platforms (e.g., Android, IOS, Windows etc.) can connect to the server and exchange information. Applications can be deployed on the server and it can be accessed any time when the client wants. We can allot storage quotas to users where users can connect to the server and access information any time they want. This is very efficient because such an 'on demand access' will enable the user to carry his files on the go as long as he has a proper internet connection. There are several APIs which allow the transmission of video and audio over the internet. Most of them have a free to use version where they provide most of the services they offer. To get the other services, the customer has to pay. All the leading APIs provide inbuilt tools which provide in depth statistics and analytics of the customers' deployed apps. This is really helpful for the customer as he can use this information for the betterment of his app. The API of choice that we have used here is OpenTok. OpenTok is a free to use API which enables the user to make use of the OpenTok library and allow the transmission of video as well as audio. They also provide in depth statistics like up time, minutes streamed (publisher, subscriber or mediator), The country from where most of the stream is uploaded and received. Such in depth analysis is not available with the other open source APIs that we have tried to implement. In this paper authors propose a framework which allows both the paramedic and doctor to login to an android application which allows the live transmission of video, compression and transmission of images.

2 LITERATURE SURVEY

In [1] the basic architectures of the cloud computing system are discussed in brief. They are hardware layer, infrastructure layer and application layer. The author has also briefly described about past and current technologies like grid computing, utility computing and virtualisation. There are brief discussions about public clouds, hybrid clouds and virtual clouds. In [2] there is information about a workshop which was conducted in the year 2008. The hot topic in LADIS 2008 was cloud computing. This laid a foundation for many researchers and scholars to come forward with their ideas and thoughts about data centric cloud storage as well as cloud computing. In [3] the author does an experiment where he creates and sets up a server environment and pushes many data sets with different delays to it. The environment of choice was OpenNebula. The experiment was conducted to test the latency and efficiency of the server environment. In [4] the author discusses about the four types of cloud namely, private clouds, public clouds, virtual clouds and hybrid clouds. This is more like a survey paper where the author has described about the evolution of cloud computing. Several advantages and disadvantages to the cloud environment are also mentioned. In [5] the author proposes an efficient method to handle the traffic and balance the workload between multiple nodes easily. This will be more applicable in a distributed environment consisting of a large number of computers. The author proposes an algorithm for this. The algorithm consists of four sub modules. The first module is coordinator algorithm which receives the status of all the nodes in the network. The second module is data placement which calculates the weights for each node after receiving the status from them. The third module is load rebalance which periodically checks if the nodes are heavily loaded with the processes or not. They identify the heavier ones and the less heavy ones. The fourth module is data migration which transfers the processes from the heavily loaded node to the node with less load in it. This will increase the performance as the loads are evenly distributed across the network. In [6] the author talks about most of the popular cloud computing platforms like Openstack, Amazon Web services etc. It is a general comparison on from the perspective of a general user. In [7] the author talks about the traffic issue that all cloud service providers face. The issue is the huge traffic while pushing voluminous amount of data. The authors propose a technique in which an abstract layer named UDS is kept between the user and the cloud server to batch files in an efficient manner.

In [8] the author proposes a system which is deployed in Firebase, uses Google maps API and MySql database to form a framework which can be used to get real time data of the check up between the doctor and the patient. Maps API is used to extract the position of the doctor. The entire database is deployed on the server and the user can access it whenever he wants to. In [9] the author considers some of the most popular cloud storage services such as Google Drive, Dropbox,

Cloud drive, Mega and some private cloud servers such as Wuala, Horizon etc. The services are analysed and benchmarked on different basis such as download, uplink, Chunking, Client-side deduplication, Compression. But as a whole, the authors have considered time as the baseline factor as most of the users won't be aware of the in depth benchmark statistics. The distance to the server is also a major factor as the ping might vary as the distance varies. The authors have concluded by saying that in most of the test cases Onedrive has shown the best results. In [10] the author compares and contrasts public and private cloud services like Google Drive, Dropbox, Microsoft Azure etc. The measurement is done on the basis of download, uplink, client side duplication, compression etc. Out of all the tests that the authors had done, Microsoft Onedrive has produced the optimal results. In [11] the author discusses and offers a few partial solutions to make WebRTC more dynamic. WebRTC is an open source and free project which allows the developer to implement video and audio calling via APIs (Application interface programmes). In short, the author proposes partial solutions to manage session across borders. In [12] the author proposes a system which binds patients and doctors together. The patients are able to login and book appointments with doctors. This system makes use of Firebase API for authentication, RDBMS for database and Opentok API has been implemented to allow remote access or to video conference with the doctor. The system is developed as an Android application. In [13] the author focuses on peer to peer (P2P) live video streaming services. When we connect two or more computers and allow the sharing of resources between them such that there is no need for a third server, we call it a peer to peer (P2P) connection. There are two types of video streaming services, they are: Live & On demand. Live streaming is a synchronous connection whereas On demand is a non synchronous connection. The author says that the quality of experience is limited compared to a traditional TV and satellite channel. In their comparison, P2P takes a longer start up and there is a considerable delay in channel. In [14] the author conducts a survey by taking 36 medical scenarios. A scenario where the patient would be treated in the clinic and via video conferencing. The authors conclude by saying that in most of the scenarios, the economical value of a hospitalization via video calling is high. In [15] the author proposes an efficient compression algorithm which is an improvement over JPEG(Joint photographic experts group). The MSE (Mean squared error), CR (Compression ratio) & PSNR (Peak signal-to-noise ratio) is calculated. Both colour and grayscale images are used as test inputs. The default quantization table(8×8) is improved. In [16] the author proposes a framework or a system which allows the efficient transmission of information (text & multimedia). The author also implements the LRJPEG compression algorithm from [15] in order to efficiently compress and send the image across the internet. The system makes use of Firebase API in order to authenticate a user and to store the information he/she sends. In [17] the paper covers a health care frame work which can merge the data at various levels and synchronize them to the Cloud engine developed. It is a integrated framework with helpful design and features that support the system working. In [18] the details covered the journal are about how cloud computing facilitates the various types of data in connection to healthcare and including storage, analytics and deduction of intelligent inferences from the data stored related to healthcare. The paper also gives the various progressive approached of cloud computing techniques in Health care sector and its future.The cloud and IoT based systems also acts as a backbone for the healthcare services which greatly helps in developing an Information system which helps in collecting, aggregating and migrating the data with proper E-health application engine [19]. The cloud based system works as risk prevention system by analyzing the data and then taking a decision using the proposed frame work. The system has a proper data processing module and data analysis module with machine learning which works as the decision support system [20] The privacy friendly approaches for the cloud frame work with the help of block chain environments are also in consideration for developing the systems where block chain ensures the privacy of the highly secure and important data. [21]. The literature survey thus covers many similar system which collects the health data and use cloud as platform to process and store it. But there arises a need for efficient transfer of the data especially when it comes to images and videos. The proposed work focuses on the compression of the images and transferring them efficiently to support the Cloud based system to process the results and make decision easily. The next section of the paper discusses on various types of tools that are used in building the proposed framework. Institutions are to be listed directly below the names of the authors.

3 TOOLS USED IN THE DEVELOPMENT

3.1 *LRJPEG*

A luminance based modification for JPEG-LRJPEG is an algorithm which helps us to compress an image without the loss in quality. The algorithm calculates the CR(Compression ratio), PSNR (Peak signal-to-noise ratio) and MSE(Mean squared error). Compression ratio(CR) is the ratio or the amount of compression that has occured. CR has to calculated in order to analyse the changes in the image sizes to analyse the efficiency of the system. PSNR and MSE are two error measures used to evaluate the quality and performance of image compression. MSE shows the cumulative squared error between the original image and the compressed image, while PSNR shows the peak error measure.

3.2 *Firebase*

Firebase can be called as a BaaS(Backend as a service). Any BaaS service enables the developers to stop worrying about managing the server or the backend API etc. This enables the developer to focus on improvising the user experience. In short, firebase allows the developer to design and develop the apps quicker. Firebase offers many functionalities like, Authentication, Cloud functions, Database, File Storage, Cloud messaging. In app messaging.All these features are integrated with the developers Gmail account. When we make use of all these features collectively, we can develop and deploy an efficient application. Firebase also gives the developer in depth analytics and statistics of their application growth.

3.3 *OpenTok SDK*

OpenTok SDK is developed and maintained by TokBox Inc. They are a PaaS(Platform as a service) based company. The server side SDKs and client side SDKs maintained by TokBox enables the user to integrate live video calling into their application without focusing much on the backend server. OpenTok'sWebRTC supports Android, Javascript and IOS. The basic services of OpenTok are made free by the company. However the advanced features like group video calling would require payment. Therefore, we are making use of the free services and performing analysis on the obtained data to check if the outcome obtained is favourable or not. The next section discusses on the Archiecture of the Proposed system.

4 ARCHITECTURE OF THE PROPOSED SYSTEM

The Architecture of the system in Figure 1 depicts the various key elements present in the proposed system such as the mobile based API in which there are 3 modules covering the Doctor, Paramedic where the details from the remote telemedicine observed are filled by the paramedic and it gets loaded in to the cloud. The Cloud engine used here is a fire base cloud which can store the heterogeneous types of data collected such as patient vitals, images and videos. The system also integrates the compression technique to make the image sharing easy and uploading faster. The compression testing of various images are also done and results are displayed. The system has a full fledged computing and processing support through the cloud platform but the focus is on the compression of the image and transfering it.

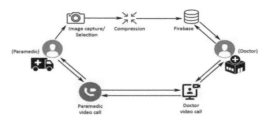

Figure 1.   Archietcture of the proposed system with cloud and mobile platforms.

5 METHODOLOGY OF THE PROPOSED SYSTEM

The proposed system has multiple modules which help in integrating the mobile application used by the paramedic and the doctor by interconnecting them through a Cloud based engine. The working of the system begins with a Paramedic module.

5.1 Paramedic module

The paramedic who will be inside the emergency vehicle will have an Android application which enables him to interact with the doctor who is sitting at the hospital. The paramedic will be able to capture and upload live images which will be reflected on the doctor's interface in a synchronous manner. The paramedic can also choose to make use of the live video calling feature which helps the doctor to give a better 'pre-hospitalisation' based on the visual references. The working and the functionalities of the Paramedic module is representer in Figure 2. Paramedic places a vital role in the prehospitalization as the images and videos needs to be send to the doctor inorder to measure the seriousness of the emergency. The Paramedic the the trained professional will follow the instructions from the doctor and record the vitals as well as images or videos based on the request that is send from the doctor or the Hospital Authorities. The paramedic part of the Mobile application also controls and process the compression of Images and data that has ro be transferred to the doctor.

5.2 Doctor side

The doctor also plays a vital role as he has to judge the patient condition based on the inputs that is transferred by Paramedic The doctor who will be at the hospital will also have an android application with which he can access the photos uploaded by each user. The photos would be stored in the order of their time of upload. The doctor can also interact with the paramedic using the video calling feature. This helps the doctor to gain visual knowledge about the patient's physical condition and can do the arrangements in the hospital accordingly. The functionalities of doctor module are depicted in Figure 3.

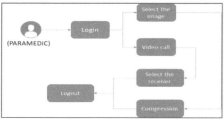

Figure 2.   Paramedic module.   Figure 3.   Doctor module.

5.3 Backend/firebase cloud and compression module

As time is a major factor during the transit of a patient, the system makes use of the LRJPEG image compression algorithm which would efficiently compress the images without losing the clarity so that the images are quickly uploaded to the server. The paramedic & doctor login is handled by the authentication module in firebase. All the information about the users are stored in a hierarchical way inside Firestore. There is almost no data loss while the data is being transmitted to the Firebase server. The user login information is being hashed by the Firebase API before storing in the server and is highly secure.

6 RESULTS AND DISCUSSION

For testing purposes intially a mobile application as discussed in the methodology is created and connected to fire base engine. Then the operation of the paramedic module is carried out and the data is collected and tested, for the purpose of testing the compression sixrandom images are chosen and then loaded. Given below are the statics of size of images before compression and after compression respectively. The testing is carried out and the observations are plotted in the Table 1 given below.

Table 1. Results of compression of the random test images, befor and after.

Image no	Before compression	After compression
Image 1	4.53 MB	534 KB
Image 2	2.62 MB	175 KB
Image 3	2.10 MB	126 KB
Image 4	6.23 MB	934.5 KB
Image 5	8.06 MB	1.01 MB
Image 6	1.45 MB	233.870 KB

The equation for calculating the compression ratio is as given below where we can see tha the ratio of uncompressed to compressed image is calculated

$$Compression\ ratio(CR) = \frac{Uncompressed\ image\ size}{Compressed\ image\ size} \tag{1}$$

The calculation of the various images selected are performed below for the ratio understanding

$$CR\ of\ image\ 1 = \frac{4.53 * 1000}{534} = 8.4831 \tag{2}$$

$$CR\ of\ image\ 2 = \frac{2.62 * 1000}{175} = 14.971 \tag{3}$$

$$CR\ of\ image\ 3 = \frac{2.10 * 1000}{126} = 16.667 \tag{4}$$

$$CR\ of\ image\ 4 = \frac{6.23 * 1000}{934.5} = 6.667 \tag{5}$$

$$CR\ of\ image\ 5 = \frac{8.06}{1.01} = 8.514 \tag{6}$$

$$CR\ of\ image\ 6 = \frac{1.45 * 1000}{233.870} = 6.2002 \tag{7}$$

As the calculation in the equation shows the Compression ratio of various images and their transfer rates it makes us understand that compression works on the images very well and this will definitely help in the faster transfer of the images.

Figure 4 shows the compression ratio (CR) of all the test image across a bar chart. Here we can see that the compression ratio always comes between 5% ~ 20%. The difference between the time to upload a compressed image vs the time to upload an uncompressed image is very huge. Considering the fact that the quality loss of the image after is very minimal and undetectable by the human eye, compressing the image makes the system extremely efficient. In depth statistics about the stream published and subscribed using OpenTok API are also available for the developer this is the main feature which helps the processing of the data, images and videos faster without queue limits.

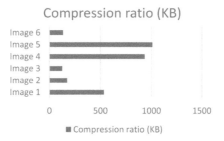

Figure 4.   Compression ratio of the various images.

The Table 2 given above shows the usage statistics based on the user interaction. Each account holder has an unique ID and they can create project which are also depicted with unique ID, each of the users will have their type of stream or route they would use in the application which is recorded in the cloud, Usage stastics represents the various users who are working in terms of using the application in hours. The Table 2 result is obtained from the cloud dashboard API which is taken in count for the calculation of the performance of the system based on usage statistics. The proposed API with the help of the image compression helps in faster uploading and downloading data which is very much requires in the case of remote telemedicine approach. Thus this system is an useful frame work for interactive data transfer.

Table 2.   Usage statics of streaming of images available from API.

Account ID	Project ID	Type	Usage
3574702	46248262	Hours	0.34166667
3574702	46248262	Streamed Subscribed Minutes	17.95
3574702	46248262	Streamed Subscribed Routed Minutes	17.95
3574702	46248262	Session Minutes	17.9

7   CONCLUSION

The pandemic and necessity for telemedicine data transfer has created need for having an efficient technique or interactive frame work for data sharing. This proposed work demonstrates a frame work which manages the data sharing with help of API and compression technique. The system interconnects paramedic who is present with patient to share the data or perform a live streaming to a doctor which helps in the pre hospitalization scenario. The system is a one stop solution for data sharing which done with a real time experimentation conducted with an android end interface Open-tok API and cloud based barracuda server API. The system thus provides an efficient interactive framework to transfer the images and live stream videos fast as there is an efficient compression frame work. The results of usage of the real time system shows that the system accelerates the data sharing which will aid for telemedicine and remote consultation. This frame work can be integrated to hospital management system to aid many difficult pre hospitalization or hospital transfer very efficiently with the user friendly application and efficient and faster storage and retrieval system.

REFERENCES

[1]   Al-Dhuraibi, Y., Paraiso, F., Djarallah, N., & Merle, P. (2018). Elasticity in Cloud Computing: State of the Art and Research Challenges. *IEEE Transactions on Services Computing*.

[2]   Birman, K., Chockler, G., & van Renesse, R. (2009). Toward A Cloud Computing Research Agenda. *ACM SIGACT News*. https://doi.org/10.1145/1556154.1556172

[3] Russkov, A. A. (2014). Shared Storage Performance of Cloud Computing: OpenNebula Case. In *2014 IEEE 3rd International Conference on Cloud Networking*, CloudNet 2014. https://doi.org/10.1109/CloudNet.2014.6969007

[4] Wu, J., Ping, L., Ge, X., Ya, W., & Fu, J. (2010). Cloud Storage as The Infrastructure of Cloud Computing. In *Proceedings – 2010 International Conference on Intelligent Computing and Cognitive Informatics*, ICICCI 2010.

[5] Prabavathy, B., Priya, K., & Babu, C. (2013). A Load Balancing Algorithm For Private Cloud Storage. In *2013 4th International Conference on Computing*, Communications and Networking Technologies, ICCCNT 2013. https://doi.org/10.1109/ICCCNT.2013.6726585

[6] Chilipirea, C., Laurentiu, G., Popescu, M., Radoveneanu, S., Cernov, V., & Dobre, C. (2016). A Comparison of Private Cloud Systems. *2016 30th International Conference on Advanced Information Networking and Applications Workshops (WAINA)*, 139–143. https://doi.org/10.1109/WAINA.2016.23

[7] Li Z. *et al.*, "Efficient Batched Synchronization in Dropbox-Like Cloud Storage Services,"*3rd International Conference on Cloud Based Systems* pp. 307–327, 2013.

[8] Rahmi, A., Piarsa, I. N., & Buana, P. W. (2017). *FinDoctor – Interactive Android Clinic Geographical Information System Using Firebase and Google Maps API*, (7), 8–12.

[9] Bocchi, E., Drago, I., & Mellia, M. (2015). Personal Cloud Storage Benchmarks and Comparison. *IEEE Transactions on Cloud Computing*, 5(4), 751–764. https://doi.org/10.1109/tcc.2015.2427191

[10] Bocchi, E., Drago, I., &Mellia, M. (2015). Personal Cloud Storage Benchmarks and Comparison. *IEEE Transactions on Cloud Computing*.

[11] Johnston, A., Yoakum, J., & Singh, K. (2013). Taking on webRTC in an Enterprise. *IEEE Communications Magazine*, 51(4), 48–54. https://doi.org/10.1109/MCOM.2013.6495760

[12] Poddar, S. R. (2018). Delivering Health Care in Rural Communities Through Telemedicine, (March).

[13] Liu, Y., Guo, Y., & Liang, C. (2008). A Survey on Peer-to-Peer Video Streaming Systems. *Peer-to-Peer Networking and Applications*, 1(1), 18–28. https://doi.org/10.1007/s12083-007-0006-y

[14] Wade V.A., Karnon J., Elshaug A.G. & Hiller J.E. (2010). A Systematic Review of Economic Analyses of Telehealth Services Using Real Time Video Communication. *BMC Health Services Research*, 10.

[15] Mukhopadhyay, A., Raj, A., & Shaji, R. P. (2018). LRJPEG: A Luminance Reduction based Modification for JPEG Algorithm to Improve Medical Image Compression. *2018 International Conference on Advances in Computing, Communications and Informatics*, ICACCI 2018. https://doi.org/10.1109/ICACCI.2018.8554763

[16] Mukhopadhyay, A., Shaji, R. P., & Raj, A. (2018). Delay Reduced Multimedia Transmission in Medical Emergencies. *2018 International Conference on Advances in Computing, Communications and Informatics*, ICACCI2018,1313–1319. https://doi.org/10.1109/ICACCI.2018.8554907

[17] Parekh M. and Saleena B. "Designing a Cloud Based Framework for HealthCare System and Applying Clustering Techniques for Region Wise Diagnosis," *Procedia Comput. Sci.*, vol. 50, pp. 537–542, Jan. 2015,

[18] Javaid M., Haleem A., Singh R.P., Rab S., Suman R. and Khan I.H., "Evolutionary Trends in Progressive Cloud Computing Based Healthcare: Ideas, Enablers, and Barriers," *Int. J. Cogn. Comput. Eng.*, vol. 3, pp. 124–135, Jun. 2022, doi: 10.1016/J.IJCCE.2022.06.001.

[19] Hussein W.N., Hussain H.N. and Humod I.M., "A Proposed Framework for Healthcare Based on Cloud Computing and IoT Applications," *Mater. Today Proc.*, vol. 60, pp. 1835–1839, Jan. 2022,

[20] Simeone A., Caggiano A., Boun L. and Grant R., "Cloud-based Platform For Intelligent Healthcare Monitoring and Risk Prevention in Hazardous Manufacturing Contexts," *Procedia CIRP*, vol. 99, pp. 50–56, Jan.

[21] Al Omar A., Bhuiyan M. Z. A., Basu A., Kiyomoto S. and Rahman M.S., "Privacy-friendly Platform for Healthcare Data in Cloud Based on Blockchain Environment," *Futur. Gener. Comput. Syst.*, vol. 95, pp. 511–521

The analysis and interpretation of higher education teachers based on student and teachers feedback

Vinita Singh
Mansarovar Global University, Sehore Vill Gadia Bilkisganj, Sehore, MP, India

Dharm Raj Singh & Vijay Kant Sharma
Department of Computer Application, Jagatpur P G College, Varanasi, UP, India

Ranjana Singh
Ghamahapur, Gangapur, Varanasi, Uttar Pradesh, India

ABSTRACT: The National Assessment and Accreditation Council (NAAC), which rates and accredits higher education institutions (HEIs), as well as the perspectives and perceptions of university/college instructors, are all subjects of our investigation in this paper. We create a mixed method to examine several components of quality indicators in higher education, if any, for estimation and acceptance. The mixed method is used to combine quantitative and qualitative data. The fundamental assumption is that integrating quantitative and qualitative approaches produces a better understanding of the research problem and subject than either method used independently. The design and implementation of the programme assessment frame of reference are used to evaluate quality assurance in higher education institutions.

Keywords: Introduction resources, Teaching and learning quality management, Result analysis

1 INTRODUCTION

Guest *et al.* explain a mixed method for collecting qualitative and quantitative data [1]. To sustain higher education quality is becoming a serious problem. Josip *et al.* examine that after primary education the role of particular and school-specific variables in the desire for higher education information of pupils in advance [2]. Jayaram highlighted all aspects of higher education in India [3]. Madhukar observed that higher education in India has grown considerably in the terms of quantity without a corresponding increase in the level of quality of instruction [4]. He stressed that the quality issue in higher education had arisen as a result of the institution's uncontrolled expansion. Gibbs examined the problems associated with educational quality, concluding that the most significant task for education in the new century was to develop coordination in terms of "quantity" and "quality" [5]. Taylor focused on an investigation into the factors contributing to the downfall of higher education quality [7]. Rodriguez discussed in his paper to examine the issues surrounding the regulation of higher education in India the institutional hurdles to the entrance have resulted in a lack of both quality and quantity in terms of offerings [6]. According to Wareing [8], the most serious issue to be concerned about is that academic staff members get preoccupied with developing and adhering to formal quality assurance procedures, which causes them to lose focus on teaching and research. Garcia *et al.* show the effects of the base on students of evaluation they already have, the evaluation of the student's perception, and evaluation types [14]. Leiber [9] observed the quality assurance in higher education, "some infrastructure and teaching/learning process related aspects are required to be monitored and assessed" to maintain a minimum standard of higher education. Sangeeta observed that the capacity and skill of the teachers determine the quality of education provided to students [10]. Lungu and Moraru present a new dimension to the analysis of student surveys in higher education [15]. Aithal *et al.* present a SWOC analysis of theory A and its application to different types of organizations [16]. Arun Jamkar [11] suggested that The NAAC had been successful in getting institutions and academia to embrace its evaluation and accreditation process as a result of an active awareness campaign and contacts with academics. Zhang *et al.* advised the NAAC for Higher Education are the basic principles and quality assurance requirements, as well as the accreditation process [12]. Ntim was a survey on Gannian teachers' skills and pay more attention to cultivating critical also with appreciable knowledge of teachers [13].

In this paper, we examine quality concerns with TQM in higher education institutions (HEIs) as assessed and certified by NAAC, as well as the perspectives and perceptions of college/university teachers. The traditional evaluation and accreditation system overlooks the influence of students' efforts, motivation, and reactions to various pedagogical attitudes on the teaching-learning process in HEIs. Because it is believed that the learning gained in HEI is the result of a wider range of factors to external institutions. On the other hand, it is a synthesis of the outcomes of instruction, practice, behavioral skills, and other elements outside of the college itself. We have designed a mixed method that is the most effective tool for quality assurance of higher education institutions. It helps to focus on the right issues and can be applied to higher education to revive the system. It is used by various instruments and hypotheses to support human resource development, imbibing a positive atmosphere, and emphasizing the importance of student-centric teaching in the study to evaluate teaching and learning to provide students with timely.

The remaining sections are arranged as follows: Section 2 explains the research methodology. Data and the result analysis are described in Section 3. The conclusion is described in section 4.

2   RESEARCH METHODOLOGY

The 'programme evaluation' frame of reference has been used in the design and execution. An open-ended survey of higher education institutions' providers (teachers) employed a qualitative approach, which include document analysis of NAAC reports and other relevant research materials. To collect quantitative data based on replies to the survey's open-ended questions. For evaluation purposes, one schedule has been created for faculty members. The schedule is based on the teaching, research, and extension missions of higher education institutions. The interview schedule was created with the six criteria based on NAAC has identified for evaluating and accrediting university/college teachers in mind: teaching learning and evaluation; curricular aspects; research, consultancy, and extension; infrastructure and learning resources; student-supported teacher's view; and innovative practice. Depending on the degree of agreement or lack of awareness/applicability, respondents have been asked to score each statement on a 3-point continuum (Agree, Disagree, or No Response).

When capturing instructors' opinions, a three-point scale is used:

- In cases of satisfaction indicating agreement (Yes).
- When there is disagreement indicating (No).
- If you can't react or it isn't applicable (No Response).

Field surveys are used to gather information about college teachers' perspectives on quality issues in higher education. 70 teachers have been chosen from college for this study. Teachers who are socially committed.

3   DATA AND RESULT ANALYSIS

A random sample of 70 university/college teachers was recruited to assess their perspectives on key quality concerns in higher education. These fundamental concerns have been chosen based on the standards of the NAAC in higher education institutions, which include some, general statements.

Apart from these questionnaires, a set of questions are sub-divided into various categories as mention:

- Teaching, learning, and assessment;
- Aspects of the Curriculum

- Extension, research, and consulting
- Resources for learning and infrastructure
- A teacher's perspective on student support
- Creative Techniques

Percentage analysis is used to examine the replies of teachers (on a three-point rating scale) to different quality concerns related to the aforementioned factors. The findings of this study have been categorized into the following sections:

3.1 *Teaching, learning, and assessment*

The observations of teachers regarding the Perceptions of Teacher's Teaching learning and evaluation in Higher Education have been given in Figure 1 based on the following question.

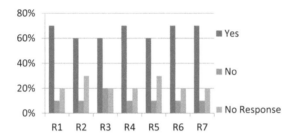

Figure 1.   Percentage of perceptions of teachers about teaching learning and evaluation in higher education.

R1: Indian education system needs digital transformation.
R2: Is the new education policy plays a significant role in the Indian education system?
R3: Higher educational institution needs to improve their policies for better cooperation among student and faculty.
R4: Higher educational institutions also include smart classes to improve their teaching experience.
R5: Indian government increases the educational budget.
R6: Is the Indian higher educational institutions need to work on their quality education programme?
R7: Educational institutes include skill development programme which is job oriented.

3.2 *Curricular aspects*

The observations of teachers regarding the quality issues related to curricular aspects at higher education have been given in Figure 2 based on the following questions.

A1: The syllabus of different courses is reviewed and updated regularly in the university to incorporate the latest knowledge.

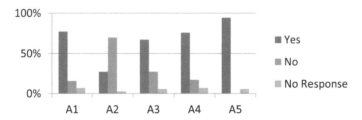

Figure 2.   Teachers' perceptions of quality of curricular aspects.

A2: The sufficient knowledge and skills in the present syllabi of different courses equipped the students with the emerging demands of the employment market.

A3: The syllabus nowadays is being revised to promote research and develop analytical thinking among students.

A4: By consulting different stakeholders like alumni, parents, and industry and social organizations while developing the curriculum would improve its quality.

A5: The contribution to the improvement of the quality of education is to move from an annual to the semester system and from the award of marks to an award of credits.

3.3 *Research, consultancy, and extension activities*

The teacher's comments on many elements of quality of research, consultancy, and extension activities based on the following questions:

B1: In the University, different departments are undertaking significant and good-quality research.

B2: The University is providing excellent research facilities in the form of laboratory facilities, library facilities, software for statistical analysis, etc. for promoting research work in all the departments.

B3: The University promotes research activities among teachers and students by organizing research development programs and workshops.

B4: The University faculty has accomplished a great deal in terms of grants, projects, publications, patents, and research awards.

B5: The University's department collaborates on a national and international level when it comes to the sharing of research knowledge, infrastructure use, and other resources.

The quality of research, consultancy, and extension activities has been assessed by percentage analysis as shown in Figure 3.

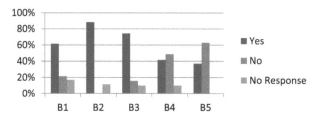

Figure 3.   Percentages of teachers' perceptions of quality of research, consultancy, and extension activities.

3.4 *Infrastructure & learning resources*

Figure 4 summarizes teachers' assessments of the quality of infrastructural and educational resources based on the following questions

C1: The University has excellent infrastructural facilities for different curricular and co-curricular activities.

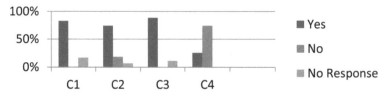

Figure 4.   Percentages of teacher's perceptions of the quality of infrastructural and learning resources.

C2: The infrastructural facilities in the university are properly maintained and are augmented from time to time.

C3: The library facilities available in the university are satisfactory in terms of regular book addition and subscription to reputed journals.

C4: The University providing good hostel facility to the students.

3.5 *Student-supported teacher's view*

Figure 5 show summarizes teachers' views on several areas of student support services in higher education based on the following questions

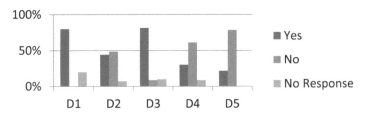

Figure 5.   Percentages of teachers' perceptions of student support teachers view.

D1: The university/college has developed a successful system for offering financial aid (scholarships/free ships) to guarantee that students from socioeconomically disadvantaged groups of society are fairly represented.

D2: In order to fulfill the unique needs of both slow and advanced learners, the university/college has been successful in adopting a few particular tactics.

D3: Effective support services are in place in the university/college for differently-abled students.

D4: In recent years, the placement record of university/college graduates has been particularly excellent.

D5: Students receive effective placement and counseling services from the college.

3.6 *Innovative practices as quality parameter*

Figure 6 provides a summary of teachers' perspectives on faculty and higher education institution introductions of innovative techniques based on the following questions:

E1: The University/College has an internal quality assurance mechanism and is effective to a considerable extent in ensuring the quality of education.

E2: Internal Quality Assurance Cell (IQAC) of the university/college attempts to institutionalize best practices and create academic benchmarks on a regular basis.

E3: To incorporate its various stakeholders in the planning, execution, and evaluation of the academic programme, the university/college has developed processes.

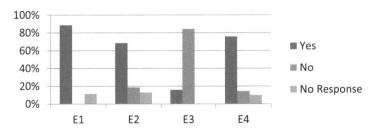

Figure 6.   Percentages of teachers' perceptions of innovative practices.

E4: The teachers regularly organize such activities which promote a sense of social responsibility and good citizenship among the students.

4 CONCLUSION

The documentary analysis of NAAC evaluation reports on the assessment and accreditation of degree colleges and universities has revealed the institutions' strengths, weaknesses, and opportunities. In this study, the mixed method is used to produce quality assurance and a comprehensive plan to improve a higher education institution. Examining all of the questions yields the conclusion that respondents' responses reflect a more positive than negative viewpoint on the process of educational excellence in the context of higher education. Responses to open-ended questions revealed that quality teachers lack some flexibility which may affect their academics and concepts. This seems to be a very promising field of research. It enables imaginative data mining and removes perception-based judgments about the success of teaching and learning.

REFERENCES

[1] Guest, G., & Fleming, P. *"Mixed Methods Research. Public Health Research Methods"*. Thousand Oaks, CA: Sage, 581–610, (2014).

[2] Šabić, J., & Jokić, B. "Elementary School Pupils' Aspirations for Higher Education: The Role of Status Attainment, Blocked Opportunities and School Context. *Educational Studies*, 47(2), 200–216, (2021).

[3] Jayaram, N. "Research on Higher Education in India. India Higher Education Report 2017": *Teaching, Learning and Quality in Higher Education*, (2018).

[4] Madhukar, B. S. "Managing Quality at Institutional Level. India Higher Education Report 2017:" *Teaching, Learning and Quality in Higher Education*, (2018).

[5] Gibbs, P. "A Pedagogy of Emergent Self-Cultivation; Why Students Should Have a "Sameness" and Why They Should Not1". *Philosophy and Theory in Higher Education*, 3(2), 43–58, (2021).

[6] Rodriguez, A. "Inequity by Design? Aligning High School Math Offerings and Public Flagship College Entrance Requirements". *The Journal of Higher Education*, 89(2), 153–183, (2018).

[7] Taylor, B. J. "Recruitment and Admission Management, Higher Education Institutions. *Journal:*" The *International Encyclopedia of Higher Education Systems and Institutions*, 2425–2430, (2020).

[8] Wareing, S. *"Staff Development and Quality Assurance. In Handbook of Quality Assurance for University Teaching"* (pp. 363–370), (2018).

[9] Leiber, T. "A General Theory of Learning and Teaching and a Related Comprehensive Set of Performance Indicators for Higher Education Institutions" *Quality in Higher Education*, (2019).

[10] Angom, S. "Role of Teachers in Academic Reforms for Quality Higher Education." In *Higher Education and Professional Ethics* (pp. 27–44), (2018).

[11] Jamkar, A. "National Assessment and Accreditation Council (NAAC)'s Accreditation of Health Sciences Institutes: Challenges and the Road Ahead." *Journal of Education Technology in Health Sciences*, 6(3), 57–58, (2019).

[12] Zhang, C., Moreira, M. R., & Sousa, P. S "A Bibliometric View on the Use of Total Quality Management in Services." *Total Quality Management & Business Excellence*, 32(13–14), 1466–1493, (2021).

[13] Ntim, S. (2017). Transforming Teaching and Learning for Quality teacher Education in Ghana: Perspectives From Selected Teacher Trainees and Stakeholders in Teacher Education. *International Journal of Education*, 9(3), 149.

[14] García, M. N. G., Escamilla, A. C., Fernández, E. M., Barrio, M. I. P., & de la Rosa García, P. "The Influence of Different Variables in Evaluation Within the Building Degree in the Polytechnic University of Madrid. *Procedia-Social and Behavioral Sciences*, 176, 691–698, (2015).

[15] Lungu, I., & Moraru, M. "New Dimensions to the Analysis of Student Survey Results in the Instructional Process in Higher Education." *Procedia-Social and Behavioral Sciences*, 180, 376–382, (2015).

[16] Aithal, S., & PM, S. K. "Theory a for Optimizing Human Productivity." 526–535, (2016).

Recent Trends in Computational Sciences – Gururaj, Pooja & Flammini (Eds)
© 2024 The Author(s), ISBN 978-1-032-42685-3

Implementing AI on microcontrollers in fog and edge architectures

Dalibor Dobrilovic
Technical Faculty "Mihajlo Pupin", Zrenjanin, University of Novi Sad, Zrenjanin, Serbia

ABSTRACT: We are witnessing the era of evolution of complex system architectures, starting from cloud computing and gradually to the edge, fog, and mist computing paradigms. The development of applications in those complex systems is equally important for companies as well as for universities. Both branches need fast, low-cost, and effective environments for rapid prototyping. These demands can be enhanced with the recent microcontroller developments. The microcontrollers grow in their performance and, simultaneously, are available for considerably lower prices. Open-source hardware movement is another important base for rapid fog and edge prototyping. This paper presents the approach of using open-source hardware development boards in edge computing architectures by investigating the possibility of implementing Artificial Intelligence on these edge devices. The performance of AI-supported edge devices is monitored and analyzed in terms of their speed, capacity, and accuracy. The Arduino, ESP8266, and ESP32 development boards are evaluated in this research.

Keywords: AI on microcontrollers, AI in edge computing, k-Nearest Neighbor, open-source hardware, intelligent sensor networks

1 INTRODUCTION

Considering the rapid evolution of complex system architectures, from cloud computing to the edge, fog, and mist computing paradigms, we should analyze their impact on the market and academic institutions. The rapid changes in the development of complex systems demand the immediate answer of companies and academies to market challenges. Both branches need fast, low-cost, and effective environments to rapidly prototype applications for the aforementioned systems.

Two major factors influenced the efforts to meet these demands. The rapid development of microcontrollers and development boards resulted in their higher processing powers and significantly decreased prices. The second facilitating feature is open-source hardware movement. The open-source movement started approximately ten years ago and rapidly spread and overflowed the market. Considering the scientific publication databases, a significant number of research papers are based on using Arduino UNO, Arduino MEGA, NodeMCU, and similar platforms.

This paper presents the approach of using open-source hardware development boards in edge computing architectures by investigating the possibility of implementing Artificial Intelligence on edge devices such as Arduino UNO and its clones. The motivation for this research goal lies in the fact that the ability of Arduino and clone boards to support AI can greatly enhance the teaching process and can enable the inclusion of AI implementation in fog and edge computing architectures in university curricula.

Besides the analyses of possible utilization of AI methods on Arduino and clone boards, the paper presents a methodology for implementing AI in the wireless sensor network based on Arduino and clone development boards. This implementation gives an example of using AI in a specially designed wireless sensor network to acquire solar radiation data.

DOI: 10.1201/9781003363781-19

The paper is structured as follows. After the introduction, the evolution of cloud architecture is briefly explained. After the cloud computing evolution, an overview of the usage of open-source hardware microcontroller boards and the possible implementation of AI is discussed as the current trends in the field. Section 4 overviews popular microcontroller development boards used in this research. Section 5 presents the wireless sensor network for solar radiation data acquisition. Section 6 describes an approach to implementing AI for solar radiation estimation. In the end, the concluding remarks and future work are discussed.

2 CLOUD COMPUTING ARCHITECTURES EVOLUTION

It is important to briefly point to the differences in various architectural approaches for introducing the topic. The Cloud computing paradigm [1] is based on remotely accessible data centers. These remote data centers offer centralized storage, computing power, and data processing. The users can use these resources. The service providers are responsible for the system's maintenance, and users usually pay for these services. The system's main feature, its centralization, is its biggest weakness. This approach requires high bandwidth demands, high power consumption, complex storage and security issues, and high latency.

Alternative approaches are required to overcome these problems [2,3]. These alternative approaches are challenged with decentralization in system architecture. The common thing for all approaches is focused on decentralizing data processing and data storage. The location of the decentralized processing and data storage can be different.

Edge computing brings data processing to end devices, allowing almost real-time processing and lower latency. The side effect of this approach is balancing the cloud pressure. Fog computing moves the processing and data storage between the edge devices and the cloud. This approach allows decentralized computing in the middle of the system. The third approach, mist computing, is the newest one. It is enabled by the recent development of microcontroller boards, followed by a significant cost drop. The computing power and overall performance of microcontroller boards allow widely decentralized processing at the extreme edge of the system.

3 OVERVIEW OF IMPLEMENTATION OF AI IN EDGE AND FOG COMPUTING

The popularity of edge, fog, and mist computing in the scientific community can be tracked by the number of scientific papers published. Two databases are searched: ScienceDirect and Springer. The results for the period 2015–2023 are presented numerically by year in Table 1. This research is based on the popularity of AI implementation topics in fog, edge, and cloud computing [4,5]. The keywords used for the search are AI and Arduino. e.g., "fog computing" AND "Artificial intelligence" AND "Arduino"; "edge computing" AND "Artificial intelligence" AND "Arduino"; "mist computing" AND "Artificial intelligence" AND "Arduino."

Table 1. Number of publications by architecture and years.

Year	2015	2016	2017	2018	2019	2020	2021	2022	2023
Springer Cloud	13	19	26	66	96	137	504	826	118
Springer Edge	0	1	4	16	33	70	249	516	66
Springer Fog	0	4	5	14	36	45	191	381	56
Springer Mist	0	0	0	0	0	2	11	15	3
ScienceDirect Cloud	1	2	11	18	41	65	78	116	4
ScienceDirect Edge	1	0	1	6	14	25	46	81	3
ScienceDirect Fog	0	0	3	7	15	26	30	59	2
ScienceDirect Mist	0	0	0	0	0	0	1	2	0

4 MICROCONTROLLER BOARDS OVERVIEW

The other interesting pillar of the research presented in this paper is the open-source hardware (OSHW) movement. OSHW movement promotes open hardware design, meaning that the designed devices are composed of low-cost and available components, with additionally available schematics, wiring diagrams, PCB layout data, documentation, source code, and code examples.

Arduino microcontroller boards are one of the most frequently used open-source hardware platforms. The numerous examples of using Arduino boards can be easily accessed as published makers and homemade projects. Also, Arduino already has a wide implementation in academic research and engineering education as a rapid prototyping platform. The wide appliance of this platform is evident in Table 1, presented in Section 3. This is the main motivating factor for investigating the possible usage of Arduino and its clones as fog/edge/mist computing devices with the implementation of AI techniques.

With this research and its findings, we will have a deeper view of the possibility of using Arduino and its clones in AI-supported edge and fog systems. The proof of usability of Arduino boards in building prototype platforms and testbeds will give us a low-cost platform for teaching edge AI-related topics within university curricula. Although today exist, a few AI friendly platforms in the Arduino family, such as Arduino Nano 33 BLE and Arduino Portenta H7, this research will be focused on ESP8266 and ESP32 Arduino development boards, such as NodeMCU ESP8266, NodeMCU ESP32, Wemos D1 R2, and Wemos D1 R32. The reason for this approach is that those boards have significantly lower costs than the aforementioned boards. Besides that, those boards have integrated communication modules which make them suitable for integration in wireless sensor networks, IoT, and edge computing systems as well.

5 WIRELESS SENSOR NETWORK FOR SOLAR RADIATION DATA ACQUISITION

An example of implementing AI-to-edge devices is given in the scenario of a wireless sensor network designed for solar radiation data acquisition. The motivation for designing this sensor network is based on a necessity for experimentation with solar-powered sensor nodes and education in that field as well. The complexity of the systems based on sensor nodes, the large and constantly growing number of nodes, and the need for their constant operation leads to the importance of deploying solar-powered sensor devices.

The sensor network presented in this Section is designed to collect solar radiation data. Collected data can optimize the energy consumption of sensor nodes, help in mapping solar radiation, and enhance planning solar-powered sensor node deployment. The sensor network is shown in Figure 1. The presented network is designed to be used as (1) the teaching platform in IT courses (IT Engineering and Software Engineering) for the development of energy-efficient solar-powered sensor nodes; (2) the teaching platform in Mechanical engineering courses to analyze solar panels behavior, their efficiency, and dependency on various ambient conditions; and (3) the platform for scientific research on solar radiation data acquisition and analyses.

The planned configuration of the sensor nodes can be like this: DHT-11 or DHT-22 sensor for ambient temperature and humidity, and for a solar panel temperature measuring, BH1750 and UV sensor for light intensity measuring; ACS712 or INA169 current sensor, MAX471 current and voltage sensor, etc. Sensor nodes have additionally the solar charger used for managing Li-Po battery 500 mAh charging and discharging. The solar panel dimensions range from 160mm x 138mm x 2.5 mm to 130 mm $\times$ 87 mm x 2.5 mm. The panel's efficiency is 16%, voltage is 5.5V, and power is 3W. The sensor node can be Arduino UNO with a communication module for Wi-Fi technology such as ESP8266 or ESP32, but

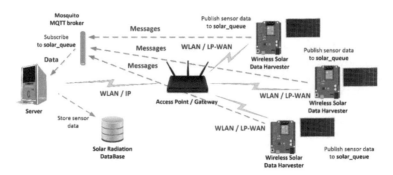

Figure 1. Sensor network for solar radiation data collection.

as an alternative, the usage of Wemos D1 or NodeMCU boards with integrated ESP8266/ESP32 modules is planned. This paper will present the methodology for using measurements of low-cost sensors (BH1750 and UV) to estimate the values of solar radiation, which will otherwise be measured with a much more expensive pyranometer.

The network is developed on the experience of designing and prototyping wireless IoT and smart technology systems [9–12]. The presented network was designed on a model for working environment monitoring in smart manufacturing systems [9] developed around Wemos D1 R2 and NodeMCU platforms. The same sensor network is built partially on the experience of designing the learning platform for Smart City Application development [10] with Wemos D1 R2 and NodeMCU platforms. Both platforms use the MQTT protocol for data transfer. The valuable authors' experiences in developing solutions based on Arduino MEGA and LoRaWAN technology applied in the urban traffic noise monitoring system [11] and investigation on long-range LoRa technology [12] also influenced the design of the presented system.

One additional aspect of edge computing architecture is certainly very important. The edge IoT security issues are addressed in [13]. The authors reviewed the edge-based IoT security research efforts analyzing the security architecture designs, distributed firewalls, intrusion detection systems, authentication and authorization algorithms, and privacy-preserving mechanisms. The strengths and weaknesses of the presented designs are discussed.

The paper [14] examines and describes security issues in accordance with a generic IoT architecture. This paper presents the main communication protocols that are used in the application, transport, network, and physical layer; and identifies and describes current security threats in IoT and current challenges and possible solutions in IoT edge security.

Although the security of AI-supported edge computing devices and the presented sensor network is equally important, security is not considered in this research phase. The reason is the purpose of the sensor network, which is primarily designed for solar data acquisition and for usage in teaching and academic research, and therefore estimated as a system without high-security risks.

6   THE APPROACH OF IMPLEMENTATION OF AI IN EDGE DEVICES

The reason for implementing AI in edge devices in the presented sensor network is to estimate solar radiation without deploying very expensive pyranometers. For example, the research effort is made to estimate solar radiation based on the measured light intensity with low-cost BH1750 light and UV sensors. In this example, the AI method is based on k-NN (k-Nearest Neighbor) algorithm. k-nearest neighbors' algorithm (k-NN) is a non-parametric supervised learning method developed by Evelyn Fix and Joseph Hodges (in 1951) and expanded by Thomas Cover. The approach of using k-NN in edge computing is investigated in [15,16]. For this implementation, the modified code [17] is used together with the experience in implementing AI techniques in data analyses [18,19].

The proposed methodology for implementing AI consists of the following steps. The first step is the initial solar radiation data collection with the wireless sensor nodes, including the pyranometer (1). The second step is analyzing and processing collected results (2). The third (3) and fourth (4) steps are a generation of the inputs (training data) for k-NN Arduino code based on collected and processed results (5). The input generation is performed externally on a PC using GNU Octave. The sixth step (6) is building Arduino code based on the analyses and collected data processing. Step (5) use GNU Octave to generate k-NN training data considering the impact of BH1750 and UV sensor measurements on the classification. For the model evaluation, generated classification inputs will be compared with the measured values (7). In the case of low accuracy (8), the process will be turned back to step 2. The process should be repeated until the highest accuracy can be reached.

This research first conducts the performance test of Arduino UNO, MEGA, ESP 8266, and ESP32 development boards. The test measures the capacity of 6 Arduino clone boards with a maximum supported number of inputs (training data) for k-NN classification. The time of loading inputs in milliseconds is additionally logged. The results of the tests are given in Table 2. The test showed that Arduino UNO could deal with a limited number of inputs (25), while Arduino MEGA could load up to 200 inputs. All boards with integrated ESP32 and ESP8266 processors support a much higher number of inputs, reaching up to 900, with the ESP8266-based Wemos D1 R2 and NodeMCU boards supporting up to 1,000 inputs. The time needed for loading a corresponding number of inputs at the start of powering development boards ranges from less than 1 millisecond to 87 milliseconds for 1,000 inputs on ESP8266-based boards.

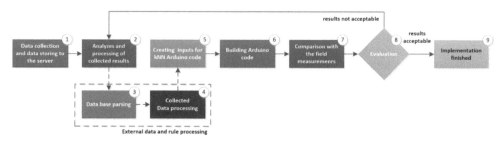

Figure 2.   Methodology of AI implementation in edge devices.

The number of supporting inputs proves that ESP32 and ESP8266 boards are both usable in prototyping AI-supported edge devices in complex scenarios.

Table 2.   Arduino and clone boards k-NN start time (ms) and capacity test.

Inputs	NodeMCU ESP8266	Wemos D1 R2	NodeMCU ESP32	Wemos D1 R32	Arduino MEGA	Arduino UNO
25	0	0	0	1	9	9
50	1	1	0	1	9	N/A
100	2	2	2.5	2	10	N/A
150	3.5	3.5	3	4	10	N/A
200	5	5.5	4.5	5	17	N/A
300	11	10	8.5	8	N/A	N/A
400	18	17	13	13	N/A	N/A

(*continued*)

Table 2.   Continued

Inputs	NodeMCU ESP8266	Wemos D1 R2	NodeMCU ESP32	Wemos D1 R32	Arduino MEGA	Arduino UNO
500	24	25	19	21	N/A	N/A
600	34	34	27	28	N/A	N/A
700	45	45	35	37	N/A	N/A
800	57	57	45.5	46	N/A	N/A
900	71	70.5	54.5	56	N/A	N/A
1000	87	86.5	N/A	N/A	N/A	N/A
1100	N/A	N/A	N/A	N/A	N/A	N/A

The second analysis investigates the accuracy of k-NN Arduino implementation for given tasks. The k-NN is used to predict the pyranometer's measurements of solar radiation. The pyranometer is an expensive device that measures the solar radiation in W/m^2 ranging from 0 to 1.300 W/m^2 in this experiment. Two significantly lower-cost sensors are used for predicting pyranometer measurements: light sensor BH1750 and UV sensor. The experimental measurement uses BH1750, UV sensor, and pyranometer simultaneously for a very short period. The 90 measurements are logged with different intensities of natural solar radiation in outdoor environments. Those 90 measurements are used for creating 90 k-NN classification inputs with the GNU Octave, and there are uploaded to development boards as a part of the Arduino code. The pyranometer's measurements are discretized by dividing the measured values by 100 and thus classifying them into 13 classes.

The accuracy of the k-NN for classification was made with 178 control measurements made with all three sensors, but only values of BH1750 and UV sensors were used for testing k-NN classification accuracy. The third value, measured with a pyranometer, was the control value. The control measurements were also made in a short period, with the different intensities of solar radiation and with the addition of an artificial light source in an indoor environment added to natural light. For the evaluation, the K value was set to 3.

The classification accuracy is presented in Table 3, with 136, or 76.40% of accurate classifications. Thirty-eight estimations (21.35%) are made with the inaccurate classification in the neighboring class (e.g., class 1 or 3 instead of class 2). Only four estimations (2.24%) are inaccurately classified in the non-neighboring class (e.g., class 4 or 5 instead of class 2).

The results presented in Table 3 justify the usage of k-NN implementation in Arduino clone devices for a given scenario. The accuracy of k-NN in this example is high, considering the short period of measurements used for training and test data, a small number of measurements, and totally different conditions of experimental measurements (indoor and outdoor, natural and artificial light combination). Additionally, no correction and filtering of training data were performed without efforts to select the best possible training data. Additionally, only 10% of the capacity of Arduino clone boards was used, considering the capacity of boards presented in Table 3.

Table 3.   Arduino and clone boards k-NN accuracy test.

No. of tests	Results	Percentage
136	Successful classification	76.40
38	Neighboring class error	21.35
4	Non-neighboring class error	2.24
178		**Total**

All these factors give excellent justification for using k-NN on AI-supported edge devices in numerous scenarios and the scenario proposed here in this paper – the wireless network for collecting and estimating solar radiation data. These findings can be used for solar radiation data mapping, by utilizing many lower-cost sensors, compared to the much more expensive pyranometers. Thus, estimating solar radiation can be possible with significantly more devices and in various wireless sensor networks covering wide areas. The lower accuracy of solar radiation estimation will be compensated by the greater number of devices used for estimation and greater data acquisition coverage.

The additional advantage of implementing this form of ML and AI in edge devices can have other benefits. For example, using the estimation presented in this paper, the solar-powered sensor nodes can detect lower solar potential and change their operating modes from full operation to modem-sleep, light-sleep, or deep-sleep modes. In that way, edge devices will use AI to manage their power consumption efficiently and to extend their battery lifecycle.

7 CONCLUSION

Moving processing and data analyses to the edge devices in the complex IoT and Smart environments relieve the core of these systems from overloading. The data processing location defines whether we deal with edge, fog, or mist computing architectures. Because Arduino and its clone boards prove their usability in research and development and rapid prototyping, it was interesting to analyze its usability in edge architectures and the implementation of AI on Arduino boards as edge nodes.

This paper gives a brief overview of the recent trends in scientific publications on using Arduino in fog, edge, and mist computing. This paper further deals with implementing AI on edge devices based on Arduino platforms. The example of integrating AI to the Arduino edge device is given for the wireless sensor network for solar radiation data collection. The k-NN (k-Nearest Neighbor) algorithm is implemented on the edge side. The k-NN estimates solar radiation using two low-cost sensors (light intensity and UV sensor) instead of a pyranometer. The test evaluated the classification accuracy and the board performances by testing their capacity and speed. Tested boards and k-NN implementation justify their utilization in the presented sensor network for solar radiation data collection. It was shown that AI could be implemented in Arduino edge devices, at least for usage in academy teaching and research. Further work will be focused on the optimization process of training data selection, together with increasing the classification accuracy. Furthermore, the k-NN will be used to optimize the sensor node operations and possibly energy efficiency.

ACKNOWLEDGMENT

This research was funded through the project "Creating laboratory conditions for research, development, and education in the field of the use of solar resources in the Internet of Things" at the Technical Faculty »Mihajlo Pupin« Zrenjanin, financed by the Provincial Secretariat for Higher Education and Scientific Research, Republic of Serbia, Autonomous Province of Vojvodina, Project number 142-451-2684/2021-01/02.

REFERENCES

[1] Marinescu D.C., *"Cloud Computing: Theory and Practice"*, Morgan Kaufmann, 2017
[2] Buyya R., Srirama S.N., *"Fog and Edge Computing: Principles and Paradigms"*, John Wiley & Sons, 2019.

[3]   Dobrilovic D., Flammini F., Gaglione A., Tokody D., "Open-Source Hardware Performance in Edge Computing", *IEEE Smart Cities Newsletter*, July 2021, https://smartcities.ieee.org/newsletter/july-2021/open-source-hardware-performance-in-edge-computing

[4]   Firouzi F., Farahani B., Marinšek A., "The Convergence and Interplay of Edge, Fog, and Cloud in the AI-driven Internet of Things (IoT)", *Information Systems*, Vol. 107, 2022, https://doi.org/10.1016/j.is.2021.101840.

[5]   Gui H., Liu J., Ma C., Li M., Wang S., "New Mist-edge-fog-cloud System Architecture for Thermal Error Prediction and Control Enabled by Deep-learning", *Engineering Applications of Artificial Intelligence*, Vol. 109, 2022, https://doi.org/10.1016/j.engappai.2021.104626.

[6]   Gill S.S., Xu M., Ottaviani C., *et al.*, "AI for Next Generation Computing: Emerging Trends and Future Directions, *Internet of Things*", Vol. 19, 2022, 100514, https://doi.org/10.1016/j.iot.2022.100514.

[7]   Caiazza C., Giordano S., Luconi V., Vecchio A., "Edge Computing vs Centralized Cloud: Impact of Communication Latency on the Energy Consumption of LTE Terminal Nodes", *Computer Communications*, Vol. 194, 2022, pp 213–225, https://doi.org/10.1016/j.comcom.2022.07.026.

[8]   Iftikhar S., Gill S.S., Song C., *et al.*, "AI-based Fog and Edge Computing: A Systematic Review, Taxonomy and Future Directions", *Internet of Things*, Vol. 21, 2023, 100674, https://doi.org/10.1016/j.iot.2022.100674.

[9]   Dobrilovic D., Brtka V., Stojanov Z., Jotanovic G., Perakovic D., Jausevac G., "A Model for Working Environment Monitoring in Smart Manufacturing", *Applied Sciences* 2021, 11, 2850 https://doi.org/10.3390/app11062850

[10]   Dobrilovic D., Malić M., Malić D., "Learning Platform for Smart City Application Development", *INDECS*, Vol. 17 No. 3-B, pp 430–437, 2019, DOI 10.7906/indecs.17.3.

[11]   Dobrilović D., Brtka V., Jotanović G., *et al.*, "The Urban Traffic Noise Monitoring System Based on LoRaWAN Technology", *Wireless Netw* 2022, 28, 441–458. https://doi.org/10.1007/s11276-021-02586-2

[12]   Dobrilović D., Malić M., Malić D., Sladojević S., "Analyses and Optimization of Lee Propagation Model for Lora 868 MHz Network Deployments in Urban Areas", *Journal of Engineering Management and Competitiveness (JEMC)* 2017, 7, 55–62.

[13]   Sha K., Yang T.A., Wei W., Davari S., "A Survey of Edge Computing-based Designs for IoT Security", *Digital Communications and Networks*, Vol. 6, Issue 2, 2020, pp 195–202, https://doi.org/10.1016/j.dcan.2019.08.006.

[14]   Gerodimos A., Maglaras L., Ferrag M.A., Ayres N., Kantzavelou I., "IoT: Communication Protocols and Security Threats", *Internet of Things and Cyber-Physical Systems*, Vol. 3, 2023, pp 1–13, https://doi.org/10.1016/j.iotcps.2022.12.003.

[15]   Xue W., Shen Y., Luo C., Xu W., Hu W., Seneviratne A., "A Differential Privacy-based Classification System for Edge Computing in IoT", *Computer Communications*, Vol. 182, pp 117–128, 2022, https://doi.org/10.1016/ j.comcom.2021.10.038.

[16]   Aazam M., Zeadally S., Flushing E.F., "Task Offloading in Edge Computing for Machine Learning-based Smart Healthcare", *Computer Networks*, Vol. 191, 2021, 108019, https://doi.org/10.1016/j.comnet.2021.108019

[17]   Arduino_KNN *Library for Arduino*, Copyright (c) 2020 Arduino SA, https://github.com/arduino-libraries/Arduino_KNN

[18]   Brtka V., Makitan V., Brtka E., Dobrilovic D., Berkovic I., "LP-WAN Performance Analysis by Semi-Linguistic Summaries", *Ad Hoc & Sensor Wireless Networks*, (2020), Vol. 48, No. 1–2, pp 145–165

[19]   Dobrilovic D., Brtka V., Berkovic I., Odadzic B., "Evaluation of the Virtual Network Laboratory Exercises Using a Method Based on the Rough Set Theory", *Computer Applications in Engineering Education*, Vol. 20, Issue 1, pp 29–37, 2012, https://doi.org/10.1002/cae.20370

Deep learning and applications

Recent Trends in Computational Sciences – Gururaj, Pooja & Flammini (Eds)
© 2024 The Author(s), ISBN 978-1-032-42685-3

Abnormality detection in chest radiograph using deep learning models

P.R. Pruthvi*, M. Manju, Megha Subramanya, R. Nidhi & C.V. Lavanya
Vidyavardhaka College of Engineering Mysuru, India
*ORCID ID: 0000-0001-6127-6033

ABSTRACT: Chest radiography is commonly used in annual health screenings to determine the health of the lungs. As a result, developing a smart system to assist clinicians or radiologists in automatically detecting possible malformations in radiographs would be preferable. This develops a new anomaly identifier method which relies on an "autoencoder" that delivers the reestablished version of the input besides the uncertainty prediction using pixel. Higher uncertainty is frequently seen in regular areas or boundaries, where reconstruction errors are relatively larger, but not at happening regions where abnormalities are present. As a result, a natural measurement for detecting abnormalities such as Pleural effusion, Consolidation, Pneumothorax, Edema, Atelectasis and Cardiomegaly in images is the normalized reconstruction error multiplied by uncertainty.

Keywords: CXR (Chest X-ray), Autoencoder, Pleural effusion, Consolidation, Pneumothorax, Edema, Atelectasis and Cardiomegaly

1 INTRODUCTION

The most often prescribed diagnostics in the medical world today are chest X-rays (CXR), which are essential for identifying pulmonary problems. An estimate shows that around 110 million Chest X-Rays are carried out yearly once in the USA, among which only 39,500–40,000 clinicians have said to provide the actual reading. Due to advancement in smartphone technology, taking photographs or pictures of CXRs and forwarding them to their colleagues for obtaining instant opinions has been a routine for all doctors. In radiology, where "turn-around time is king," radiologists are estimated using the reverse time rather than using the "quality of their reports". Providing more importance to reverse time can result in confusion, subpar reports, mis-diagnosis, and lack of communication with primary care physicians, specifically in rustic areas where doctors depends on "teleradiology" for CXR interpretation. All of these factors have a significant negative impact on treatment and might result in life-altering incidents for patients. Here, abnormality chest X-Ray images or photographs imported from ChestXRay14, which has the largest database, is applied with convolutional neural network (CNN) which is taken as the "test data" later after being pre-processed, to further classify them into normal images, abnormal images and its types. Although, the feasibility of using smartphones for automation is yet to be evaluated, "image-based teleconsultation" using smartphones has gradually been incremented with respect to its popularity where specific deep learning models have been implemented to read anomalies in chest radiographs.

1.1 *Image processing*

Image Processing is widely used by Radiologists for detecting diseases or abnormalities in human body. Abnormalities in chest region, can be diagnosed by using image processing

DOI: 10.1201/9781003363781-20

133

[28]. An image processing takes poor-quality images as the input and gives improved quality of an image as an output. Image processing includes the following: image enhancement, restoration, image acquisition, pre-processing along with encoding and compression.

1.2 *Deep learning*

A part of machine learning algorithms that uses ANNs is called deep learning. The different deep learning network architectures Alex Net, Google Net, VGG16Net and Quoc Net can be used for classification [46]. Alex net is a successful deep learning architecture and google Net is deeper than Alex Net [2].

1.3 *Convolutional neural network*

ConvNet, also known as a convolutional neural network, is a multi-layer neural network that has been used to extract information from input images. CNN does not require a lot of pre-processing. Convolutions and pooling can be used to fit an image into its basic features.

From the Radioscopy report, CXR is diagnosed. The arrow in the image represents the location of abnormal parts. (b) and (c) shows focus of Model-RECA and the Model-ORIG examined by CXR original. (e) and (f) shows the anchor of the Model-RECA and the Model-ORIG which are examined on CXR pictures. Each image above shows the strengths of the contribution from every nook and cranny for predicting consolidation is denoted by the colours from blue to red.

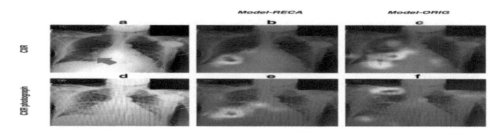

Figure 1.   Diagnostic focus of two models and their visualisation.

2   RELATED WORK

2.1 *Internal validation*

MIMIC-CXR database was developed first, and models were tested using this. MIMIC_CXR database derives the training images, the Photo-MMC, and testing photographs. From the standard AUROC of 0.86 to 0.77 (p < 0.0001) performance decrease is shown by Model-ORIG comparatively as shown in the Figure 2a. From the standard AUROC of 0.77 to 0.84 (p < 0.0001), performance recovery is shown by ModelRECA after model recalibration. The Receiver Operating Characteristic (ROC) curves are displayed in Figure 3. Blue lines represent the contrast references. The performance of Model-RECA's on the Chest X Ray images is indicated by the green lines, while the performance Model-ORIG's when calculated from the Chest X Ray images is shown by the yellow lines. The results underscore the importance of this study and reiterates the two insights. First, original CXRs trained by model was not able to maintain its performance on CXR photographs. Secondly, improvement in the model presentation transferred image-based models' diagnosing was increased by recalibration process.

2.2 External validation

CheXpert are external database photographs these models were tested and progressed from the MIMIC-CXR database to look over whether the two insights specified above can be generalised to another database. Based on the on the Photo-CXP, the models were tested. As in Figure 2b, from an averaged AUROC of 0.75 to 0.67, Model-ORIG decreases its performance when tested on the Chest X Ray photographs ($p < 0.0001$) across the six radiological findings. AUROC from 0.67 to 0.75, again captures the performance loss ($p < 0.0001$) significantly by Model-RECA. The results here were more consistent compared to the internal validation. Later, the transfer learning model (Model-TRNS) had better performance compared with ModelPHOT (0.71 vs. 0.59, $p < 0.0001$) and Model-ORIG (0.71 vs. 0.67, $p < 0.0001$), the Model-RECA still defeat Model-TRNS (0.75 vs. 0.71, $p < 0.0001$). The recalibration process bring forth a better solution for diagnosis of Chest X Ray photographs were implied by the results but still domain shift problems are solved by using transfer learning strategy. Compared to the figures, The MIMIC-Chest X Ray and CheXpert databases' differences, shown in Figure 2a and 2b, led to AUROC reductions for all labels and models with the exception of the pleural effusion.

2.3 End-user scenario

To evaluate the models and simulate the model performance for usage in actual clinical practise, the Photo-MED dataset was employed. Pictures from their own mobilephones and desktop monitors were taken by nine medical residents, and they were asked to redirect them to their colleagues for the next discussion on the photographs. As shown in Figure 2c, the comparison results can be seen. As with internal or external validation, the same results were obtained. Among the four models examined, ModelRECA, the recalibrated model, performed exceptionally well (0.75 vs. 0.72, p 0.0001; 0.75 vs. 0.70, p 0.0001; and 0.75 vs. 0.61, p 0.0001). The achievement has the same comparative reference (0.75 vs. 0.75) as the comparison. For the real clinical works recalibrated model has the best potential method to be deployed and results are also demonstrated the models improvement is not user-dependent.

2.4 Diagnostic visualization

For demonstration of explainability, activation maps were created in this study. The user is provided some insight by recognising the segment of the Chest X Ray photographs that challenfing with respect to the algorithm output. Dependability of an algorithm's classification can be determined particularly using this. The Fig illustrates Model-adaptability RECA's to noise disruption in a scenario known as consolidation using gradient-weighted class activation mapping. When applied to the original Chest X Ray pictures, both models correctly detected the abnormal opacification at the right lower lobe. ModelORIG was confused by photographic noise when applied to the Chest X-ray picture and decided to consolidate labelling based on the right clavicle. Conversely, making model stability apparent The Model-RECA discovered the same focus point when it was tested on the original Chest X Ray photos, despite the fact that there was apparent noise in the image. However, if the Chest radiography or the smartphone photo quality is poor, noise may be introduced into the algorithm.

2.5 Cross-database validation

To re-evaluate the stability of the recalibration method, more CheXpert-based models (Model-ORIG, Model-TRNS, Model-RECA, and Model-PHOT) were built. For training and testing, the roles of the MIMICCXR and CheXpert databases were switched. To ensure that our results were consistent, all methods were reviewed once again. The results and related results are displayed in the Supplementary Figure 2 below.

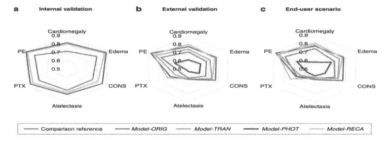

Figure 2.   Related results of review – efficiency of radiography.

According to AUROCs, Figure 2 illustrates the efficiency of radiography detection for six labels, including cardiomyopathy, edoema, consolidation, atelectasis, pneumothorax, and pleural effusion.

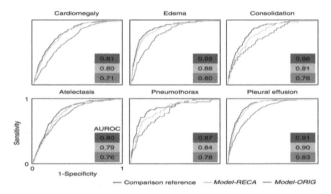

Figure 3.   ROC curves for the detection of edoema, cardiomegaly, atelectasis, pneumothorax, pleural effusion, and consolidation.

Figure 3 shows the ROC curves for the detection of edoema, cardiomegaly, atelectasis, pneumothorax, pleural effusion, and consolidation in Chest X Ray images. The ROC curves produced by the uncalibrated model are shown by blue lines. Yellow and green lines, respectively, represent how the Model-ORIG and the recalibrated model Model-RECA interpret CXR images. For each disease ($p < 0.0001$), The AUC of Model-RECA is less than that of Model-ORIG.

2.6   *Generalized methods for disease detection*

Initially, collect X-Rays Chest from two repositories and create a CXR image database by taking digital CXR photographic images. Create a host of additional functions to enlarge training data sets into CXR images, instead of accumulating multiple images. By comparing

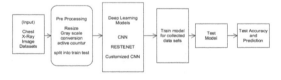

Figure 4.   Methodology in detection of chest radiograph.

the similarities between the actual 175 images and additional effects, "Hyperparameters for extension functions" are tuned. CXR images that were eventually acquired in enlargement were used to train the reprinted model. Three other models (CNN, Rest net and custom cnn) were built to compare. Finally, the models were tested in a database of four CXR images comprising four tests (internal validation, external verification, end user status, and device variability testing.

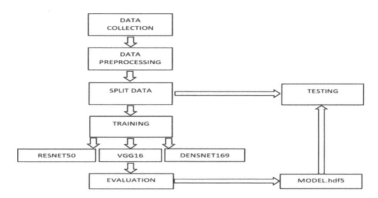

Figure 5.   Steps in collecting, training, and testing the model.

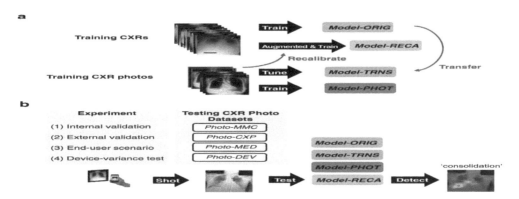

Figure 6.   Shows an example of training the CXRs and converting it from one format to other.

3   COMPARATIVE ANALYSIS

TITLE	DESCRIPTION	RESULT
Recalibration of profound learning models for irregularity location.	It fostered a strategy for recalibration to foster profound learning models to recognize discoveries on CXR pictures. Mirror CXR and CheXpert are the	This model after recalibration accomplished regions under the collector trademark bend for around 80 to 90%, to identify cardiomegaly edema, cardiomegaly, atelectasis,

(continued)

137

TITLE	DESCRIPTION	RESULT
Profound Learning for irregularity discovery in Chest X-Ray pictures	These issues would mitigate using computer-aided triage system by directly communicating to the physician when tests are necessary and more. It is also a technique that has significant growth that allows for classifying the heterogeneous images in a huge dataset.	1. Foster a straightforward pre-handling pipeline by master radiologist guidance and furthermore by picture handling. 2. Make a pipeline that can apply three models of brain network in order of assignments InceptionNet, ResNet and, GoogLeNet of CXR pictures. 3. Use representation of brain network techniques to comprehend which models would weigh intensely and their elements.
Irregularity Detection in Chest X-Ray Images Using Uncertainty Prediction Autoencoders	Gives one of the new anomaly discovery strategies which depends on an autoencoder.	Yields are the vulnerability expectations utilizing pixel-wise alongside the remade typical variant of the information picture Larger the remaking mistake, higher is the vulnerability at ordinary district limits.
CXNet-m1: Anomaly Detection on Chest X-Rays with Image-Based Deep Learning	This mainly develops VGG model with few architecture layers. It also brings about ambiguous diagnosis To overcome the insufficient differentiation of chest X-beam pictures	This model diminishes the boundaries by 51.92% contrasted with DenseNet121, 97.51% than VGG-16, 85.86% than Res-50, 83.94% than Xception, however expanded contrasted with MobileNet by 4%.
Irregularity Detection in Chest X-Ray Images Using Uncertainty Prediction Autoencoders	Gives one of the new anomaly discovery strategies which depends on an autoencoder.	Yields are the vulnerability expectations utilizing pixel-wise alongside the remade typical variant of the information picture Larger the remaking mistake, higher is the vulnerability at ordinary district limits.
Reproducibility of anomaly location on chest radiographs utilizing convolutional brain network in matched radiographs acquired inside a transient span	The assessment of the reproducibility of PC supported discoveries (CADs) with a CNN (convolutional brain organization) on CXRs (chest radiographs) of examples of unusual aspiratory in	From 2010–2016 anonymized CXRs (n = 9792) were obtained and it comprised of five types of patterns of diseases which includes the pleural effusion (PLE), nodule (N), consolidation (C),
Reproducibility of anomaly location on chest radiographs utilizing convolutional brain network in matched radiographs acquired inside a transient span	The assessment of the reproducibility of PC supported discoveries (CADs) with a CNN (convolutional brain organization) on CXRs (chest radiographs) of examples of unusual aspiratory in patients, with an exceptionally short time frame.	From 2010–2016 anonymized CXRs (n = 9792) were obtained and it comprised of five types of patterns of diseases which includes the pleural effusion (PLE), nodule (N), consolidation (C),

4 CONCLUSION

The complexity of CNN's well-designed designs is sufficient to discriminate between normal and pathological chest x-ray pictures with an accuracy that is far above that of random sections. The neural network architecture utilised and the model are resilient for the two image sizes examined (224×224 and 512×512 pixels). Additionally, network data show that the model successfully studies macroscopic aspects. The model's collection of typical Chest X Ray images appears to emphasise symmetry as a key component. Since the performance of deep network design has not altered, the system anticipates that this functionality will also appear in the development of InceptionV3 and Residual Network. There are several approaches to improve the model mentioned above: 1. more preoperative processing, such as pulmonary imaging in the images or margins of CXR images to emphasise lung regions 2. To direct clinical application, measurement of instances like sensitivity with clarity is stressed. 3. Including uncertainty in diagnostic procedures 4 integrating the division so that the network learns simple, tiny aspects as opposed to simply complex ones; and 5. the use of odd but very natural examples uncommon and distinctive. Although this model is not yet ready for clinical discovery, it holds out the promise of a workable future diagnostic network that can differentiate between normal and abnormal chest x-rays and give primary care doctors and radiologists the knowledge they need to drastically cut down on diagnostic time and further enhance the current standard of care.

REFERENCES

[1]   *Recalibration of Deep Learning Models for Abnormality Detection in Smartphone-captured Chest Radiograph*

[2]   *Deep Learning for Abnormality Detection in Chest X-Ray Images*

[3]   *Abnormality Detection in Chest X-Ray Images Using Uncertainty Prediction Autoencoders*

[4]   *CXNet-m1: Anomaly Detection on Chest X-Rays With Image-Based Deep Learning*

[5]   *Pneumonia Detection from Chest X-ray Images Based on Convolutional Neural Network*

[6]   *Reproducibility of Abnormality Detection on Chest Radiographs Using Convolutional Neural Network in Paired Radiographs Obtained Within a Short-term Interval*

[7]   *Unsupervised Deep Anomaly Detection in Chest Radiographs*

[8]   *Anomaly Detection using Machine Learning Techniques*

[9]   Po-Chih Kuo, Cheng Che Tsai, Diego M. López Alexandros Karargyris, Tom J. Pollard, Alistair E. W. Johnson, *Deep learning Models for Abnormality Detection in Chest Radiograph, 2020–2021*

[10] Christine Tataru, Darvin Yi, Archana Shenoyas, Anthony Ma, *Deep Learning for Abnormality Detection in Chest X-Ray Images*, 2017

[11] Yifan Mao, Fei-Fei Xue, Ruixuan Wang, Jianguo Zhang, Wei-Shi Zheng1, and Hongmei Liu, *Abnormality Detection in Chest X-Ray Images Using Uncertainty Prediction Autoencoders, 2018–2019*

[12] Shuaijing Xu, Hao Wu, Rongfang Bie, CXNet-m1: *Anomaly Detection on Chest X-Rays with Image-Based Deep Learning*, 2018

[13] Dejun Zhang, Fuquan Ren, Yushuang Li, Lei Na and Yue Ma, *CXNet-m1: Anomaly Detection on Chest X-Rays with Image-Based Deep Learning*, 2018

[14] Takahiro Nakao, Shouhei Hanaoka, Yukihiro Nomura, Masaki Murata, Tomomi Takenaga, *Unsupervised Deep Anomaly Detection in Chest Radiographs*, 2021

[15] Yongwon Cho, Young-Gon Kim, Sang Min Lee, Joon Beom Seo & Namkug Kim, *Reproducibility of Abnormality Detection on Chest Radiographs Using Convolutional Neural Network in Paired Radiographs Obtained Within a Short-term Interval*, 2020

[16] Julia Amann, Alessandro Blasimme, Effy Vayena, Dietmar Frey, Vince I. Madai, *Explainable AI in Healthcare*, 2020.

[17] Mettler, F. A. Jr. *et al.* Patient Exposure from Radiologic and Nuclear Medicine Procedures in the United States: Procedure Volume and Effective Dose for the Period 2006–2016. *Radiology* 295, 418–427 (2020).

[18]   Rosenkrantz, A. B., Hughes, D. R. & Richard Duszak, J. The U.S. Radiologist Workforce: An Snalysis of Temporal and Geographic Variation by Using Large National Datasets. *Radiology* 279, 175–184 (2016).

[19]   Boissin, C., Blom, L., Wallis, L. & Laflamme, L. Image-based Teleconsultation Using Smartphones or Tablets: Qualitative Assessment of Medical Experts. *Emerg. Med. J.* 34, 95–99 (2017).

[20]   Giansanti, D. WhatsApp in mHealth: an Overview on the Potentialities and the Opportunities in Medical Imaging. *Mhealth* 6, 19–19 (2020).

[21]   Auffermann, W. F., Gozansky, E. K. & Tridandapani, S. Artificial Intelligence in Cardiothoracic Radiology. *Am. J. Roentgenol.* 212, 997–1001 (2019).

[22]   McBee, M. P. *et al.* Deep Learning in Radiology. *Academic Radiol.* 25, 1472–1480 (2018).

[23]   Rajpurkar, P. *et al. Chexnet: Radiologist-level Pneumonia Detection on Chest x-rays with Deep Learning.* Preprint at https://arxiv.org/abs/1711.05225 (2017).

[24]   Baltruschat, I. M., Nickisch, H., Grass, M., Knopp, T. & Saalbach, A. Comparison of Deep Learning Approaches for Multi-label Chest X-ray Classification. *Sci. Rep.* 9, 1–10 (2019).

[25]   Taylor, A. G., Mielke, C. & Mongan, J. Automated Detection of Moderate and Large Pneumothorax on Frontal Chest X-rays Using Deep Convolutional Neural Networks: A Retrospective Study. *PLoS Medicine* 15, https://doi.org/10.1371/journal.pmed.1002697 (2018).

[26]   Annarumma, M. *et al.* Automated Triaging of Adult Chest Radiographs with Deep Artificial Neural Networks. *Radiology* 291, 196–202 (2019)

[27]   Rajpurkar, P. *et al.* Deep Learning for Chest Radiograph Diagnosis: A Retrospective Comparison of the CheXNeXt Algorithm to Practicing Radiologists. *PLoS Med.* 15, e1002686 (2018).

[28]   Wang, X. *et al.* ChestX-ray8: Hospital-scale Chest X-ray Database and Benchmarks on Weakly-supervised Classification and Localization of Common Thorax Diseases. *Proc. IEEE Conf. Computer Vis. Pattern Recognit.* 2017, 2097–2106 (2017).

[29]   Majkowska, A. *et al.* Chest Radiograph Interpretation with Deep Learning Models: Assessment with Radiologist-adjudicated Reference Standards and Populationadjusted Evaluation. *Radiology* 294, 421–431 (2020).

[30]   Nam, J. G. *et al.* Development and Validation of Deep Learning–based Automatic Detection Algorithm for Malignant Pulmonary Nodules on Chest Radiographs. *Radiology* 290, 218–228 (2019).

[31]   Lakhani, P. & Sundaram, B. Deep Learning at Chest Radiography: Automated Classification of Pulmonary Tuberculosis by Using Convolutional Neural Networks. *Radiology* 284, 574–582 (2017).

[32]   Tang, Y.-X. *et al.* Automated Abnormality Classification of Chest Radiographs using Deep Convolutional Neural Networks. *npj Digital Med.* 3, 70 (2020).

Recent Trends in Computational Sciences – Gururaj, Pooja & Flammini (Eds)
© 2024 The Author(s), ISBN 978-1-032-42685-3

A comprehensive review on hate speech recognition utilising natural language processing and machine learning

S. Shilpashree
Department of CSE-JSSATEB, Visvesveraya Technological University, Belagavi

D.V. Ashoka
Department of ISE, JSS Academy of Technical Education, Visvesveraya Technological University, Belagavi, Bengaluru
ORCID ID: 0000-0002-0701-3221, 0000-0003-1326-2387

ABSTRACT: Many social media sites offer a low-cost mode of communication that makes it possible for anybody to immediately connect with countless individuals. Because of this, anybody can produce anything on these networks, and everyone who is intrigued by the material can acquire it, which represents a revolutionary transformation in our community. On the other hand, it has resulted in the emergence of a number of problems, one of which is the propagation and distribution of materials that contain hateful rhetoric. The recognition and surveillance of inciting hatred are becoming an increasingly difficult societal problem as a whole, as well as for individuals, strategists, and scholars. In spite of attempts to leverage automated processes for identification, the capabilities of these methods remain a long way from being adequate, which consistently demands further study on the subject. This study presents a comprehensive assessment of the previous research that has been done in this area, with a particular interest in natural language processing (NLP) algorithms. Our goal is to gain an understanding of the prevalence of offensive speech in social media platforms found online, the most popular manifestations of hatred, as well as the impact obscurity has on hateful texts.

Keywords: Natural Language Processing (NLP), Machine Learning (ML), Hate speech recognition, Computational Linguistics

1   INTRODUCTION

In today's age of social technology, people's interactions with one another, particularly on online chat rooms and social networks, stand out far more than ever. Even though social networking sites offer users the opportunity to speak freely, disrespectful and damaging information is all too common, and it may have a significant influence not only on the customer experience but also on the decorum of a whole society [1]. Verbal abuse is one example of this type of damaging material. Hate speech is defined as language that either explicitly targets or encourages hatred toward a community or a person due to the actual or apparent components of their identification, including their race, faith, or sexual preference [2,3]. These vitriolic remarks that are released publicly ultimately lead to criminal acts. Large social media businesses are conscious of the detrimental impact of hateful speech, and as a result, they have regulations in place concerning the filtering of posts containing such speech. On the other hand, the most frequently utilised systems have many drawbacks [4]. This work adds to previous surveys by updating the literature search on automated linguistic hateful speech identification, focusing on NLP techniques to help achieve this objective.

2 KEY CHALLENGES

About five billion people are using the internet across the globe. As of the beginning of the year 2022, China held the top spot globally in terms of the number of people who used the internet. The world's most popular country, China, had 1.02 billion internet consumers, which is more than three times as many as the United States, which was placed third and had approximately 307 million internet users. Each BRIC country generally had over 100 million users, making them responsible for four of the eight nations with over 100 million users [5]. Figure 1 represents some of the most internet-using countries in the world at the contemporary time.

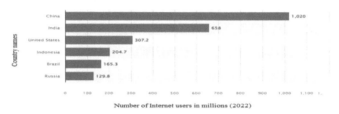

Figure 1. Top 6 internet-using countries in 2022 (*Source: Statistica* [50]).

The proliferation of social media platforms has made it much simpler for individuals to communicate with one another and voice their opinions. Nonetheless, the use of social media is shifting away from sharing or blogging about societal, economic, political, and ideological conflicts to the dissemination of inciting hatred [6]. This trend is occurring throughout every nation. The proliferation of offensive speech and false online information can interfere with democratic national conventions and practices, make it easier for people to violate fundamental rights [7–9], and later lead to isolating minority populations [10]. It is becoming progressively vital to investigate the manner in which detrimental speech promotes thoughts and actions in actual life, such as racist violence, in light of the rise in the amount of hate speech that is being posted online. Nevertheless, there has been a lot of difficulty in experimentally finding the linkages connecting internet speech vitriols and material world repercussions [11]. In contrast, professional censors and consumer complaints do not work very well. It indicates that when a hostile message is identified and removed from the website, it has already begun to have an adverse effect on users [12]. These issues, together with the prevalence of material that can be interpreted as online-hatred, provide a strong drive for the development of automated hateful speech identification. The various issues and challenges in hate speech detection using NLP are Grammatical and lexical deviations,Scarce and heavily skewed data,Sample selection biases,Conflicts of interest in data collection and analysis,Implicity, Multilingual aspects,and Romanization of different languages.The challenge of automatically detecting instances of such utterances is a difficult one due to varying interpretations of what constitutes hateful speech [13,14]. Because of this, the same piece of content could inspire hatred in some people's hearts while having no such effect on others, depending on how each person defines the term [15,16]. The social media nature of hate speech detection faces numerous difficulties, including difficulties related to syntax and language that are more likely to be encountered. A user's idiosyncratic choice of words presents problems with generalizability right at the origin, primarily because it hinders the application of standard model training methods. The issue becomes prominent in the case of dialects [17]. Social media texts can include various slang, codewords, and euphemisms [18]. This makes it difficult for a system to detect the hateful instigation residing in a text [19].

3 RESEARCH MOTIVATION AND METHODOLOGY

Research can be done for a wide variety of causes and motives. The methods, approach, and ultimate objective of this study are founded on a number of fundamental questions. The reasons why this subject is being worked on are addressed here.

3.1 Research queries

The report as a whole addressed, analysed, and provided answers to the following questions like,what is hate speech, what are the challenges in detection of hate speech, how NLP and deep learning techniques used to, from which platform data sets collected and utilized.

3.2 The sources of related works

The selection of academic articles on hate speech detection was made to address the study's queries. Publications from relevant works that neither discussed nor even endorsed the research areas were disqualified for consideration. The following repositories house our key sources for conducting the research article searches after they have been released are, ScienceDirect, IEEE Xplore, Springer MDPI, and Various other publication sources.

3.3 Searching procedure

We used 'Google Scholar' to search for this study's related works and datasets. We combed for keywords linked to this review work, and they are: 'Hate speech recognition, 'Abusive language identification', 'sexism', 'cyberbullying', 'religious hate detection', 'covid hate speech' etc. After the search was complete, the data collected were saved into an Excel file. These data are analysed later.

3.4 Information assortment

The procedure of selecting the proper data source and kind and the proper means to gather the data is known as the data collection process. There are few significant points for picking up the data resources. By considering acdemic research paper published from 2017–22 time range, articles published in prominent journals and covers the thoughts about hate speech-related concerns [49].

4 HATE SPEECH AND NLP

4.1 Hate speech

Using language in such a way that incites hatred is a nuanced topic that is intricately tied to the interactions among diverse groups of people. Various groups and people have attempted to establish criteria for identifying hateful speech. Hate speech, as the term is often understood, is objectionable communication directed towards a group or a person on the basis of personal traits (like ethnicity, faith, or gender) that compromise societal harmony. According to United Nations, the definition of hate speech is any interaction in statement, text, or attitudes that damage or utilises derogatory term or hateful speech with regard to an individual or group simply because of who they are; this includes religious references, ethnic background, national origin, racial group, appearance, descendance, sex, as well as other identity factors [21]. It is possible for hateful speech to be communicated across any expression, such as pictures, comics, jokes, items, actions, and symbolism, as well as being distributed either offline or online. Communication that is motivated by hatred might be "discriminating," which means it is opinionated, discriminatory, or inflexible, or it can be "judgemental," which means it is prejudicial, scornful, or disparaging of a person or group. In contrast to conventional media, hate speech published and disseminated on the internet can be done quickly, cheaply, and discreetly [22]. It offers the capacity to communicate in real-time with a public that is both wide and international. The comparative longevity of hostile material on the web is another source of concern, given that it has the potential to reappear as well as gain prominence with the longer it stays. To effectively shape new reactions, it is essential first to comprehend and monitor instances of hate speech that occur along various online groups and channels. However, these attempts are frequently thwarted because of the massive magnitude of the phenomena, the technical limits of automatic surveillance systems, and the absence of openness displayed by businesses.

4.2 *NLP*

The field of computer engineering and, more particularly, the subfield of AI known as NLP focuses on teaching machines how to comprehend spoken and written languages in a manner similar to that of humans. NLP is the application of statistics-based ML and deep learning algorithms to the discipline of computational linguistics (the regulated modelling of natural speech). When combined, such tools provide machines with the capability to "comprehend" natural communication or written language in all its nuance, including the subject's or author's purpose and emotion. Because of the inherent inconsistencies in natural speech, developing software that can reliably decipher spoken or written information content is an extremely challenging problem [23–25]. Coders must educate usual language-based systems to identify as well as interpret effectively from the outset the idiosyncrasies of natural speech, such as homonyms, humour, metaphors, parables, grammatical and terminology anomalies, and differences in sentence construction. The usage of NLP in detecting hate speech is quite effective. Although, more research advancements in this area make this technology more efficient day by day. Still, it needs further improvements.

5 LITERATURE REVIEW AND COMPARATIONS

We found various information from the list of data we accumulated from the investigative search. The related works published in this domain lead us to various statistics around hate speech detection.

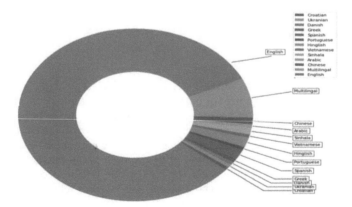

Figure 2.   Statistics (presented in doughnut chart) of hate speech recognition researches on different languages.

However, non-English HS identification toolkits have witnessed a continuous increase due to the development of international parsers and AI technologies. Kocijan *et al.* presented a technology to identify and classify a variety of smears, taunts, and, eventually, hateful speech in the Croatian language [26]. Melnyk *et al.* (2021) use a semi-automated ML method that employs K-means clustering and Word embeddings to identify COVID-19-related hateful speech in the Ukrainian and Russian language remarks and their subjects [27]. Various technologies like these developed in recent years for hate speech recognition in different languages are presented in detail in Table 1. Table 1 represents the Hate Speech Recognition Technologies in Different Languages, most of the authors have explored twitter data and applied traditional machine learning and advance deep learning. The languages explored are Arabic and Spanish. Table 2 represents the NLP-based Hate Speech Recognition Technologies inMultiple languages. Arabic, Italian, Portuguese, German, Indonesian, English languages are targeted to find the hate speech and less NLP techniques are explored.

Table 1. Hate Speech Recognition Technologies in Different Languages.

Paper and Author	Year	Targeted Language	Source	Used Technology Details
Kocijan et al. [26]	2019	Croatian	Facebook news pages	NooJ
Melnyk et al. [27]	2021	Ukrainian, Russian	News Websites	K-means clustering, Word embeddings
Bick et al. [28]	2019	Danish	–	A method for annotating adjectives semantically
Perifanos et al. [29]	2021	Greek	Twitter	BERT, ResNet
Qu et al. [30]	2021	Mexican Spanish	Twitter	XLM, CNN
García-Díaz et al. [31]	2022	Spanish	Media, Twitter	BERT, BETO
Chavez et al. [32]	2020	Peruvian Spanish	Twitter	SVM, MLR, RF, NB
Silva et al. [33]	2020	Portuguese	Twitter	NB, SVM, RF, MLP
Biradar et al. [34]	2021	Hinglish	Social Media	TIF-DNN
Kumar et al. [35]	2022	Hinglish	Social Media	CNN-BiLSTM
Vo et al. [36]	2022	Vietnamese	Social Media	Transfer learning
Wimalasena et al. [37]	2021	Sinhala	Facebook	LR
Hettiarachchi et al. [38]	2020	Romanized Sinhala	Facebook	NB, RF
Al-Hassan et al. [39]	2021	Arabic	Twitter	LTSM, CNN + LTSM, GRU, CNN + GRU
Althobaiti et al. [40]	2022	Arabic	Twitter	BERT
Masadeh et al. [41]	2022	Arabic	Social Media	Ara-BERT, BERT-AJGT
Tang et al. [42]	2020	Chinese	Online world	Hierarchical attention capsule network

Table 2. NLP-based Hate Speech Recognition Technologies in Multiple languages.

Paper and Author	Year	Targeted Languages	Source	Technology Details
Elouali et al.	2019	Arabic, Italian, Portuguese, Indonesian, English, German, Hinglish	Twitter	CNN
Vitiugin et al. [43]	2021	English, Spanish	Twitter	Multilingual Interactive Attention Network
Corazza et al. [44]	2020	English, Italian, German	Twitter, Facebook	LSTM, GRU, Bi-LSTM
Song et al. [45]	2021	Portuguese, Spanish, French, Italian, Russian, Turkish, English	Civil Comments, Wikipedia	MBERT, XLM-R

(continued)

Table 2. Continued

Paper and Author	Year	Targeted Languages	Source	Technology Details
Huang *et al.* [46]	2022	Italian, English, Spanish, Portuguese	Twitter	RNN
Zhang *et al.* [47]	2022	46 Languages	Reddit	MT5, DialoGPT
Vashistha *et al.* [48]	2020	English, Hindi	Various sources	LR, CNN-LSTM, BERT

6 CONCLUSION AND FUTURE DIRECTIONS

In this piece of review work, we offered a critical assessment of how the automatic identification of hateful speech in the material has grown over the past several years. Additional types of inciting hatred, such as cyberbullying, verbal abuse, derogatory language, and unpleasant situations, were taken into consideration in this research. Through this study, we found several languages, although some of the world's top spoken languages are not considered for research in this field. One of the reasons for this may be the lack of resources and interest of scholars in those languages. Dialects and usage of abbreviated terms in a language can make the speech more complex, leading towards low efficiency of hate speech detection models. We also saw that Twitter is a significant breeding ground for hateful speech among all social platforms. Romanisation of a language can be an advantage, as well as a disadvantage for an NLP approach. This review also indicates that the industry of hate speech identification requires advanced skilled annotators.

CONFLICTS OF INTEREST

The authors declare that there is no conflict of interest regarding the publication of this article.

ACKNOWLEDGEMENTS

The paper reflects only the authors' views. The authority is not responsible for any use that may be made of the information it contains.

REFERENCES

[1] Nobata C., Tetreault J., Thomas A., Mehdad Y., and Chang Y., "Abusive Language Detection in Online User Content," in *Proceedings of the 25th international conference on world wide web*, 2016, pp. 145–153.

[2] Talat Z., Thorne J., and Bingel J., "Bridging the Gaps: Multi Task Learning for Domain Transfer of Hate Speech Detection," in *Online Harassment*: Springer, 2018, pp. 29–55.

[3] Sharma S., Agrawal S., and Shrivastava M., "*Degree Based Classification of Harmful Speech Using Twitter Data,*" 2018.

[4] Gao L., Kuppersmith A., and Huang R., "*Recognizing Explicit and Implicit Hate Speech Using a Weakly Supervised Two-path Bootstrapping Approach,*" 2017.

[5] Statistica, "Countries with the Largest Digital Populations in the World as of January 2022," *Statista Research Department*, Web Jul 26 2022.

[6] Negussie N., Ketema G. J. A. J. o. H., and Sciences S., "Relationship Between Facebook Practice and Academic Performance of University Students," vol. 2, no. 2, pp. 1–7, 2014.

146

[7]   Duzha A., Casadei C., Tosi M., and Celli F., "Hate Versus Politics: Detection of Hate Against Policy Makers in Italian Tweets," *SN Social Sciences*, vol. 1, no. 9, pp. 1–15, 2021.

[8]   Fuchs T. and Schäfer F., "Normalizing Misogyny: Hate Speech and Verbal Abuse of Female Politicians on Japanese Twitter," in *Japan Forum*, 2021, vol. 33, no. 4, pp. 553–579: Taylor & Francis.

[9]   Brown A. and Sinclair A., *The Politics of Hate Speech Laws*. Routledge, 2019.

[10]   Pelzer B., Kaati L., and Akrami N., "Directed Digital Hate," in *2018 IEEE International Conference on Intelligence and Security Informatics (ISI)*, 2018, pp. 205–210: IEEE.

[11]   Modha S., Mandl T., Majumder P., and Patel D., "Tracking Hate in Social Media: Evaluation, Challenges and Approaches," SN Computer Science, vol. 1, no. 2, pp. 1–16, 2020.

[12]   Chen H., McKeever S., and Delany S. J., "The Use of Deep Learning Distributed Representations in the Identification of Abusive Text," in *Proceedings of the International AAAI Conference on Web and Social Media*, 2019, vol. 13, pp. 125–133.

[13]   Millar S., "Hate Speech: Conceptualisations, Interpretations and Reactions," in *The Routledge Handbook of Language in Conflict*: Routledge, 2019, pp. 145–162.

[14]   Karppinen K., "Beyond Positive and Negative Conceptions of Free Speech," *Blurring the Lines: Market-Driven Democracy-Driven Freedom of Expression*, pp. 41–50, 2016.

[15]   Balkin J. M., "Digital Speech and Democratic Culture: A Theory of Freedom of Expression for the Information Society," in *Law and Society Approaches to Cyberspace*: Routledge, 2017, pp. 325–382.

[16]   Young S. J., Russell N., and Thornton J., *Token Passing: A Simple Conceptual Model for Connected Speech Recognition Systems*. Cambridge University Engineering Department Cambridge, UK, 1989.

[17]   Blodgett S. L. and O'Connor B., "*Racial Disparity in Natural Language Processing: A Case Study of Social Media African-american English*," 2017.

[18]   Taylor J., Peignon M., and Chen Y.-S., "*Surfacing Contextual Hate Speech Words Within Social Media*," 2017.

[19]   Bodapati S. B., Gella S., Bhattacharjee K., and Al-Onaizan Y., "*Neural Word Decomposition Models for Abusive Language Detection*," 2019.

[20]   Mulcaire P., Kasai J., and Smith N. A., "*Low-resource Parsing with Crosslingual Contextualized Representations*," 2019.

[21]   Strossen N., "United Nations Free Speech Standards as the Global Benchmark for Online Platforms' Hate Speech Policies," *Mich. St. Int'l L. Rev.*, vol. 29, p. 307, 2021.

[22]   George C., "Hate Speech Law and Policy," *The International Encyclopedia of Digital Communication Society*, pp. 1–10, 2015.

[23]   Chaudhary M., Saxena C., and Meng H., "*Countering Online Hate Speech: An nlp Perspective*," 2021.

[24]   Nadkarni P. M., Ohno-Machado L., and Chapman W. W., "Natural Language Processing: An Introduction," *Journal of the American Medical Informatics Association*, vol. 18, no. 5, pp. 544–551, 2011.

[25]   Chowdhary K., "Natural Language Processing," *Fundamentals of Artificial Intelligence*, pp. 603–649, 2020.

[26]   Kocijan K., Košković L., and Bajac P., "Detecting Hate Speech Online: A Case of Croatian," in *International Conference on Automatic Processing of Natural-Language Electronic Texts with NooJ*, 2019, pp. 185–197: Springer.

[27]   Melnyk L., "Hate Speech Targets in COVID-19 Related Comments on Ukrainian News Websites," *Journal of Computer-Assisted Linguistic Research*, vol. 5, no. 1, pp. 47–75, 2021.

[28]   Bick E., "A Semantic Ontology of Danish Adjectives," in *Proceedings of the 13th International Conference on Computational Semantics-Long Papers*, 2019, pp. 71–78.

[29]   Perifanos K. and Goutsos D., "Multimodal Hate Speech Detection in Greek Social Media," *Multimodal Technologies Interaction*, vol. 5, no. 7, p. 34, 2021.

[30]   Jia S., "*Non-contextual Binary Classification for Mexican Spanish with XLM and CNN*," 2021.

[31]   García J. A.-Díaz, Jiménez-Zafra S. M., García M. A.-Cumbreras, and Valencia-García R., "Evaluating Feature Combination Strategies for Hate-speech Detection in Spanish Using Linguistic Features and Transformers," *Complex Intelligent Systems*, pp. 1–22, 2022.

[32]   Cuzcano X. M. Chavez and Ayma V. H. Quirita, "*A Comparison of Classification Models to Detect Cyberbullying in the Peruvian Spanish Language on Twitter*," 2020.

[33]   da Silva and Roman N. T., "*Hate Speech Detection in Portuguese with Naive Bayes, SVM, MLP and Logistic Regression.*"

[34]   Biradar S., Saumya S., and Chauhan A., "Hate or Non-hate: Translation Based Hate Speech Identification in Code-mixed Hinglish Data Set," in *2021 IEEE International Conference on Big Data (Big Data)*, 2021, pp. 2470–2475: IEEE.

147

[35] Kumar A., Kumar M., and Yadav D., *"Hate Speech Recognition in Multilingual Text: Hinglish Documents,"* 2022.

[36] Hong-Phuc Vo H., Nguyen H. H., and Do T.-H., "Online Hate Speech Detection on Vietnamese Social Media Texts In Streaming Data," in *International Conference on Artificial Intelligence and Big Data in Digital Era*, 2022, pp. 315–325: Springer.

[37] Wimalasena K., *"Detecting Sinhala Language Based Racial and Religious Offensive Statements in Social Media,"* 2022.

[38] Hettiarachchi N., Weerasinghe R., and Pushpanda R., "Detecting Hate Speech in Social Media Articles in Romanized Sinhala," in *2020 20th International Conference on Advances in ICT for Emerging Regions (ICTer)*, 2020, pp. 250–255: IEEE.

[39] Al-Hassan A. and Al-Dossari H., "Detection of Hate Speech in Arabic Tweets Using Deep Learning," *Multimedia Systems*, pp. 1–12, 2021.

[40] Althobaiti M. J., "BERT-based Approach to Arabic Hate Speech and Offensive Language Detection in Twitter: Exploiting Emojis and Sentiment Analysis," *International Journal of Advanced Computer Science Applications*, vol. 13, no. 5, 2022.

[41] Masadeh M., Davanager H. J., and Muaad A. Y., *"A Novel Machine Learning-Based Framework for Detecting Religious Arabic Hatred Speech in Social Networks."*

[42] Tang X., Shen X., Wang Y., and Yang Y., "Categorizing Offensive Language in Social Networks: A Chinese Corpus, Systems and an Explanation Tool," in *China National Conference on Chinese Computational Linguistics*, 2020, pp. 300–315: Springer.

[43] Vitiugin F., Senarath Y., and Purohit H., "Efficient Detection of Multilingual Hate Speech by Using Interactive Attention Network with Minimal Human Feedback," in *13th ACM Web Science Conference 2021*, 2021, pp. 130–138.

[44] Corazza M., Menini S., Cabrio E., Tonelli S., and Villata S., "A Multilingual Evaluation for Online Hate Speech Detection," *ACM Transactions on Internet Technology*, vol. 20, no. 2, pp. 1–22, 2020.

[45] Song G., Huang D., and Xiao Z., "A Study of Multilingual Toxic Text Detection Approaches Under Imbalanced Sample Distribution," *Information*, vol. 12, no. 5, p. 205, 2021.

[46] Huang X., *"Easy Adaptation to Mitigate Gender Bias in Multilingual Text Classification,"* 2022.

[47] Zhang Q., Shen X., Chang E., Ge J., and Chen P., *"MDIA: A Benchmark for Multilingual Dialogue Generation in 46 Languages,"* p. arXiv: 2208.13078, 2022.

[48] Vashistha N., Zubiaga A., and Sharma S., *"An Online Multilingual Hate Speech Recognition System,"* 2020.

[49] Malagi, Sanjan S., Rachana Radhakrishnan, R. Monisha, S. Keerthana, and D. V. Ashoka. "Content Modelling Intelligence System Based on Automatic Text Summarization." *International Journal of Advanced Networking and Applications* 11, no. 6 (2020): 4458–4467

[50] Statistica, "Countries With the Largest Digital Populations in the World as of January 2022," *Statista Research Department*, Web Jul 26 2022

Prevalence of migraine among collegiate students in greater Noida

Usman Yola Doka Yakubu, Jyoti Sharma & Shahiduz Zafar
Department of Physiotherapy, Galgotias University, India
ORCID ID: 0000-0005-8278-1190, 0000-0002-6510-3903, 0000-0003-2764-7755

ABSTRACT: Migraine is a prevalent headache illness that affects 9.4 percent of men and 20.2 percent of women on average. The third most common condition and the sixth most common particular cause of impairment worldwide. However, studies aimed at prevalence and effects of migraine among Indian and International college students have not been conducted. The present study sought to better understand the prevalence and effects of migraine among university students in Greater Noida, India. Methods: University students of Greater Noida participated in the study between the period of 1st April 2022 and the 21st of June 2022. A questionnaire-based study was undertaken. Undergraduates, postgraduate, and PhD students from various universities in Greater Noida, India, were the target population irrespective of the gender. The study included students from first to last year students who confirmed having migraine at least twice in the previous three months. Having provided informed consent to participate in the study and being able to respond to the survey questions. The students were given access to the questionnaire using an online portal (Mail and What app). A total of 200 responses were received and analysed. Analysis of data was done through SPSS version 22, the tool used was multivariate linear regression to find the association of independent variables (age, gender, nationality and course) with dependent variable migraine headache. MNNOVA tool was further used to find the comparison of each question to independent variables (age, nationality, gender and course).Conclusion: Regression analysis resulted into the significant impact of gender on migraine among college students. In gender, female is more prone to have migraine and feel likely to lie down during the episode of migraine. This limits the activities of daily living among females. Rest other variables (age, nationality and course) doesn't have any significant impact on migraine among college students in Greater Noida.

Keywords: Migraine, college students, Prevalence, Greater Noida

1 INTRODUCTION

The primary headache illness known as migraine is characterized by repeated bouts of varying severity unilaterally throbbing pain, frequently followed by nausea and/or phobic reactions to light and sound. There are two main types: migraine without aura (MO) and migraine with aura (MA) both of which have concomitant migraine symptoms as well as visual, sensory, or other CNS symptoms prior to the headache [1]. Because of its frequency range, accompanying symptoms of disability, and decreased performance, migraine, one of several neurologic illnesses, creates a substantial public health issue. Additionally, migraines might hinder one's ability to interact socially, work, or study when they are at their most productive [2]. The prevalence of migraine is thought to exceed 1 billion people worldwide. The second-leading cause of impairment worldwide in 2016 was found to be migraine, which is also the most disabling illness for people between the ages of 15 and 49 [3]. In a study of

500 students in a medical college in South India, the proportion of headaches was found to be 28 percent overall, but migraine accounted for 42 percent of the headache group [4].

According to a study, a large percentage of medical students at the undergraduate level suffer from the migraine, with sleep apnea being the primary culprit [5]. In addition to higher occurrences of fatigue, irritability, headache pain severity, and comorbidities, migraine is linked to greater familial load, elevated direct and indirect medical costs, and all of these factors [6]. The four categories of primary headaches are migraine, tension-type headache, trigeminal autonomic cephalgia's, and other primary headache diseases, according to the International Headache Society (IHS) (IHS 2013, 2018). Migraine with and without aura is one of the two primary kinds of the headache condition. Symptoms of the nervous system that appear just before a migraine attack are referred to as an aura. Chronic migraine, which differs from episodic migraine in that it occurs 15 days a month or more for at least 3 months and seems to have migraines complaints at least eight days per month, is another classification of migraine (IHS 2013) [7]. Nearly 12% of Americans experience migraines each year1, and those who experience them suffer gravely lessening of life quality (QoL), diminished functioning, and co-existing psychiatric illnesses are some of the consequences of migraine [8].

As a result, it is challenging to address the effects of migraine disorders due to the inherent characteristics of migraine headaches as well as problems related to daily tasks. Patients may actually be restricted in their daily life working effectively during ictal phases and able to perform daily tasks with greater capacity during interictal ones [9]. A disability score of 0.7 from the World Health Organization places migraine among the most incapacitating illnesses (WHO). Most students have been observed to self-medicate, which can result in improper management and occasionally overusing analgesics, which can lead to treatment resistance, in the current climate of rising headache prevalence [10]. Stress and inconsistent sleep are two common migraine causes in the lives of medical students. They are frequently exposed to pressure because to their deferred responsibility for their coursework, the requirement for improved performance, their numerous exams, and their lengthy educational careers.

Additionally significant migraine triggers include diet and nutrition, particularly for children and teenagers. The primary alleged triggers include fasting, chocolate, cheese, and alcohol. Food triggers can have an impact by both directly stimulating brain stem and cortical neural pathways, which causes vasoconstriction and the release of serotonin and norepinephrine [11]. In addition to weather and environment, skipping meals, loud sounds, and menstrual cycles, university students also cited other triggers like poor sleep and study-related stress as common trigger factors [12]. However, studies aimed at prevalence and effects of migraine among Indian and International college students have not been conducted. The present study sought to better understand the prevalence and effects of migraine among university students in Greater Noida, India.

2 MATERIALS AND METHODS

Various university students of Greater Noida participated in the study, between the period of 1st April 2022 and the 21st June 2022 through a questionnaire-based study design. Both male and female undergraduate, postgraduate, and PhD students from various universities in Greater Noida, India, were the target population. Indians and foreign students from all different racial backgrounds were enrolled in the subjects with the age categories from 18 to 32 years or older. All students received an HIT-6 (headache impact test) questionnaire, and those who answered "yes" to the question about experiencing a headache were chosen. Having provided informed consent to participate in the study and being able to respond to the survey questions. The students were given access to the questionnaire using an online portal (Mail and What Sapp).

3 DATA PROCESSING AND ANALYSIS

Analysis of data was done through SPSS version 21, tool used was multivariate linear regression to find the association of independent variables (age, gender, nationality and course) with dependent variable migraine headache. MNNOVA tool was further used to find the comparison of each question to independent variables (age, nationality, gender and course).

4 RESULTS

A total of 441 questionnaire forms were circulated among the student, out of which 308 students had fully filled the questionnaire. But, only 208 participants confirmed of having headache by reporting "yes" to the inclusion criteria question of "Did you have headache 2 or more times in the past 3 months". Eight students were excluded, as they have filled the questionnaire multiple times. Finally, the total participants of 200 students were included in the study. The demographic variables undertook for study were age group, gender, nationality and courses undertaken. Age group variable was categorized into four categories; 1st category consisted of age group range of 18–22years, included 57.5%(N = 115) of participants, 2nd category consisted of age group range of 23–27years, included 35.5%(N = 71) of responses, 3rd category consisted of age group range of 28–32years, included 5%(N = 10) of participants, and the 4th category consisted of 32years and above, included 2%(N = 4) of participants. The female participants' response were 55%(110) and male participants were 45%(N = 90). Nationality; 51.5%(110) response were Indians and 48.55%(90) were internationals. All UG, PG and PhD students participated in the study and the responses received included 67%(134) UG response, 31.5%(63) PG response and 1.5%(3) PhD responses. As depicted in Figure 1.

Table 1.

Age	Number	Percentage
18-22	115	57.5%
23-27	71	35.5%
28-32	10	5%
32 and above	4	2%
Gender		
Male	90	45%
Female	110	55%
Nationality		
India	103	51.5%
Others	97	48.55
Course		
Undergraduate	134	67%
Postgraduate	63	31.5%
PhD scholar	3	1.5%

Figure 1.   Percentages Of Demographic Variables.

Further to analyse the relationship of the demographic variables with migraine through HIT-6 questionnaire on college students, the multivariate regression analysis was done through SPSS version 21. Regression analysis of the first hypothesis, stated that the gender carries significant impact on migraine on college students. The dependent variable migraine was regressed on predicting variable gender to test the hypothesis. H_{01}. Gender significantly predicted that migraine among college students is significantly affected by gender (R = .170, p = .016). These results clearly direct the positive effect of the gender. For analysing the second hypothesis, age carries no significant impact on migraine on college students. The dependent variable migraine was regressed on predicting variable age to test hypothesis. H_{02}. Age (R = .191a, p = .121) which indicates that age have no significant role in migraine among college students.) As, level of significance, (p>.05) which clearly indicate that courses have no significant role in migraine among college students.

For third hypothesis testing, the demographic variable, nationality have no significant impact on migraine among college students. The dependent variable migraine was regressed on predicting variable gender to test hypothesis. H_{03} Nationality (R = .000, p = .999) which indicate that nationality did not have significant role in migraine among college students.To analyse, fourth hypothesis, course do not have significant impact of migraine among college students. The dependent variable migraine was regressed on predicting variable course to test hypothesis. H_{04}. Course (R = .065, p = .359) which clearly indicate that courses does not have significant role in migraine among college students.Overall gender has significant on migraine (p < 0.5) but there is no significant effect of age, nationality and course on migraine. HIT-6 migraine score for university students suffering from migraine were analysed by multivariate linear regressions, the result is depicted in Figure 2 below:

Table 2. Regression Analysis

Hypothesis	(Regression) R	Beta coefficient	R^2	F	t- value	p- value	Hypothesis supported
H_{a1}(gender)	.170a	.170	.029	5.884	2.426	.016b	Yes
H_{a2}(age)	.191a	.026	.037	1.849	.323	.121b	Yes
H_{a}^{3}(nationality).000		-.002	.000	.000	-.001	.999b	Yes
H_{a}^{4}(course)	.065a	-.980	.004	.844	-.919	.359	Yes

The regression analysis showed: At the reference p < 0.05

Figure 2.

5 DISCUSSION

Our study findings gender is significantly associated with migraine, which reveals that females have more prevalence of migraine compared to men. The similar were highlighted in study done by Gorjizadeh et al. (2021) which focused on the causes of migraine headaches in an Iranian community, revealed that women were more likely than males to experience migraines. The reason supposedly may have physiologically basis, as female hormones may contribute to the development of female sufferers of migraines. According to reports, migraine attacks rise when oestrogen levels fall and vice versa. Nearly 50% of women say that menstruation may be a migraine trigger, making it the most important risk factor for the development and maintenance of headaches and migraines [13]. Also supported by Panigrahi et al. (2019) study on prevalence, pattern and associated psychosocial factors of headache among undergraduate students of health profession. Stated that the result in Female students experienced headaches more frequently, and these headaches were strongly linked to lifestyle factors such sleep disruption, daily soft drink intake, and health dissatisfaction [14]. More so in HIT-6, question no. 3: When you have a headache, how often do you wish you could lie down? Female students were more likely to lie down which is significant at .016. Also in question 2: How often do headaches limit your ability to do usual daily activities including household work, work, school, or social activities? Females tends to be more affected our result is partially significant at .055. Age group between 18–22 are more prone to migraine, mean value 3.00 followed by 28–32 and 32 and above, mean values each 2.50. It is crucial to study headaches in university students since, due to factors associated to academic activity such worry, stress, lack of sleep, and poor eating habits, perhaps compared to the general population, this group is more likely to get headaches. Students studying to become health professionals are also exposed to these headache-causing elements. The current study found that students in the health professions experienced a high incidence of headaches, with an overall frequency of 73.1 percent in the previous year. According to Birru *et al.* 2016. 67.2 percent of undergraduate students studying medicine and health sciences reported having headaches the previous year [14]. Others have also noted a sizable relationship between gender, the presence of a family history of headaches, and the frequency of migraine in both male and female students [13]. However, in our study level of education is not significantly associated with migraine this was supported by a Canadian article stated there were no significant differences in marital status, education levels or income bracket in migraineurs [15]. In our study nationality is not significantly associated with migraine which is in contrast by a study stated that differences in prevalence can be ascribed to racial, environmental, dietary, psychological, and social factors that affect a particular group and cause headaches [10]. In the future study empirical study can be done in factors that may result in migraine. Also, individual nation wise or ethnic can be done. And more study is encouraged to minimize the occurrence of migraine among college students.In our study nationality is not significantly associated with migraine which is supported by a Chinese study stated that even though these distinctions were insignificant, America had the highest pooled prevalence, followed by Asia, Europe, and Africa [16]. Which is in contrast with a study that stated that differences in prevalence can be ascribed to racial, environmental, dietary, psychological, and social factors that affect a particular group and cause headaches [10]. Age group between 18–22 are more prone to migraine, mean value 3.00 followed by 28–32 and 32 and above, mean values each 2.50. Similar to this, several research have discovered that the peak age for female migraine incidence is between 20 and 24 [17].

6 CONCLUSION

Finally, through the study the conclusion done found gender to be more significantly associated with migraine and female are more likely to lie down when having migraine may be due to multitasking. However, females tend to be more affected by migraine in limiting the activities of daily living. The study was limited to private universities in Greater Noida. Similarly large sample size can be used. Future study can be done on ethnicity and nationality to understand the effect of migraine. However, the study was limited to students' only future study can be done on lecturers and workers.

REFERENCES

[1] Sutherland, H. G., "Advances in Genetics of Migraine", *The Journal of Headache and Pain*, Vol.20, pp 72, 2019.
[2] Ibrahim, N. K., "Migraine Among Students from The Faculty of Applied Medical Sciences, King Abdulaziz University, Jeddah, Saudi Arabia", *Journal of Advances in Medicine and Medical Research*, Vol.22, pp 31–14, 2018.
[3] Houts, C. R., "Content Validity of HIT-6 as a Measure of Headache Impact in People with Migraine: A Narrative Review", *The Journal of Head and Face Pain*, 2019.
[4] Alharbi, A. A., "Migraine among Medical and Non-Medical Students of Hail University", *The Egyptian Journal of Hospital Medicine*, pp. 3343–3350, 2018.
[5] Irfan, K.J. (2014). "A Cross Sectional Analysis for Prevalence of Migraine Among Undergraduate Medical Students: An Institutional based Study", *Indian Journal of Basic and Applied Medical*.
[6] Houts, C. R., "Reliability and Validity of the 6 Item Headache Impact Test in Chronic Migraine from the PROMISE 2 Study", *Quality of Life Research*, pp.931–943, 2021.
[7] Sharpe, L., "Psychological Therapies for the Prevention of Migraine in Adults", *Cochrane Library*, 2019. DOI: 10.1002/14651858.CD012295.pub2.
[8] Smitherman, T. A., "Negative Impact of Episodic Migraine on a University Population: Quality of Life, Functional Impairment, and Comorbid Psychiatric Symptoms", *American Headache Society*, Vol.5, pp.581–589, 2011.
[9] Leonardi, M., "A Narrative Review on the Burden of Migraine: When the Burden is the Impact on People's Life", *The Journal of Headache and Pain*, Vol.20, pp.41, 2019.

[10]   Nandha, R., "Prevalence and Clinical Characteristics of Headache in Dental Students of a Tertiary Care Teaching Dental Hospital in Northern India", *International Journal of Basic & Clinical Pharmacology*, Vol.22, pp 79–80, 2012.

[11]   Oraby, M. I., "Migraine Prevalence, Clinical Characteristics, and Health Care-seeking Practice in a Sample of Medical Students in Egypt", *The Egyptian Journal of Neurology, Psychiatry and Neurosurgery*, Vol. 57, pp 26, 2021.

[12]   Gu, X., "Migraine Attacks Among Medical Students in Soochow University, Southeast China: A Cross-sectional Study", *Journal of Pain Research*, 2018.

[13]   Gorjizadeh, N., "Trigger Factors Associated with Migraine Headache Among Northern Iranian Population", *Romanian Journal of Neurology*, 2021.

[14]   Panigrahi, A., *"Prevalence, Pattern, and Associated Psychosocial Factors of Headache Among Undergraduate Students of Health Profession"* Elsevier, 2019.

[15]   Brna, P., "Health-related Quality of Life Among Canadians with Migraine", *Headache Pain*, Vol8, pp 43–48, 2006.

[16]   Wang, X., "The Prevalence of Migraine in University Students: A Systematic Review and Meta-analysis", *European Journal of Neurology*, pp 1–12, 2015.

[17]   Smitherman, T. A., "Negative Impact of Episodic Migraine on a University Population: Quality of Life, Functional Impairment, and Comorbid Psychiatric Symptoms", *American Headache Society*, Vol.51, pp 581–589, 2011.

Recent Trends in Computational Sciences – Gururaj, Pooja & Flammini (Eds)
© 2024 The Author(s), ISBN 978-1-032-42685-3

Indoor navigation using BLE beacons

H.A. Chaya Kumari, Spoorthi S. Bhat, M. Sucharith, T.S. Pooja & B.C. Thanmayi
Vidyavardhaka College of Engineering, Mysuru, India
ORCID ID: 0000-0002-5757-4430

ABSTRACT: Institutes/Hospitals usually provide a directory to their available services, but these directories are most of the time static and do not provide any interactivity features to the visitors/ clients. With the rise in smart devices and technology, we see a remarkable demand for real-time location-based services. In this paper, we present a mobile application-grounded navigator. Various indoor localization techniques are studied, out which the beacons based wireless Indoor Positioning System (IPS) is an efficient approach. In an indoor space, Bluetooth Low Energy (BLE) beacons are installed to track the user's position. Parameters such as Received Signal Strength Indicator (RSSI) are considered to evaluate the distance between the smart device and the installed BLE tags. An android application is developed as the user interface to interact with the system. The main purpose behind the conceptual idea of this project is that when a user wants to reach a certain locat ion in an indoor environment, they should easily be able to traverse along with a seamless navigation experience.

Keywords: Indoor Positioning System (IPS), Bluetooth Low Energy (BLE) beacons, Received Signal Strength Indicator (RSSI), Android Application

1 INTRODUCTION

For as long as people have had the urge to detect new places and find their way, navigation has been a important key to future in those areas. In the initial days, navigation by the stars gradually developed into using compasses and today the Global Position System (GPS) can give us a very accurate position almost everywhere in the world. However, it does not support sufficient positioning accurateness for the inside of buildings. Signals are attenuated inside of buildings or opaque high building regions because various materials between satellites and receivers establish interference to the signal propagation. Indoor positioning technologies have began to cover this problem.

With users approval applications to catch snippets of entertainment, streamline calendars and grocery lists, while on the go and make banking more convenient, it is evident that the number of mobile users is greater today than the number of computer users [1]. As smart-phones come equipped with a GPS that helps in finding a way for travelers, it has brought advanced portable location-aware systems into our day-to-day life. Thus, we can provide a solution using a mobile application as it is convenient, faster, and easier to browse. Back in time, the use of Bluetooth for Indoor Positioning Systems was not entertained as it suffered high energy consumption and communication was restricted from point to point.

With the introduction of Bluetooth BLE/Bluetooth Smart/Bluetooth 4.0, the energy consumption was lesser in comparison with the standard Bluetooth. The BLE beacon is a small wireless device transmitting constant radio signals. The signal is detected by nearby smartphones through Bluetooth Smart. The signal passes the ID number to the smart device. The smart device, in turn, delivers the ID number to the cloud server where it checks the action to be performed. BLE is companionable with iPhone, Android phones, Small PCBs, USBs, MAC laptops, Apple TVs [2]. The promising long battery life and lower maintenance requirements make it highly economical and efficient. The drawback of the system includes installation of extensive number of BLE devices as attenuation of the radio signal is reasonably large. Unlike Wi-Fi network, it does not transit widely.

2 LITERATURE SURVEY

R. Abhilash and P. Asha adapts web browsers for implementing an Indoor Navigation System. The user opt for the source and the destination places. Using this data, the route map displays all the possible paths to reach the destination from the source [3]. The shortest path among all the possible routes and important landmarks is emphasized. Whenever a user request is sent with source and destination data, the record generates the floor map and creates the route with the use of the coordinates. The floor map and the route created are matched with the site map of the given area, then the routing directions are created which assists the user [3].

The advantage of this system is that there is no use of internet technologies to create the map and there is no use of the Global Positioning System (GPS) to locate the user's location. Also, the system makes navigation simple by designing it to work in offline mode [3]. It is difficult for the user to track where they are a t a given instance without knowing his/her real time location. Thus, this system fails to offer a smooth navigating experience.

As described by Mila n Herrera Vargas [4], another way to implement a n indoor positioning system is a combination of wireless, positioning and android technologies. An application based on the Wi-Fi manager API of android is developed. It is based on the Received Signal Strength Indicator (RSSI) and Basic Service Set Identifier (BSSID) of each wireless access point (AP). Fingerprinting is a technique used to compare this radio map to the strength of the Wi-Fi signal received at a random point to assume a pattern of mesh that most closely resembles your location.

The advantage of finger printing is that it is an efficient and effortless way to get location information from your existing Wi-fi infrastructure. The infrastructure can be scaled up quickly. However, when equipment is replaced or relocated, radio mapping/surveying is no longer valid and eventually becomes an ongoing maintenance issue. In parts of a building where access points have less coverage (especially at the end of a hallway), there is not enough data to determine their location. In these areas, errors often occur when finding a room or floor. This approach consumes a lot of electricity, and it is not cost – effective.

M. Segura implements indoor navigation using Ultra wideband [5]. Ultra-wideband (UWB) is a short-range radio technology which can be used for indoor navigation. This technology requires an precision of 10-30 centimeters. This short-range radio technology uses an extremely wide frequency range with a bandwidth of at least 500 MHz. UWB works with a low transmission power (0.5 mW / -41.3 dBm/MHz) to avoid disturbing already occupied frequency ranges. To detect the signals, we use unique UWB receivers. To measure the running time of light between a transmitter (UWB tag) and number of receivers are determined based on their location. Currently, an efficient UWB system is not in the picture. The basic requirement involves installation of antennas in every corner of the venue before indoor positioning with UWB can be enabled. For accurate positioning several tags are required. We cannot send enormous amount of data as it is identified by low-speed data transmission. UWB signals can hinder the existing lines of communication as it uses a wider bandwidth.

3 IMPLEMENTATION

A system architecture is the conceptual model that specifies the structure, behavior, and more viewpoints of a system. An architecture portrayal is a formal description and representation of a system, arranged in a way that supports reasoning about the structures and behaviors of the system.

The construction of the proposed system is shown in the Figure 1 the proposed system architecture can be spited into three parts.

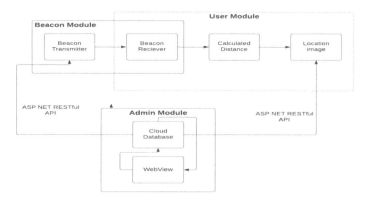

Figure 1.   Indoor navigation system architecture.

3.1   *User module*

This includes the operations that users can perform with the system. It helps to know what all tasks a user can perform and view. It includes the beacon receiver, view calculated distance and viewing the location image.

- *Beacon Receiver:*
  This is the application installed in the user's device which acts as a receiver of the beacon signa ls. When a transmitter is recognized, this application displays the necessary details to the user.
- *Calculated Distance and Location Image:*
  This is a block in the user interface displaying the distance calculated from the transmitter to the receiver and the image for the location that the user is in. A map is displayed on the user interface.

3.2   *Beacon module*

This module includes all the Bluetooth devices involved in the indoor positioning system. It consists of the beacon receiver and the beacon transmitter devices.

- *Beacon Transmitter:*
  The beacon transmitter is used to transmit radio signals so that the receiver receives the signal and displays the wanted information. Here, we use android devices as the beacon transmitter.
- *Beacon Receiver:*
  The beacon receiver is the one which is detects the signals from the transmitter. The bcacon receiver is present in the user module as well as the beacon module.

3.3   *Admin module*

The admin module is a vital part of the system. The admin details are present in the database and authenticated individuals can only access the portal. This includes a web view where the admin is allowed to enter beacon's ID, location URL and description. The user is also allowed to edit and delete the beacon entries. The beacon transmitting signal will be recognized by the receiver if and only if the transmitter ID is entered by the admin and present in the cloud database.

3.4   *Activity model*

- The Receiver device is initialized as shown in the Figure 2.
- Permission to connect is requested if the device is not already linked to the transmitter.

157

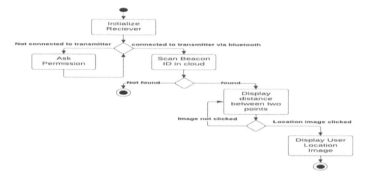

Figure 2.   Activity diagram of indoor navigation system.

- If the device is connected to the Transmitter via Bluetooth, the Beacon ID is scanned and retrieved from the cloud.
- If the appropriate Beacon ID is not found, it can be registered in the database.
- If a Beacon ID is found, the distance between the source and destination is displayed in a list format.
- The user can see the destination image by tapping the list item.

3.5   The indoor navigation system

The Indoor Navigation system. The system's first component is the Cloud Database where the beacon's information such as UUID, Description, URL and Admin Credentials are stored.

3.5.1   Admin module

- The admin module fetches the username and password from the cloud database.
- This is required to perform a validation check on the user interface and allow authenticated admin to log into the web application.
- After the login, the admin can create, edit, delete, and view beacon entries which are then stored and updated in the database.

3.5.2   Beacon module

- The beacon module uses cloud database to make API calls and fetch the required destination image. The beacon transmitter and android device are connected via Bluetooth.
- The transmitter broadcasts the required destination image as an advertising message. The response is received by the beacon receiver and displayed on the screen.
- The receiver can receive multiple advertised messages from multiple transmitters. The respective transmitter distance from the user's location is displayed on the screen as a list view.

3.6   Web application

Web Application of the system contains of two components Login Module and Beacon Entries View.

The **LoginModule** includes the class Login Component. It consists of the member's username and password. The attributes are the admin details stored in the cloud database. This module also includes validation checks for the members.

The class contains the function Login Verify. The function includes,

- Verification of the fields that were entered. When the status is successful, the admin session begins, and the admin is directed to the beacon entry page.
- When the status is unsuccessful, appropriate errors are notified.

3.7 *Beacon entries view*

This view's prime operations are created, read, update and delete by the admin. It includes the Beacon entry Component class having the members – ID, UUID, Description and URL describing the beacons.

The following functions are defined and invoked to perform the admin operations.

- **GetEntries:** This function is invoked to display the updated list of beacon entries present in the database on the admin web view.
- **showForEdit:** This function is called when the admin wishes to edit the details of any beacon entries. The edit flag is made true. The alterable attributes are ID, UUID, Description and URL describing the beacon.
- **showForDelete**: This function deletes the selected beacon entry from the database and updates the list of beacon view with the updated list of beacon entries from the database.
- **onSubmition:** On the beacon entry form submission, the function checks for any empty or invalid fields and shows error messages accordingly. If the edit flag is true then the existing beacon entry is updated, else a new entry is created.

3.8 *Android application*

Android Application of the system comprises of two components Transmitter and Receiver. The Transmitter component contains the following functions:

- **onRequestPermissionResult:** Requests Bluetooth connection to execute the application.
- **initObjects:** The Beacon ID View and Shared Preferences is initialized. The UUID of transmitter beacon is displayed in Beacon ID View component. Shared Preferences is utilized to obtain Bluetooth related information from the mobile device.
- **processTransmitBeacon:** A condition check is made to see if the Bluetooth adapter is enabled. If not, the Bluetooth Settings screen is displayed. A function call is performed through Shared Preferences, to obtain an existing Beacon ID. If ID is empty, a random id is generated and set to Beacon ID View.
- **transmitBeacon:** A beacon object is created from Beacon library. The value of UUID is set as beacon Id which is obtained from Shared Preferences. The format of the advertisement message is required to broadcast it. Therefore, the beacon layout for All Beacon is set by providing the required specifications.
- **startAdvertising:** Instructs the beacon transmitter object to transmit the advertising message. The advertising image which the transmitter beacon has to broadcast is obtained from the cloud database.

The **Receiver** component contains the following functions:

- **onRequestPermissionResult**: To operate the application, permission to enable Bluetooth connection is requested.
- **initObjects:** A Recycler View is initialized to display a list of transmitter beacons. Text View Message is set to alert the user that the application process is fetching beacons.
- **initRecyclerView:** The Adapter for Recycler View is configured here. The Recycler View displays the data which is fetched from the API calls and automatically updates it on the screen.
- **initRefresh:** When user refreshes the application, the API calls are executed again, and the screen is updated if any incorrect response is received.
- **getBeaconEntries:** The Transmitter Beacon information such as description and advertisement message are fetched through an API call using transmitter ID as a parameter. If an empty response is received, appropriate message is displayed on the screen.
- **setLoading:** Until a response from the required API call is obtained, a loading page is displayed on the screen to keep the user informed.
- **monitorBeacons:** Bluetooth Settings screen is displayed if Bluetooth adapter is not enabled. If there is no Transmitter beacon advertising any messages, an appropriate message such

as 'Scanning for beacons' is displayed on the screen. If found, Text View is set to null, and the response is shown using Recycler View. Each list item in the Recycler View contains distance from the transmitter, advertising image and other transmitter information.

- **setOnClickListener:** The destination image or the advertising image received from a particular transmitter beacon is displayed on screen when a list item is tapped by the user.

4 RESULTS

The system includes two android applications, one which acts as a beacon (transmitter), the other which acts as a user application (receiver). It also includes one web application for the admin and a cloud database. The receiver captures the signa l only if the transmitter beacon information is already present, the UUID is retrieved from the database. If not, a randomly generated UUID is produced and displayed on the screen, which the admin can subsequently add to the database. The transmitter beacon broadcasts the advertising message. The receiver application receives the advertised information from the transmitter. The location image with the calculated distance is shown in the form of a list view. On tapping the list item, the advertising message which was broadcasted by transmitter is displayed as the desired destination image by the receiver application. The web application provides a login option through which admin can log into the application. the login page has two fields: username, password. Both the fields are mandatory. After the user submits the form, a validation check is performed, and if any errors are made by the user, an appropriate message is displayed. The admin/user dashboard for Beacon Entries contains a form. In the form field provided, admin can create, read, update, delete beacon entries to and from the database. A dashboard containing details of the Registered Beacons is displayed. In the table, the UUID, Description, URL of each beacon entries is shown, with an option to edit and delete the respective beacon entry.

5 CONCLUSIONS

In this work, we developed an android app and a web app that utilizes Bluetooth Low Energy beacons to provide users with basic indoor navigation. The Android application receives the advertised map from the beacon and shows it to the user. Here in our project, we have implemented the Bluetooth Low Energy beacon using an Android application that transmits an image of the database-stored indoor map. To obtain the advertised image and show it to the user, another Android application is used. The map data and beacon device information can both be created, read, updated, or deleted by the campus or indoor space administrator via the web application. The key feature of our project is the usage of Bluetooth Low Energy beacons to implement a basic indoor 4.navigation system. Utilizing most use of our sources and information we have developed this application which takes the present yet another step closer to the future.

REFERENCES

[1] Estel M. and Fischer L., "Feasibility of Bluetooth Ibeacons for Indoor Localization," *in Digital Enterprise Computing 2015.*

[2] Maria Fazio, Alina Buzachis, Antonino Galletta, Antonio Celesti Massimo VillariA "Proximity-based Indoor Navigation System Tacklingthe COVID-19 Social Distancing Measures", *IEEE 2020.*

[3] Abhilash R., Asha P. "Indoor Navigation System", *International Journal of Applied Engineering Research ISSN* 0973- 4562 Volume 10, Number 4 (2015) pp. 10515–10524 (27 January 2016).

[4] Milan Herrera Vargas *"Indoor Navigation Using Bluetooth Low Energy Beacons (BLE)".*

[5] Segura M., Mut V., Sisterna C. "Ultra-widebrand Indoor Navigation System", *IET Radar Sonar Navig.,* 2012, Vol. 6, Iss. 5, pp 402 – 411.

[6] Ankush A. Kalbandhe1, Shailaja. C. Patil "Indoor Navigation with Bluetooth Low Energy in Crowded Places" 978-1-50901338-8/16/$31.00 ©*2016 IEEE.*

Recent Trends in Computational Sciences – Gururaj, Pooja & Flammini (Eds)
© 2024 The Author(s), ISBN 978-1-032-42685-3

Strategic health planner and exercise suggester

Deepa Angadi
CMRIT

C. Kavyashree
New Horizon College

Veena V. Kangralkar
GIT, Belgaum

ABSTRACT: The year is 2022 and it's a high time for the people to take extra care of their health. As per the survey from WHO 13%of the world's adult population is obese which consisted of 11% male and 15% female. Our body is a temple and we must take good care of it to survive in longer race. Strategic Health Planner is the key to unlock the secrets of a healthy body. It provides the proper diet plan according to your body mass index. It reminds you your time to time calorie intake and also requests you to stay hydrated. It has a special feature of smart snap which provides you the nutritional information of the food just by taking its picture from camera. This application not only takes care about your nutrients but also becomes your free personal trainer by giving you access to varieties of exercises to keep your body fit and flexible. The Strategic Health Planner also has Google maps embedded in it to show you nearby gyms, parks and grocery stores. This application is a complete package for your healthy life. The application uses artificial intelligence to plan a diet according to your body mass index.

Keywords: Health Planner, Exercise Suggester, BMI

1 INTRODUCTION

There are many units of application in market that claim to be the foremost effective health planner. but invariably they're going to have one or two choices missing Associate in Nursing if they have all the choices then the appliance area unit a paid one which could charge an honest amount of money for its membership. Strategic Health Planner may be a shot to form the health observance casier by turning into acadcmic degree one-stop declare a healthy life. It not only exclusively monitors your calories but in addition becomes a non-public trainer and takes care of your fitness and suppleness. This application in addition provides the location of handy gyms for serious exercise or Shut Park if the user loves a brisk walk. It in addition shows you the location of shut grocery stores thus making a healthy meal be further easier. There are many units of structures on the market to develop application. There are many structure on the market to develop a cell phone apps. It provides a native frame supported Java language and iOS provides a native frame work supported Objective-C / Swift language. However, to develop academic degree application supporting every OSs, we wish to code in two altogether totally different languages practice two different anatomy [2]. There exists mobile structures supporting every OS. These frameworks vary from easily understandable hybrid mobile app framework (which uses language for program and JavaScript for application logic) to difficult language specific framework (which do the work of fixing code to native code). Not with standing their simplicity or quality, these frameworks invariably have many disadvantages, one in each of the foremost disadvantage being their low reliability. Throughout this case, Flutter an easy and high rated platform supported Dart language provides high contribution towards production by rendering the UI directly among the processes rather than through traditional approach.

2 EXISTING SYSTEM

There are many health planner like healthify, daily routine planner but none provides a complete one stop solution. If the application provides one stop solution then it usually charges you a good amount for its membership. Very few application runs on both mobile and the desktops. The active feedback mechanism and continuous improvement lacks in most of the application. Its drawbacks are although the coder designs an optimized code, the code readability is very inadequate and cannot be modified by the user even if required. Any change to the code requires change in the model itself. The amount of complexity and requirement requested by the code has led to the reduced usage of this application and manual code conversion is preferred. Studies performed on general populations showed that meal planning was positively associated with frequencies of home food preparation and family meal, as well as the presence of fruits for dinner. To our knowledge, only one study in the literature has evaluated the potential link between meal planning and food consumption. It focused on fruit and vegetables specifically, and showed that planning meal ahead was associated with higher fruit and vegetable intakes. However, the latter presented weakness in the dietary intake assessment method since it consisted only of questions on the number of servings eaten per day. Additionally, meal planning was evaluated, among various practices, as a tool to maintain weight among successful weight losers but no data exists on the potential relationship with weight status in the general population. In the present study, we hypothesize that meal planning might encourage home meal preparation, and therefore have beneficial effects on dietary quality and consequently on weight status. Thus, we first described meal planning practices among a large sample of individuals. Then, we investigated the relationships between meal planning and diet quality, based on adherence to nutritional guidelines, energy, macro nutrients and food group intakes, as well as food variety. Finally, we evaluated the association between meal planning and weight status.

3 PROPOSED SYSTEM

The goal of introducing Strategic Health Planner is to create a one-stop solution for a healthy life. The application is made on flutter which is a leading open-source software which gives the application capability to run on IOS, Android and also Windows. The application uses artificial intelligence to plan a diet according to your body mass index. It provides us instructions to perform exercises at home. It has Google maps embedded in it to provide location of nearby gym, parks and grocery stores. Regular reminder is given for breakfast, lunch, snacks and dinner. Also the application makes sure that you stay hydrated [9].

4 ANALYSIS

Technical analysis could be a commercialism discipline utilized to gauge investments and establish commercialism opportunities by analyzing applied math trends from commercialism activity, like value movement and volume.

In contrast to basic analysis, that makes an attempt to gauge a security's worth supported business results like sales and earnings, technical analysis focuses on the study of value and volume. For running this technique, we would like not have any routers that square measure extraordinarily economical. That the system is economically potential enough [7]. Economic analysis is performed to evaluate the development cost weighed against the ultimate income or benefits derived from the developed system. For running this system, we need not have any routers which are highly economical. So the system is economically feasible enough.

5 SYSTEM DESIGN

Software style could be a method through that wants square measure translated into an illustration of software. Style could be a place wherever style is fostered in software package Engineering. Style is that the common thanks to accurately translate a customer's demand within the finished product. Style creates an illustration or model, provides details about software organization, design, interfaces and parts that square measure necessary to implement a system. The system of logic style got wind of as a results of systems0analysis is regenerate into physical system style.

5.1 *Data flow diagram*

The DFD is also called as bubble chart. It is a simple graphical representation that can be used to represent a system in terms of the input data, processing carried out on these data, and the output data is interpreted by the system.

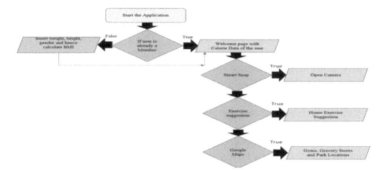

Figure 1.   Data flow diagram.

Figure 1 essentially shows how the data control flows from one module to another. Unless the input filenames are correctly given the program cannot proceed to the next module. Once the correct input filenames are given by the user parsing is done individually for each file. [2] The required information is taken in parsing and an adjacency matrix is generated for that. From the adjacency matrix, a lookup table is generated giving paths for blocks. And the final sequence is computed with the lookup table and the final required code is generated in

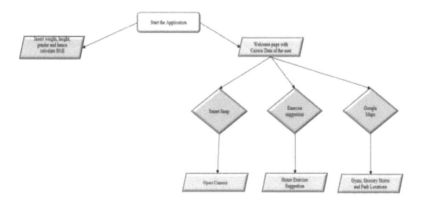

Figure 2.   Component diagram.

an output file. In case of multiple file inputs, the code for each is generated and combined together.

The component diagram for the decentralized system ideally consists of different modules that are represented together via a common module for the user. The user is required to have the input files in the current folder where the application is being used. It is interesting to note that all the sequence of activities that are taking place are via this module itself, i.e. the parsing and the process of computing the final sequence. The parsing redirects across the other modules till the final code is generated.

6 IMPLEMENTATION

Perpetration has form of motifs that deserves to bandy and conjointly punctuate a number of their limitation that encourage happening chancing result additionally as high lights a number of their benefits that reason these motifs and their options square measure utilized in this paper. During this script, Flutter – a straightforward and high performance frame grounded on Dart language, provides high performance by rendering the UI directly within the in operation system's Flutter contraptions conjointly supports Strategic Health Planner hardiness and gestures. The operation sense is grounded on reactive programming. By ever-changing the state of the contrivance, Flutter can mechanically compare the contrivance's state (previous and new) and render the contrivance with solely the mandatory changes rather of re-rendering the entire contrivance. [6]

The goal of this project was to use the largest publicly available collection of recipe data (Recipe1M+) to build a recommendation system for ingredients and recipes. Train, evaluate and test a model able to predict cuisines from sets of ingredients [22]. Predict the probability of negative recipe-drug interactions based on the predicted cuisine. Finally, to build a mobile application as a step forward in building a reccommendation system which accounts for the user taste preferences maximizes the number of healthy compounds and minimizes the unhealthy ones in the food.

Tensor Flow may be a software system library or frame platoon to use machine skill and deep skill generalities within the best manner. It combines the procedure pure mathematics of improvement ways in which for simple computation of diverse fine expressions. Allow us to currently take into account the subsequent necessary options of Tensor Flow one [6]. It includes a degree that defines, optimizes and performs accurate evaluation with the assistance of multi-dimensional arrays referred to as tensors. It includes study of neural network skill. It includes a high climbable purpose of calculation with impressive knowledge sets. Tensor Flow uses GPU computing, automating operation.

It conjointly includes a singular purpose of improvement of same memory and therefore the knowledge used Nutritionix API Nutritionix may be a hunt machine and info for nutrition info. Druggies will browse or search by cuffs and food to search out nutrition info. Druggies may also partake their nutrition knowledge with Nutritionix. [23]. The Nutritionix API permits inventors to pierce and integrate the practicality and knowledge of Nutritionix with different operations and to provide new operations. Strategic Health Planner Google Charts API may be a free internet mapping service by Google that gives varieties of geographical info. Exploitation Google Charts, one can. 1. Look for places or get directions from one place to a different. 2. Read and navigate through vertical and perpendicular bird's-eye road position pictures of colorful metropolises round the world. 3. Get specific info like business at a selected purpose. Google provides associate degree API exploitation that you'll customize the charts and therefore the info displayed on them. This chapter explains a way to load a straightforward Google Map on your internet runner exploitation HTML and JavaScript.

7 RESULTS

Figure 3.   Screenshots of application.

8   CONCLUSION

With the event of information analytics and mobile computing, this method is ready to produce a lot of sensible and convenient applications and services. Moreover, assisted by machine learning, data processing, AI, and alternative advanced techniques, health planner systems might conjointly play a crucial role as a guide of healthy lifestyles, as a tool to support higher cognitive process, and as a supply of innovation within the evolving health planner system. This project presents the strategic health planner systems assisted by information analytics and mobile computing, in this consisting of the info assortment layer, the info management layer, and also the service layer. This project conjointly introduces some representative applications supported the planned theme that are verified or incontestable to be able to give a lot of strategic, skilled, and personalized health planner services [4].

Though this project presents a comprehensive system style for strategic health planner systems assisted by information analytics and mobile computing, a lot of advanced techniques ought to be enclosed in our future work, like psychological feature computing, deep learning, and effective computing, to any improve the standard of service and user expertise. Train, value and check a model able to predict cuisines from sets of ingredients. Predict the chance of negative formula – drug interactions supported the anticipated cooking. Finally, to make a mobile application as a revolution in building a recommendation system that accounts for the user style preferences maximizes the quantity of healthy compounds and minimizes the unhealthy ones within the food. (For example, don't differentiate among departments of an equivalent organization) [14].

REFERENCES

[1]   *Meal Developing With is Said to Food Choice*, Diet Quality and Weight by *Int J Behav Nutr Phys Act.* – missionary Ducrot, Caroline Méjean https://www.ncbi.nlm.nih.gov/pmc/arti-cles/PMC5288891/

[2]   *SmartDiet: A Personal Diet Authority for Healthy Meal Developing with* – Tim A., Samantha H. https://www.researchgate.net/publication/221030630_SmartDiet_A_personal_diet_consultant_for_healthy_meal_planning

[3]   *Recommending Healthy Meal Plans by Optimising Nature-inspired Many-objective Diet Disadvantage* – Connor J.K. https://jour-nals.sagepub.com/doi/full/10.1177/1460458220976719

[4]   *set up|meal plan|design|plan} (Main plan, Evidence, Analysis, Link)* – Tom Hardy https://www.thinksrsd.com/wp-content/up-loads/2016/12/MEAL-Mnenomic.pdf

[5]   Kanungo T, Mount D. M., Netanyahu N. S., Piatko C. D., Silvermin R., and Wu A. Y., "An Economical k-means Clump Algorithm: Analysis and Implementation," *IEEE Trans. Pattern Anal. Mach. Intell.*, vol. 24, no. 7, pp. 881–892, Jul. 2002

[6]   *Nutritionix* https://developer.nutritionix.com/docs/v2

[7]   *Tensorflow* https://www.tensorflow.org/resources/learn-ml?gcli

[8]   *Exercise* https://www.healthline.com/health/at-home-workouts

[9]   *GoogleMaps* https://developers.google.com/maps/documentation

[10]   W. Xu, J. He, and Y. Shu, "DeepHealth: Deep Illustration Learning with Autoencoders for Health care Prediction," in *Proc. IEEE Symp. Ser. Comput. Intell. (SSCI)*, Dec. 2020, pp. 42–49.

[11]   *GoogleFirebase* https://firebase.google.com/docs?gclid=sAFp Netron https://github.com/lutzroeder/netro

[12]   Deng L., Hinton G., and Kingsbury B., "New Sorts of Deep Neuralnetwork Learning for Speech Recognition and Connected Applications: Aatop Level View," in *Proc. IEEE Int. Conf. Acoust.*, Speech Signal technique,8599–8603.

[13]   *Instruction* https://www.delish.com/cooking/g2150/comfort-food/

[14]   Singh, S.; Mamatha, K.; Ragothaman, S.; Raj, K.D.; Anusha, N.; SusmiZacharia. Waste Segregation System Using Artificial Neural Networks. *HELIX* 2017, 7, 2053–2058.

[15]   Ranada, P. Why PH is the World's 3rd Biggest Dumper of Plastics in the Ocean. *Rappler Blog*. Retrieved October 6, 2015, 2015.

[16]   Sousa, J.; Rebelo, A.; Cardoso, J.S. Automation of Waste Sorting with Deep Learning. 2019 *XVWorkshop de Visão Computacional (WVC)*. IEEE, 2019, pp. 43–48.

[17]   Devi, R.S.; Vijaykumar, V.; Muthumeena, M. *Waste Segregation using Deep Learning Algorithm*.

[18]   Bircanoğlu, C.; Atay, M.; Beşer, F.; Genç, Ö.; Kızrak, M.A. Recyclenet: Intelligent Waste Sorting Using Deep Neural Networks. *2018 Innovations in Intelligent Systems and Applications (INISTA)*. IEEE, 2018, pp. 1–7.

[19]   Yang, M.; Thung, G. *Classification of Trash for Recyclability Status*. CS229 Project Report 2016, 2016.

[20]   Lowe, D.G. Distinctive Image Features From Scale-invariant Keypoints. *International Journal of Computer Vision* 2004, 60, 91–110.

[21]   Krizhevsky, A.; Sutskever, I.; Hinton, G.E. Imagenet Classification with Deep Convolutional Neural Networks. *Communications of the ACM* 2017, 60, 84–90.

[22]   Sakr, G.E.; Mokbel, M.; Darwich, A.; Khneisser, M.N.; Hadi, A. Comparing Deep Learning and Support Vector Machines for Autonomous Waste Sorting. 2016 *IEEE International Multidisciplinary Conference on Engineering Technology (IMCET)*. IEEE, 2016, pp. 207–212.

[23]   Arebey, M.; Hannan, M.; Begum, R.; Basri, H. Solid Waste Bin Level Detection Using Gray Level Co-occurrence Matrix Feature Extraction Approach. *Journal of Environmental Management* 2012, 104, 9–18.

[24]   Costa, B.S.; Bernardes, A.C.; Pereira, J.V.; Zampa, V.H.; Pereira, V.A.; Matos, G.F.; Soares, E.A.; Soares, C.L.; Silva, A.F. Artificial Intelligence in Automated Sorting in Trash Recycling. *Anais do XV Encontro Nacional de Inteligência Artificial e Computacional*. SBC, 2018, pp.198–205.

Recent Trends in Computational Sciences – Gururaj, Pooja & Flammini (Eds)
© 2024 The Author(s), ISBN 978-1-032-42685-3

Perspective of deep learning strategies for analysis of 1D biomedical signals

K.R. Vishwanath
VTU
ORCID ID: 0009-0007-4756-1656

V. Rohith, D.C. Vinutha & C.P. Vijay
VVCE
ORCID ID: 0000-0002-5206-858X, 0000-0003-3096-2967, 0000-0002-0525-2368

Soundarya Govinda Rao & N. Sriraam
VTU

ABSTRACT: Artificial Intelligence and Machine learning encompass a broader range of technologies, of which Deep learning is a specific subset and it is expeditiously evolving in the medical field. Most medical devices or imaging techniques use deep learning as a critical tool for analysing the data. The proposed study is a literature review on the methodologies that are used for the sleep -wake cycle detection. The importance of deep learning in the field of ECG, EEG and analysis of infant cry signals is also demonstrated in the sections below. Further analytical studies explain about the facial pattern recognition of the new-born in NICU and also discuss the benefits of analysing the data especially in the case of infants.

Keywords: ECG, EEG, Cry signal, CNN, Deep learning, Infant, NICU

1 INTRODUCTION

Data Science (Machine learning and Deep learning) has its presence in various fields. When it comes to the biomedical domain only few open source texts are publically available. Presently biomedical research is the hottest topic in the research and in that human body is the most interesting topic to research in that. In this paper an application of the same is discussed where in the beginners can deep dive and explores the Biomedical Signal processing with the aid of Data Science [1].

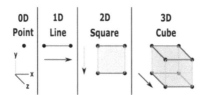

Figure 1. Different types of dimensions in data.

1.1 *One-dimension*

A point can be used as the origin to draw a straight line on either the horizontal or the vertical axis. This will give a line that is one dimension in nature. [1]

Figure (4) The data that can be collected from a human body for medical analysis are EEG, ECG, EMG, voice signal, temperature, PPG, EDA, facial features, etc. The prominent data which we have used are ECG, EEG, voice signals, and facial features from an infant.

The data collected is processed and implemented using a deep learning model. By utilizing examples, deep learning enables computers to learn, making it a machine learning technique.

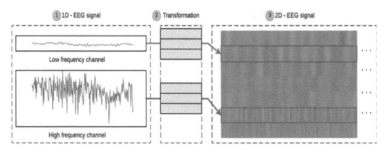

Figure 2. 2D EEG signals are transformed from 1D EEG signals. A smooth texture block is formed by the low frequency channels and a rough texture block by high frequency channels block. The high-frequency channels undergo a transformation that results in the creation of a block with a rough texture.

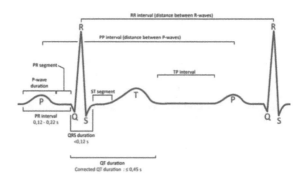

Figure 3. ECG signal breakdown. Some of the terms are explained as, P Waves – Atrial Depolarization (electrical shift between negative and positive ions). PR Interval – Start of the P wave and ends at the beginning of the Q wave. QRS Complex – Ventricle Depolarization. ST-Segment – Starts at the end of the S wave and ends at the beginning of the T wave. T Wave-Ventricular Repolarization. RR Interval – Between QRS complex. QT Interval – Starts at the beginning of the QRS complex and ends at T wave.

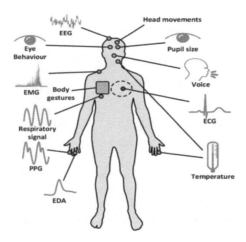

Figure 4. Data collected from a human body.

168

Following the one dimensional convolution block, there are three fully connected layers in the deep convolutional neural network. From each one dimensional convolutional block the data is flattened by bringing together the sensor output from all the sensors (Figure 5). The data that is flattened is link together into a single vector and is transported to a fully connected neural network layer which consist of 32 units. ReLU activation function is used by each of these 32 units. The sub layers consist of 16 hidden units, with the second layer following the first layer. The output layer is the final layer in the network and for applications such as binary stress detection, it consists of a single hidden unit that utilizes the sigmoid function to activate the layer's functions. [2]

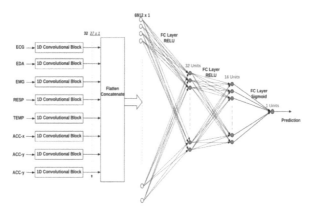

Figure 5. The diagram of the proposed deep one dimensional convolutional neural network.

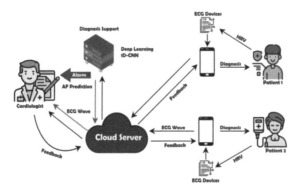

Figure 6. Flow diagram of deep learning model for AF diagnosis system [3].

Figure 6 depicts the structure for acquiring and estimating the ECG data from a server which is in the cloud, upon collecting the data the data is transmitted to the mobile terminal with the aid of bluetooth and this will also get displayed in real time. These HRV signals are verified and interpreted by the DL architecture. If the output of the neural network layer is detected with an Atrial Fibrillation (AF) the cardiologist is informed immediately. The HRV trace that is doubtful can be analysed further as a beat or ECG rhythm to arrive at a conclusion. The diagnosis can then be communicated to the patient using layman terms at a later time.

2 PROCESSING 1-D BIOMEDICAL SIGNALS

When computing and comparing biomedical signals as time series functions, a pre-determined number of steps must be followed. A habitual front facia is the acquisition of the signals using a sensory subsystem. Pre-amplifiers are buffers are used in the adjacent sub-systems for the pre-processing of the signals in the chain. The next stage is filtering and proximate to that is the analog to digital conversion (A/D) stage. The final stages comprises of the removal of the artefacts, detection of the events and interpreting the same and finally extracting the features. [4]

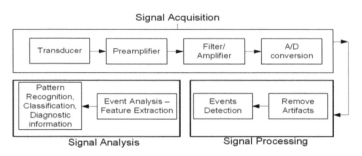

Figure 7.   Biomedical signal processing.

The initial step in measuring, identifying parameters, and characterizing bio signals is to acquire diagnostic data in the form of images or time series that contain relevant information about the underlying physical processes. Analog signals are typically amplified and filtered, and then converted to digital form for ease of processing using digital methods. Signal processing techniques, such as filtering, smoothing, feature extraction, and classification, can then be applied to further analyse the bio signal. The accuracy of the acquired data and subsequent processing steps is crucial for obtaining reliable diagnostic information. Therefore, it is important to carefully design, implement, and validate these steps. The measurement, parameter identification, and characterization of bio signals have significant implications for a range of medical and biological applications, including clinical diagnosis, medical imaging, physiological monitoring, and biofeedback.

Signal processing is typically performed on digital data using software algorithms, but analog signal processing may also be necessary in some cases. Noise is an inevitable factor in measurement systems and can limit the performance of medical devices. Various signal processing techniques are used to minimize noise in measurements. Biomedical measure-ments are subject to four types of variability, including ambient noise or interference, transducer artefacts, electronic noise, and physiological variability. Physiological variability arises because the detected biomedical signal can be affected by biological factors other than the one being studied. Environmental noise can originate from sources both inside and outside the body. An example commonly used to illustrate the need for adaptive noise reduction techniques is the measurement of foetal electrocardiograms. In this scenario, the desired signal is often obscured by the maternal ECG, which creates ambient noise. Traditional noise reduction techniques can be partially effective, but because the source of the ambient noise is unknown, adaptive techniques typically perform better at filtering out unwanted noise. [4]

3   MODEL

Artificial Neural Networks are models that basically learn from experience. ANN models are very useful in pattern recognition problems as they can model complex functions and are easy to use. [6]

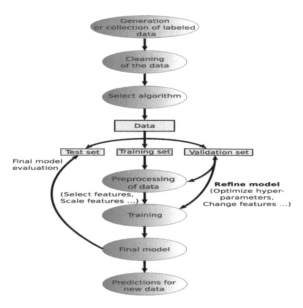

Figure 8.   Model flow for deep learning model.

To evaluate the accuracy of infant expression detection, the VOC2007 metrics (Everingham *et al.* 2010) are utilized. The VOC2007 metrics are primarily designed for detection and classification tasks. Average Precision indicates the detection accuracy and mean Average Precision shows the performance of seven expressions of interest. The frequency of the False Expression Changes (FEC) is computed for each dataset to analyse the consistency of the output. A false term replacement refers to the change of a term from one frame to the next adjacent frame, which does not correspond to the interchange indicated by the ground truth. False term replacements may occur due to false classification of expression and localization of bounding boxes with false detection. In addition to computing the frequency of the facial expression changes (FEC), a reduction rate is introduced to analyse the temporal stability. [7] This analysis is only applicable to systems that use R-CNN and the proposed temporal consistency. Here, F0 represents the frequency of occurrence of false changes when temporal analysis is used, while F denotes the count of false expression changes without temporal processing. [7]

4   LITERATURE REVIEW

Sl. No	Author-Year	Evaluation algorithms	Results
[1]	Premanand S July 19, 2021	Artificial Neural Network	–
[2]	Russell Li & Zhandong Liu December 30, 2020	Deep convolutional neural network, Deep	The deep convolutional neural network was able to achieve accuracy rates of 99.80% and 99.55% for binary and 3-class classification, respectively. Similarly, the deep multilayer perceptron neural network achieved accuracy rates of

(*continued*)

Continued

Sl. No	Author-Year	Evaluation algorithms	Results
		multilayer perceptron neural network	99.65% and 98.38% for binary and 3-class classification, respectively
[3]	Bambang Tutuko, et al. 14 July, 2021	Convolutional neural network	The evaluation results of a deep learning model for AF detection using the DL Cloud System that achieved high accuracy, sensitivity, and specificity when tested with unseen data at a sample period of 0.02 seconds. The accuracy was 98.94% and the sensitivity and specificity were both 98.97%. Additionally, when the number of instances was increased for the three-class classification problem, the accuracy was found to be 96.36%, sensitivity was 93.65%, and specificity was 96.92%
[4]	Alexandros Karagiannis, et al. August 23, 2011	Emperical Mode Decomposition	The implemented algorithm is able to estimate the level of white Gaussian noise present in ECG time series data and provide a 95% confidence interval for that estimation. This can be useful in evaluating the quality of the ECG signal and identifying any sources of interference or noise that may be affecting the accuracy of any analyses or diagnoses based on that data
[5]	YUE SUN, et al. 3 December 2019	Convolutional neural network	The model achieved a high overall accuracy of 91% using two-fold cross-validation on a dataset of 6,534 comfort and 260 10,303 discomfort video frames from 24 infants. Additionally, the model was able to accurately detect comfort frames at a rate of 90% and discomfort frames at a rate of 92%. This is promising and suggests that the model may be useful in accurately detecting and distinguishing between comfort and discomfort in infants
[6]	S.E. Barajas-Montiel, et al. 28–30 November 2005	Neural networks and support vector machines	The study evaluated the performance of ensembles of neural networks (NN) and support vector machines (SVM) for classifying pain and hunger cries in infants. The results indicate that the NN ensemble achieved 96.41% accuracy for pain and no-pain cry classification, while the SVM ensemble achieved 85.16% accuracy. For hunger and no-hunger classification, the NN ensemble achieved 87.61% accuracy, while the SVM ensemble achieved 76.41% accuracy. It also seems that the SVM ensemble had a variation of 2% to 3% in correct classification in each loop. Overall, the study suggests that NN ensembles perform better than SVM ensembles for this task.
[7]	Cheng Li, et al. January 2021	Faster RCNN and HMM	The high mean average precision indicates that the system is accurately able to classify different expressions and detect discomfort in both clinical and daily settings. This suggests that the system could potentially be useful for a variety of applications in healthcare and beyond.

5 DATA AUGMENTATION TECHNIQUES

Data augmentation methods for audio include several techniques such as cropping out a portion of data, noise injection, shifting time, speed tuning, changing pitch, mixing background noise and masking frequency.

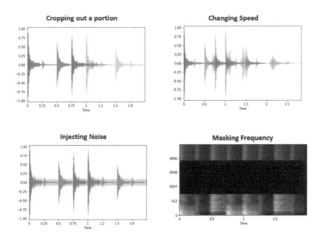

Figure 9.   Audio data augmentation of Cry signals.

6   CONCLUSION

To conclude we can understand the importance and research which needs to be done on developing different perspectives using deep learning strategies of 1D, Biomedical signals.

REFERENCES

[1]   Analyticsvidhya-   https://www.analyticsvidhya.com/blog/2021/07/artificial-neural-network-simplified-with-1-d-ecg-biomedical-data/

[2]   Li, R., Liu, Z. Stress Detection Using Deep Neural Networks, *BMC Med Inform Decis Mak* 20 (Suppl 11), 285 (2020), DOI: 10.1186/s12911-020-01299-4

[3]   Tutuko, B., Nurmaini, S., Tondas, A.E. *et al.* AFibNet: An Implementation of Atrial Fibrillation Detection with Convolutional Neural Network, BMC Med Inform Decis Mak *21*, 216 (2021). DOI: 10.1186/s12911-021-01571-1

[4]   Karagiannis, Alexandros & Constantinou, Philip & Vouyioukas, Demosthenes. (2011). *Biomedical Time Series Processing and Analysis Methods: The Case of Empirical Mode Decomposition* DOI: 10.5772/20906.

[5]   Yue Sun 1, Caifeng Shan, Tao Tan, Tong Tong, Wenjin Wang, Arash Pourtaherian, Peter H N de With Detecting Discomfort in Infants Through Facial Expressions, (2019) DOI 10.1088/1361-6579/ab55b3

[6]   S. E. Barajas-Montiel and C. A. Reyes-Garcia, *Identifying Pain and Hunger in Infant Cry with Classifiers Ensembles International Conference on Computational Intelligence for Modelling, Control and Automation and International Conference on Intelligent Agents*, Web Technologies and Internet Commerce (CIMCA-IAWTIC'06), 2005, pp. 770–775, DOI: 10.1109/CIMCA.2005.1631561.

[7]   Li, Cheng & Pourtaherian, Arash & Onzenoort, L. & With, Peter. (2021). *Automated Infant Monitoring Based on R-CNN and HMM* 553–560. 10.5220/0010299605530560.

Artificial intelligence driven applications

Recent Trends in Computational Sciences – Gururaj, Pooja & Flammini (Eds)
© 2024 The Author(s), ISBN 978-1-032-42685-3

Revolution in agriculture sector using blockchain technology

Abdul Latif Saleem, K. Paramesha & M.G. Vinay
Department of Computer Science and Engineering, Vidyavardhaka College of Engineering, Mysore, Karnataka

ABSTRACT: In recent years, blockchain technology has gotten a lot of attention. The introduction of blockchain smart contracts which enable the implementation of decentralized apps in trust-free contexts is one explanation for this new trend. As smart contracts become more widely used attacks exploiting their flaws will surely increase. Several ways have been investigated to mitigate these assaults and avoid breaches such as documenting vulnerabilities or model testing using formal verification. These methods however fall short of capturing the blockchain and user behaviour features.

Agriculture is gradually becoming into a significant financial industry with global implications. Real-time monitoring of the weather and the condition of the soil, along with improved food quality, are what are driving the technological, ecological, and economical trend. The study explores the exciting notion of fusing smart agriculture with smart contract technology to produce not only better agricultural products but also better supply chains and agricultural logistics, resulting in a variety of advantages for all parties involved. With the use of numerical and possibly categorical data, the emphasis is on calculating similarity metrics for tuples that reflect soil and climate conditions. In addition, Solidity a high-level language for creating smart contracts meant for the Ethereum Virtual Machine is provided with a sample implementation of one such measure. The final stage for smart contracts that depend on physical objects is to highlight elements of agricultural asset digitalization.

Keywords: blockchain, traceability, smart agriculture, smart farming, smart contracts, solidity, ethereum virtual machine (evm)

1   INTRODUCTION

About 17.4% of the Indian economy comes from agriculture, which also employs 54.6% of the workforce. India is second in the world for producing fruits and vegetables. The largest producer of milk is also India. The technological revolution has significantly altered farming practices in India.

Higher-yield grain types have been created as a result of the Green Revolution, which has also altered outdated and ineffective farming practices. However, because the majority of Indian farmers lack formal education, they are unable to keep up with the rapid changes. They are being tricked into spending money on farming necessities. For instance, many farmers are unable to distinguish between a greater yield seed and a regular seed, and the seller exploits this in order to deceive the farmer. This also holds true while selling their grains to the Government [1]. The existing procedure by which the Government buys crops from farmers is not well understood by many farmers. They still don't understand the meaning of the Minimum Support Price (MSP) or how the price can vary depending on the grain quality. Due to this, many farmers have fallen victim to fraud and are now compelled to sell their produce for considerably less money. Many different solutions are being put out to solve. In order to address this issue, our project introduces the usage of blockchain technology and smart contracts [2]. India is an agricultural country. Indians spend the most of their time farming. Selling produce to the government is a difficult task, just as is growing and maintaining crops. The approaches used now are outdated and ineffective. Government price fixing rarely occurs for farmers, and there are several frauds. blockchain can aid in the resolution of these issues. It is a technique that renders transaction recording significantly more transparent and efficient than the current system. The purpose of this article is to illustrate how blockchain technology and smart contracts can be utilized to enhance the trading of agricultural products between Indian farmers and the government. It also describes how blockchain might reduce fraud and speed up the payment process for farmers on time [3]. India is a very populous nation, and about 70% of the workforce is employed in agriculture. Government of India allots funds for farmers as part of its annual budget [4]. However the shortage of agricultural machinery, labour, and resources has resulted in significant financial losses for farmers. Due to these issues, an increasing number of farmers are leaving agriculture for new occupations each year. If farmers continue to leave their jobs, there will be a shortage of food goods and an increase in demand, which would lead to higher prices. The adoption of this suggested technique will help farmers continue their work in a more cost-effective manner in a variety of ways [5,6]. For the purpose of guaranteeing food safety, agricultural goods' provenance (tracing) systems are crucial. However, the stakeholders (growers, farmers, sellers, etc.) are many and geographically distributed, making centralized data management challenging [7]. As a result, trust cannot easily be established and the production process stays opaque. In order to address the trust crisis in the supply chain for goods, we suggest a blockchain-based agricultural provenance system that is characterized by decentralization, collaborative maintenance, consensus trust, and accurate data. The management operations (such as fertilization, irrigation, etc.) with a certain data structure are included in the recorded information. The origin of agricultural products may be tracked using blockchain technology, which not only expands its range of possible uses but also helps different stakeholders create a trustworthy community [8,9]. The structure of this paper is as follows: the Section II Literature Survey, the Section III Problem Statement, the Section IV Proposed System, the Section V Architectural Design, the Section VI Usecase Diagram, the Section VII Results, the Section VIII Conclusion and Future Work.

2 LITERATURE SURVEY

The purpose of an agriculture traceability platform is to save information for the production supply chain, which includes the production, processing, storage, transportation, and distribution of agricultural goods, so that the entire procedure may be observed by outside parties (customers, insurance companies, etc.). As was said in the introduction, the features of blockchain technology are ideal for the requirements of an agriculture traceability platform. The goal of developing a blockchain based platform for agriculture traceability is to store all relevant data on blockchain architecture. This refers to bringing together various businesses and organizations [10]. It is necessary to define a generalized organized representation of the data. We define and introduce the terms used in the agricultural traceability system in this section. The relatively straightforward transaction history is what is kept on the bitcoin blockchain [11,12]. The content of the agricultural traceability system, which includes businesses, seeds, fertilizers, pesticides, time, agricultural operations, and residue testing, is substantially more sophisticated. If all of this information is presented in a single, unified format, there will unavoidably be a lot of repetition. Therefore, a provenance record and fundamental planting information were built as two linked components for the agriculture traceability system [12]. Think about a system that can run on the blockchain but without mining, using the example of a bank with its branches and clients (i.e., without tokens). However, mining and tokens are absent because only the bank is able to create new blocks. For instance, Peter must transfer money through the bank or a branch of the bank in order for it to be confirmed, noted in the block, and a new transaction to be formed [13].

The system under consideration's block creation algorithm is presented. The bank is the sole owner of this system, and as such, he has the authority to represent any number of nodes anywhere in the world. While the degree of decentralization is decreased and total trust is required for the structure under consideration, the system's territorial distribution and use of blockchain remain important. A platform for digital property rights may also be used as an illustration of the use of such a system. This value accounting structure necessitates total faith in the company, in this example, the bank [14].

The system owner can designate a number of nodes with auditor rights to check that all modifications are implemented successfully and alert the auditors if anything goes awry. Such a structure has the benefit of not requiring a lot of computational resources [15]. Smart contracts rely on dynamic provisions, which are created using information from a smart agriculture as well as the talks between the parties involved. Remember that in a blockchain context, all smart contract conversations and their much iteration are appended to the original contract in accordance with the specifications stated [16]. In any case, due to the delicate nature of agricultural products, caution must be used when drafting agricultural smart contracts. From a systems perspective, security rules are given to thwart malicious software agents. Furthermore, asset digitization is a significant concern as has been emphasized among others in case of as any physical industry that includes smart contracts. The primary objective is to study smart contracts using formal approaches [17].

Digital contract provisions must be updated, as previously said, using hard facts. Factors related to the soil or the climate can be expressed as tuples that can include either categorical or numerical data. Similarity measures to be constructed so that they can be the foundation in smart contracts negotiation depending on how various aspects change over [18]. Four parties are involved in the vehicular networks' blockchain-based ad dissemination framework: the Register Authority (RA), numerous advertisers, numerous vehicles, and numerous roadside units (RSUs) [18].

Key generation and identity management are the responsibilities of Register Authority, often known as RA, a dependable registration hub. It may inherit the legitimate institutions that have long existed in the actual world, such government agencies. In particular, a vehicle's private key is produced in two pieces. The car generates one portion, while the RA generates the other. The RA binds the identification with the vehicle-identification-number to assign each car a distinct digital certificate (VIN) [19,24].

V stands for "vehicles," which stands for the entities using a communication device to send or receive advertisements (on-board units, OBUs). Each vehicle has trusted platform module (TPM, as defined in [21]), which is used to hold private keys and perform predefined calculations on sensitive data. Through DSRC, vehicles can communicate with the nearest RSU or other vehicles. In an advertisement dissemination situation, they can play either the sender or the recipient role. Vehicles can forward advertisements to following vehicles in the sender position in an effort to earn the forwarding incentive. Vehicles can receive advertisements in the capacity of the receiver and reply with a signature to the sender. Without communicating with the RA, vehicles can independently create and update the blockchain addresses that are used to complete transactions [20–22]. RSUs, or roadside units, use wireless networks to connect with moving vehicles. They perform the role of consensus nodes in the upkeep of the blockchain network, which may be a PoS-based blockchain. Each RSU maintains a copy of the whole blockchain ledger, allowing it to validate blocks and transactions. RSUs in particular act as access points to transmit an advertisement to passing vehicles. As incentives for promoting RSUs, the blockchain network's transaction fees and mining rewards are offered [23].

3 PROBLEM STATEMENT

The main issue with the current farmers is suffering significant financial losses as a result of a shortage of agricultural machinery, labour, and natural resources. Due to these issues, an increasing number of farmers are leaving agriculture for new occupations each year. The use of this suggested approach would help farmers continue their work in farming in a cost-effective and efficient manner in a variety of ways.

3.1 *Proposed system*

The proposed system has the following components based on blockchain smart contract:

3.2 *Pool farming*

Pool Farming is a platform based on shared economy facilitating exchange of three different types of services: Hand, Machine and Storage.

Hand: Meant for exchanging Manpower.

Machine: Meant for exchanging tools and machinery.

Storage: Meant for exchanging the use of storage facilities.

3.3 *Farmer local banking system*

Rural banking has historically been chaotic. Farmers in India have traditionally relied on the country's indigenous banking system, which consists of shroffs, money lenders, and traders who charge exorbitant interest rates, to cover their short- and long-term credit needs. Lack of collateral security is one of the primary reasons organized banking hasn't reached rural India.

There is now a monopoly of money lenders who are squeezing the average farmer dry and reducing them to miserable conditions due to increasing operational costs and reduced profits that precluded these institutions from participating. Farmer Bank developed a smart contract to manage a fund given by a group of members and process loan requests from the members in an effort to address the issue. At the moment, Farmer Bank evaluates loan requests in accordance with the following standards: Only members may add money or make loan requests.

A member may ask for twice the amount that was contributed.

The maximum loan amount is equal to half the pool's total value.

This serves to both prevent the pool from being emptied and to encourage nutrition. Like the majority of blockchain based solutions, Farmer Bank gets its strength from a sizable user base.

Members can contribute funds to the pool.

A moderator or owner is required to add a member.

4   ARCHITECTURAL DESIGN

4.1 *Architecture of blockchain network with respect to algorithm*

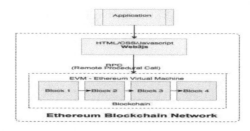

Figure 1.   Blockchain network with respect to algorithm.

180

Architecture of a blockchain network contains three layers as shown in the diagram. They are Application layer, Interface layer and the Network layer.

Application layer: The user interface of the application, or application layer, is where the user enters data. Some requested functionality is implemented by the application code.Some requested functionality is implemented by the application code. Overlays, APIs, applications, etc. are permitted by the application layer.

Interface layer: The layer that has web3js, html, CSS, and java script is known as the interface layer. For communication between the application and network layer, Web3js is utilized. JavaScript: Java Script is a dynamic computer programming language. It is most frequently utilized as a part of web pages because of how client-side script can interact with users and build dynamic pages thanks to its implementations. It is an interpretable object-oriented programming language. Originally known as Live Script, JavaScript was renamed to JavaScript by Netscape, maybe in response to the buzz that Java was causing. Live Script, the predecessor to JavaScript, had its debut in Netscape 2.0 in 1995. Netscape, Internet Explorer, and other web browsers all contain the general-purpose language's core.

Web3js: The libraries in web3.js allow you to interact with a nearby or remote ethereum node over an HTTP or IPC connection. The web3 JavaScript library is used to interface with the Ethereum blockchain. It can interact with smart contracts, send transactions, and retrieve user accounts, among other things.

Remote Procedure Call (RPC): It is a method of interposes communication. Client-server programmers use it. When computer software invokes a procedure or subroutine in a different address space by coding it as a regular procedure call without particularly coding for the remote interaction, RPC techniques are utilized. This method call also controls low-level transport protocols like User Datagram Protocol, Transmission Control Protocol/Internet Protocol, etc. It is used by programmers to convey message data.

Network layer: EVM (Ethereum Virtual Machine), which is used to build contracts, is present in the network layer. The hashing algorithm SHA256, which uses encryption keys with a bit rate of 256, is the one that is employed.

EVM: Each full Ethereum node has the Ethereum Virtual Machine (EVM), a potent virtual stack that is sandboxed and in charge of running contract bytecode. Usually, contracts are created using higher level languages like Solidity, which are then translated into EVM bytecode.

5 ARCHITECTURE WITH RESPECT TO THE APPLICATION

The above architecture contains four layers they are the Application layer or front end layer, contract layer, blockchain virtual network layer and transaction layer.

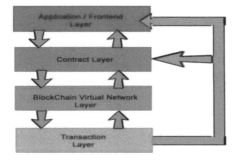

Figure 2.   Architecture with respect to application.

5.1 Application layer

Smart contracts make up the application layer. The application layer can be further broken down into the execution layer and the application layer. The applications in the application layer are what end users utilize to communicate with the blockchain network. It consists of user interfaces, frameworks, APIs, and scripts. The blockchain network serves as the back-end system for various applications, which frequently communicate with it using APIs. The sub layer known as the execution layer is made up of chaincode, smart contracts, and underlying rules. The actual code and rules that are executed are located in this sub layer.

A transaction is validated and executed at the semantic layer even though it propagates from the application layer to the execution layer (smart contracts and rules). Applications send commands to the execution layer, or chaincode in the case of the Hyperledger fabric, which carries out transactions and upholds the determinism of the blockchain (such as per-missioned blockchain like hyperledger fabric). Contract layer It consists of contracts, each of which has a certain job to do. Smart contracts cannot be changed. Written in Solidity, smart contracts for the Ethereum runtime engine. For the code to be syntactically proven, a compiler is required. The bytecode is smaller and executes more quickly on EVM because it has been compiled. Code running on EVM is completely isolated and doesn't interact with the disc or network in any way.

Smart contract: A code that contains business logic is located on the EVM and is recognised by a special address. When a transaction is made against one of the functions of a smart contract, those functions are executed. A transaction may cause a change in the status of the smart contract, depending on its logic. To create a smart contract, developers can use any language, including Python or Solidity. They can then use a specialised compiler to turn their code into bytecodes, which they can then upload to a blockchain. A distinct address is given to the smart contract once it has been implemented. On the blockchain, any user has the ability to carry out a transaction against that smart contract. For the steps of a trans-action on an electronic device, see the transaction flow below.

5.2 Transaction layer

The transaction layer stores transactions that occur in the application layer and the contract layer, and the information in the transaction layer is accessible to the other layers.

The things that give a blockchain purpose are transactions. They are a blockchain sys-tem's tiniest building blocks. An address for the recipient, an address for the sender, and a value make up most transactions. This isn't all that unlike from a typical transaction you could see on a credit card bill.

5.3 Usecase diagram

5.3.1 Usecase diagram for machinary pool

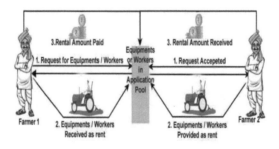

Figure 3.   Usecase diagram for machinery pool.

182

All the farmers must be the member of the pool.

Farmer 1 requests for equipment or workers.

The equipment or workers are received as rent from the equipment of workers in the application pool.

The rental amount is paid after receiving the equipment.

5.3.2 *Usecase diagram for banking transactions*

The following are the steps for banking transactions

Farmer 1 requests for loan.

Loan is given to the farmer 1 from the funds of the other farmers in smart contract pool.

The loan is paid back with the interest.

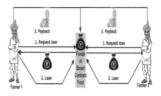  

Figure 4. Results farmer utility pooling.

6   CONCLUSION AND FUTURE WORK

Millions of metric tons of grains are purchased annually from farmers across the nation. An outline of how blockchain can be utilized to speed up the procurement process was provided in this project. Additionally, the effectiveness of smart contracts compared to the existing system and their practical use were investigated in this study. Now it's up to the Indian government to determine whether and when to incorporate blockchain technology. In order to determine whether blockchain is appropriate for any project-related components that may be lacking, a cost-benefit analysis should be created.

Additionally, this project demonstrates how blockchain can be used to decrease frauds and scams against innocent farmers, permanently store records without the risk of losing them again, accurately record daily transactions, retrieve records instantly, provide traceability for grains purchased, improve food safety and security, and guarantee that farmers receive what they are promised. This project also demonstrates how the intermediary, or the man at the procurement centre, can have less authority and the process can be automated.

Scaling up this application from the state level to the national level, where the states and union territories collectively form a national level blockchain, could be the next task. In addition, rather than creating a blockchain from scratch, the use of HyperLedger and Ethereum can be researched.

REFERENCES

[1]   Drakopoulos G., Liapakis X., Tzimas G., and Mylonas P., "A Graph Resilience Metric Based on Paths: Higher Order Analytics with GPU," in *ICTAI*. IEEE, 2018, pp. 884–891.

[2]   Lottes P., Khanna R., Pfeifer J., Siegwart R., and Stachniss C., "UAVbased Crop and Weed Classification for Smart Farming," in *ICRA*. IEEE, 2017, pp. 3024–3031.

[3] Yang Y. and Jia Z., "Application and Challenge of Blockchain Technology in the Field of Agricultural Internet of Things," *Information Technology*, vol. 258, pp. 24–26, 2017.

[4] Yuan Y. and Wang F. Y., "Blockchain: The State of the Art and Future Trends," *Acta Automatica Sinica*, 2016.

[5] Yuan Y., Zhou T., Zhou A. Y., Duan Y. C., and Wang F. Y., "Blockchain Technology: From Data Intelligence to Knowledge Automation," *Zidonghua Xuebaolacta Automatica Sinica*, vol. 43, pp. 1485–1490, 2017.

[6] Chiasserini C., Malandrino F., and Sereno M., "Advertisement Delivery and Display in Vehicular Networks: Using v2v Communications for Targeted Ads," *IEEE Veh. Technol. Mag.*, vol. 12, no. 3, pp. 65–72, Sep. 2017.

[7] Sharma P. K., Moon S. Y., and Park J. H., "Block-vn: A Distributed Blockchain Based Vehicular Network Architecture in Smart City," *Journal of Information Processing Systems*, vol. 13, no. 1, p. 84, 2017.

[8] Lv F. and Chen S., "Research on Establishing a Traceability System of Quality and Safety of Agricultural Products Based on Blockchain Technology," *Rural Finance Research*, vol. 12, pp. 22–26, 2016.

[9] Yang Y. and Jia Z., "Application and Challenge of Blockchain Technology in the Field of Agricultural Internet of Things," *Information Technology*, vol. 258, pp. 24–26, 2017.

[10] Nakamoto S., *"Bitcoin: A Peer-to-peer Electronic Cash System,"* Consulted, 2008.

[11] Yuan Y. and Wang F. Y., "Blockchain: The State of the Art and Future Trends," *Acta Automatica Sinica*, 2016.

[12] Yuan Y., Zhou T., Zhou A. Y., Duan Y. C., and Wang F. Y., "Blockchain Technology: From Data Intelligence to Knowledge Automation," *Zidonghua Xuebao/Acta Automatica Sinica*, vol. 43, pp. 1485–1490, 2017.

[13] Hong T., "Accelerating the Application of Blockchain in the Field of Agricultural Products E – commerce in China," *Journal of Agricultural Information*, pp. 18–20, 2016.

[14] Yuan Y. and Wang F.-Y., "Parallel Blockchain: Concept, Methods and Issues," *IEEE Acta Automatica Sinica*, vol. 43, pp. 1703–1712, 2017.

[15] Andreas M A. *Mastering Bitcoin: Unlocking Digital Cryptocurrencies.* O'Reilly Media, 2014.

[16] Drakopoulos G., Kanavos A., Mylonas P., Sioutas S., and Tsolis D., "Towards a Framework for Tensor Ontologies Over Neo4j: Representations and Operations," in *IISA*, 2017, pp. 1–6.

[17] Drakopoulos G., Liapakis X., Tzimas G., and Mylonas P., "A Graph Resilience Metric Based on Paths: Higher Order Analytics with GPU," in *ICTAI*. IEEE, 2018, pp. 884–891.

[18] Lottes P., Khanna R., Pfeifer J., Siegwart R., and Stachniss C., "UAV-based Crop and Weed Classification for Smart Farming," in *ICRA*. IEEE, 2017, pp. 3024–3031.

[19] Sa, Z. Chen, M. Popović, R. Khanna, F. Liebisch, J. Nieto, and R. Siegwart, "Weednet: Dense Semantic Weed Classification Using Multispectral Images and MAV for Smart Farming," *IEEE Robotics and Automation Letters*, vol. 3, no. 1, pp. 588–595, 2017.

[20] Salamí E., Barrado C., and Pastor E., "UAV Flight Experiments Applied to the Remote Sensing of Vegetated Areas," *Remote Sensing*, vol. 6, no. 11, pp. 11051–11081, 2014.

[21] Park J.-I., Lee M.-H., Grossberg M. D., and Nayar S. K., "Multispectral Imaging Using Multiplexed Illumination," in *International Conference on Computer Vision*. IEEE, 2007, pp. 1–8.

[22] Drakopoulos G., Kanavos A., and Tsakalidis K., "Fuzzy Random Walkers with Second Order Bounds: An Asymmetric Analysis," *Algorithms*, vol. 10, no. 2, p. 40, 2017.

[23] Drakopoulos G., Kanavos A., Mylonas P., and Sioutas S., "Defining and Evaluating Twitter Influence Metrics: A Higher-order Approach in Neo4j," *SNAM*, vol. 7, no. 1, pp. 52:1–52:14, 2017.

Customer churn prediction using ensemble learning with neural networks

Kannaiah Karthikeya & Vikram Neerugatti

Department of Computer Science and Engineering, Jain (Deemed to be) University, Bangalore, Karnataka, India

ORCID ID: 0000-0002-2203-7002

ABSTRACT: Modern-day companies have to keep in mind their immensely large customer base. Their businesses strictly deal with the masses. Companies have to maintain a record of how their product or service is reaching the people. And to know Customer Behavior around the product or service provided. This vital piece of information will give understanding about customer's interests aligned with respect to the product or the service provided. Companies have to keep an eye on customers leaving. Churn Rate is a measure vital to understand the growth or fall of a company. Predicting customers who are willing to churn even before they are going to churn will be quite remarkable. Based on prediction, companies now can target and work better on customers predicted to be leaving. This paper suggests an Ensemble Learning based Model using Neural Networks. Three models of Neural Networks will share their prediction and most occurred prediction is selected to be the ftnal prediction. Proposed model scored 84% accuracy which is a decent improvement over individual model.

Keywords: customer churn, churn rate, ensemble learning, neural networks & prediction

1 INTRODUCTION

Churn depicts turnover of customers. Customers are assets to the growth of companies. Losing an existing customer is a situation that a company should not be in. Getting new customers means a lot of customer acquisition costs. Getting in good terms with an unsatisfied existing customer by providing valuable service. Customers are a very valuable asset especially to the companies which are customer dependent such as the Telecom industry [1]. Customer Churn is a situation where customers are willing to change from services of one company to another [2].With advancements in the field of big data, customer-centric organizations like banks face unexpected novel challenges and competition [3]. Not only banks but this can be related to any sector that deals with immensely large customers. Organizations from these kinds of sectors usually have a methodology/system to manage customers. Customer Relationship management systems play a role in both acquiring new customers and also retention of existing customers. And this system does by developing and maintaining a healthy relationship with the customers. Churn prediction plays a vital role in this instance as it gives predicted potential churners on whom this management system can work upon and retain them [4].

Data is crucial in churn prediction. When dealing with prediction problems we may think that more the data the merrier. But it is actually more the data is relevant to the cause the merrier it will be to solve the problem. Careful inspection of data can filter useful features that have the benefit of being useful in terms of having to be an independent feature but

DOI: 10.1201/9781003363781-27

dependable for the problem. Usually in this particular field of customer churn, we have very less data on churners. In this paper, we will see some work put up related to data. This paper proposes an ensemble learning model using only neural networks. This model is used to predict potential churners by taking votes and the most voted solution wins and becomes the final prediction by ensemble model. Our model predicts if a customer churns or not. It is a binary classification problem. The remaining parts of this paper will feature 5 more sections. In Section II, features Literature Review of relevant work related to customer churn. Section III is context about the Proposed Model. In Section IV, Implementation methodology is given. In Section V, Results will be featured along with commentary. Finally in Section VI, Conclusion and Future Scope have been discussed.

2 LITERATURE REVIEW

This Section features findings of relevant works carried out by enthusiasts relevant to this problem.

A paper done by Sanket Agrawal et al. [5] features a multi-layer neural network model. Deep learning techniques have been used on a telecom dataset. Their model scored a promising accuracy of 80.03%. Their paper describes Customer behavior analysis and features which are to be kept in mind in analysis. Analysis draws correlations and helps understanding dependent features.

Seyed Jamal Haddadi et al. [6] This paper uses the concept of time-series DNNs based customer retention approach. Processed data is fed to Bi-LSTM neural network. This proposed model showed significant improvement in comparison to other models based on traditional ML techniques. This proposed model scored a promising accuracy of 84%. This paper is useful for the banking sector.

Shrisha Bharadwaj et al. [7] features two models for customer churn prediction on a telecom dataset. One is a Logistic Regression model with sigmoid activation function scores 87.52% accuracy and second model is Multi-Layer Perceptron (MLP) scores an 94.19% accuracy. MLP features three hidden layer architectures with a learning rate of 0.01. Predictive models proposed in this paper are useful in conducting market research.

Jesmi Latheef and Vineetha S did a paper on customer churn prediction which features a LSTM model and SMOTE preprocessing for an imbalanced dataset. Their model scored an accuracy of 88%. Accuracy found to be better after SMOTE technique. Framework provided by the paper is useful for organizations to discover clients who are likely to churn [4]. This paper done by Pushkar Bhuse et al. [8] describes various classifiers put up against Random Forest Classifiers. In comparison, Random Forest fared well over other models. Scoring an accuracy of 90.96%. Other models they explored are Ridge Classifier, XGBoost, KNN Classifier, Support Vector Classifier (SVC), and Deep Neural Network. This paper is based on a telecom dataset.

A comparative study done by Manpreet Singh et al. [9] explores models used in around ten papers. Their findings say that Random Forest was best in comparison to other basic classification models and Importance of feature engineering in helping performance of the model. Amany Zaky et al. [10] attempts a single hidden layer Artificial Neural network. Bank dataset is used in this paper. The model scored an accuracy 87%. The proposed network is said to be a cost effective method for bank organizations to maintain bank customers and help retain them. Paper by Aisyah Larasati et al. [11] discusses optimizing the ANN model to predict better. In results it found that the churn percentage is 36.4% and an accuracy of 76.35%.

Xin Hu et al. [12] features a model that combines prediction results and confidence of Decision Tree and Neural network. This model takes prediction results as variable and confidence of both Decision Tree and Neural network as weights. This paper uses customer information of a supermarket as a dataset. And scores an amazing accuracy of 98.87%.

3 PROPOSED MODEL

3.1 *Neural networks*

Neural Networks are inspired by the Human brain. They try to mimic the function of neurons in the human brain. Deep learning is a concept where neural network models with more than one hidden layer can be constructed. This is because of scope and advancements in the neural networks field [11].

Neural network is a black box kind of method where we don't know how it came to that particular understanding. This is because of the presence of hidden layers. Neural networks have significant advantages such as they can improve prediction and interpretability of churn detection [4].

This paper introduces three models. Three models are different from each because of how they are structured and used. First model (MODEL-1) is trained on an imbalanced dataset. The Second (MODEL-2) and Third model (MODEL-3) are trained on a balanced dataset. Section IV has information about how the dataset is balanced. While the Third model is a single hidden layer Neural network. First and Second models have the same structure of three hidden layers. This paper implements neural networks using TensorFlow and Keras library [13,14].

3.2 *Ensemble learning*

Ensemble Learning is a machine learning concept dealing with use of multiple learning algorithms. Ensemble models obtained better performance than any individual models used. Better predictive performance over constituent models is observed in ensemble learning methods which is why it is better to choose ensemble models over single constituent models. In ensemble learning, predictions generated by all constituent models are considered in final decision making. Opinions are taken to make final predictions. Every model has its own hypothesis for solving the problem. All of these hypotheses are combined to form hopefully a better hypothesis [15]. In this paper, predictions of all three models are considered. As we are dealing with a binary classification problem where prediction is either "0" or "1", suppose we have two models predicted to be "0" and another model predicted it to be "1". Then as prediction "0" is the most popular vote so "0" is chosen to be a prediction by the ensemble model. We achieve this system by adding all three predictions and then if the sum is greater than 1 (2 or 3) then "1" will be a prediction. If the sum is either 0 or 1 then prediction is said to be "0". Here "0" means customer will not churn and "1" means customer will churn.

3.3 *Evaluation*

Evaluating models is a process where one can understand how the model is performing. Evaluation measures help us understand this. This paper uses four such measures. Those measures are Accuracy(A), Precision(P), Recall(R) and F1-score (F1). You can refer to these measures in this paper [10]. A confusion matrix can be used to help calculate these measures as shown in Table 1.

Table 1.   Confusion matrix.

Type	Predicted (0)	Predicted (1)
Actual (0)	True Positive (TP)	False Negative (FN)
Actual (1)	False Positive (FP)	True Negative (TN)

4 IMPLEMENTATION

4.1 *Dataset*

Dataset is collected from Kaggle [16]. This dataset is used for Telco customer churn. This dataset has 21 features includes Customer ID, Gender, Senior Citizen, Partner, Dependents, Tenure, Phone Service, Multiple Lines, Internet Service, Online Security, Online Backup, Device Protection, Tech Support, Streaming TV, Streaming Movies, Contract, Paperless Billing, Payment Method, Monthly Charges, Total Charges, Churn. Here Churn is the prediction class.

4.2 *Data preprocessing*

Our dataset is categorical so it should be converted to numeric to be able to process. First, we remove the Customer ID column as it is not useful because all records have unique values. We remove records with empty fields. The field values "No internet service", "No phone service" into simple "No". All the columns with "Yes" or "No" values can be converted into binary values. "Yes" value can be changed to "1". In the gender column, female is changed to "1" and male to "0".

The columns Internet Service, Contract, and Payment Method which have three class values are manipulated and turned into dummy or indicator variables for the same. Tenure, Monthly Charges, and Total Charges are scaled using a Minmax Scaler. After preprocessing, a dataset with a total of 7032 records with 27 listed features is generated. But this preprocessed data is imbalanced. Records with churners are very less compared to non-churners. This dataset can't be used for training our neural network models as it leads to overfitting problems.

Table 2. Data manipulation.

Features	Dummy variables after Manipulation			
Internet Service	Internet Service DSL	Internet Service Fiber optic	Internet Service No	(Not needed)
Contract	Contract Month-to-month	Contract One year	Contract Two year	(Not needed)
Payment Method	Payment Method Bank transfer (automatic)	Payment Method Credit card (automatic)	Payment Method Electronic check	Payment Method Mailed check

4.3 *Handling imbalanced data*

To tackle imbalanced datasets there are techniques like oversampling and under sampling. In this case, an oversampling technique called Synthetic Minority Oversampling Technique (SMOTE). The SMOTE algorithm can be applied to solve overfitting problems. SMOTE generates more samples to the dataset [4]. This paper implements SMOTE by imbalanced learning library (imblearn) [17]. After application of SMOTE, the dataset will have a total of 10326 records.

Table 3. Data imbalance.

Number of records having	
Churn is equal to 0 (records of customers who are Not churning)	*Churn is equal to 1 (records of customers who are churning)*
5163	1869

Table 4.   After data balancing using smote.

Number of records having	
Churn is equal to 0 (records of customers who are not churning)	*Churn is equal to 1 (records of customers who are churning)*
5163	5163

4.4   *Training and testing*

Our first model is trained by using imbalanced data. Then our second and third models use a SMOTE balanced dataset. Balanced dataset is split into training and testing sets in the ratio of 4:1 (80% and 20%). This testing set from a balanced dataset is used to test all the models.

First and Second neural network models have the same structure. They have three hidden layers. Rectified Linear Unit (ReLU) activation function is used by Input and hidden layers. While Sigmoid activation function is used in the output layer. We use a Dense layer from Keras. Dense layer has an activation function applied to the linear prediction function as the output [14]. Third model has only one hidden layer with 50 neurons. Input and hidden layers use Rectified Linear Unit (ReLU) as activation function. The Output layer uses sigmoid activation function.

5   RESULTS AND DISCUSSIONS

All three models have been trained for 100 epochs. And then all three models are given a testing set to predict. The predicted sets from all three models are taken and considered in the ensemble model. The predictions from all three models are used as votes and the most popular vote is considered as the prediction of the ensemble model.

Table 5.   First and second models.

Layer	Input shape	Activation function
Input layer	(None, 26)	ReLU
Hidden layer 1	(None, 50)	ReLU
Hidden layer 2	(None, 20)	ReLU
Hidden layer 3	(None, 5)	ReLU
Output layer	(None, 1)	Sigmoid
Total params = 3183		

First model which was trained by an imbalanced dataset scored an accuracy of 78% and loss of 0.5492. Second Model which is trained by a balanced dataset, was able to score a better accuracy of 81% and loss of 0.5560. Third Model which is similarly trained as second has also scored 80% accuracy and loss of 0.4299. But the proposed ensemble model has scored an accuracy of 84% which is a decent improvement over three models. The Table 1 shows the predicted type positive or negative, Table 2 shows the data manipulation done, Table 7 shows the difference models with the accuracy and Table 8 shoes the confusion matric for the prediction types positive or negative.

189

Table 6.　Third model.

Layer	Input shape	Activation function
Input layer	(None, 26)	ReLU
Hidden layer	(None, 50)	ReLU
Output layer	(None, 1)	Sigmoid
Total params = 2103		

Table 7.　Comparison.

Model	Accuracy
MODEL-1	78%
MODEL-2	81%
MODEL-3	81%
ENSEMBLE MODEL	**84%**

Table 8.　Resultant confusion matrix.

Type	Predicted (0)	Predicted (1)
Actual (0)	872	161
Actual (1)	172	861

Table 9.　Classification report of ensemble model.

Churn Value	Precision	Recall	F1-score
0	0.84	0.84	0.84
1	0.84	0.83	0.84
Accuracy		**84%**	

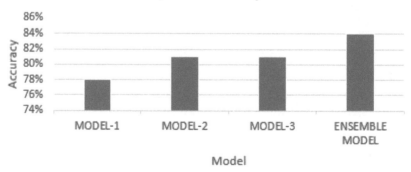

Figure 1.　Comparative analysis of models.

6 CONCLUSION AND FUTURE SCOPE

In conclusion, this paper aimed to perform customer churn prediction and used an ensemble model of three neural network models. Even though the dataset used is an telecom customer churn dataset. The proposed model can be used over any churn dataset. So any sector can use this in their customer relation or behavior analyses. Ensemble Learning considers opinion of every constituent model. These constituent models are important. In this paper, all the constituent models are based on neural networks. This paper shown results over telecom churn dataset. Accuracy found to be 84% and the least difference between accuracies of ensemble model and constituent models is 3%. This 3% difference is the improvement shown by proposed ensemble model. In Future, Neural networks used can be optimized and can use more models to form ensemble models. Data is very important in churn prediction. Working with proper datasets can provide a positive level of results.

REFERENCES

[1] Agrawal, S *et al.* (2018). "Customer Churn Prediction Modelling Based on Behavioural Patterns Analysis using Deep Learning". *2018 International Conference on Smart Computing and Electronic Enterprise (ICSCEE)*, pp. 1–6.

[2] Bharadwaj, S et al. Bhuse, P et al. (2020). "Machine Learning Based Telecom-Customer Churn Prediction". *2020 3rd International Conference on Intelligent Sustainable Systems (ICISS)*, pp. 1297–1301.

[3] Butgereit, L (2020). "Work Towards Using Micro-services to Build a Data Pipeline for Machine Learning Applications: A Case Study in Predicting Customer Churn". *2020 International Conference on Innovative Trends in Communication and Computer Engineering (ITCE)*, pp. 87–91.

[4] Contributors, Wikipedia (2019). URL: https://en.wikipedia.org/wiki/Ensemble_learning. Feng, L. "Research on Customer Churn Intelligent Prediction Model based on Borderline-SMOTE and Random Forest". *2022 IEEE 4th International Conference on Power, Intelligent Computing and Systems (ICPICS), 2022*, pp. 803–807.

[5] Haddadi S.J. *et al.* (2022). "Customer Churn Prediction in the Iranian Banking Sector". *2022 International Conference on Applied Artificial Intelligence (ICAPAI)*, pp. 1–6.

[6] Hammoudeh A., Fraihat M. and Almomani M. (2019). "Selective Ensemble Model for Telecom Churn Prediction". *2019 IEEE Jordan International Joint Conference on Electrical Engineering and Information Technology (JEEIT)*, pp. 485–487.

[7] Hu X. *et al.* (2020). "Research on a Customer Churn Combination Prediction Model Based on Decision Tree and Neural Network". *2020 IEEE 5th International Conference on Cloud Computing and Big Data Analytics (ICCCBDA)*, pp. 129–132.

[8] Larasati A. *et al.* (2021). "Optimizing Deep Learning ANN Model to Predict Customer Churn". *2021 7th International Conference on Electrical, Electronics and Information Engineering (ICEEIE)*, pp. 1–5.

[9] Latheef J. and S Vineetha (2021). "LSTM Model to Predict Customer Churn in Banking Sector with SMOTE Data Preprocessing". *2021 2nd International Conference on Advances in Computing, Communication, Embedded and Secure Systems (ACCESS)*, pp. 86–90.

[10] Singh, M *et al.* (2018). "Comparison of learning techniques for prediction of customer churn in telecommunication". *28th International Telecommunication Networks and Applications Conference (ITNAC)*, pp. 1–5.

[11] Zaky, A, S Ouf, and M Roushdy. "Predicting Banking Customer Churn based on Artificial Neural Network". *2022 5th International Conference on Computing and Informatics (ICCI), 2022*, pp. 132–139.

Recent Trends in Computational Sciences – Gururaj, Pooja & Flammini (Eds)
© *2024 The Author(s), ISBN 978-1-032-42685-3*

Securing crime case summary and E-FIR using blockchain concept

Tanuja Kayarga
VVCE, Mysore, India

C.V. Kavitha, P. Lokamathe, M. Yamuna & M.S. Meghana
Department of CSE, VVCE, Mysore, India
ORCID ID: 0000-0001-5818-3055

ABSTRACT: Electronic First Information Report (e-FIR) is a basic document filed to the police stations by a victim or someone on his/her behalf when a cognizable offense such as murder, kidnapping, rape, theft, etc. is committed. In the e-FIR database, the offense's record can be compromised due to its centralized nature, and further the intentional registration of false e-FIR can occur. Thus, data integrity and transparency are key concerns in e-FIR database. In this paper, e-FIR data integrity and false registration appended with police stations in a centralized database are addressed via a consensus-based distributed blockchain solution, as an integral part of a smart city environment. Specifically, a smart contract based intelligent framework has been utilized to explore the potential of Ethereum blockchain in providing integrity to e-FIR data stored in a police station's database. Local database is interfaced with Ethereum blockchain using Web3 Remote Procedure Call (RPC) protocol. Multiple simulations have been performed to evaluate the performance of the proposed framework. Our results show a trade-off between different hashing algorithm security level for the offenses data and number of transactions stored in a single block on blockchain ledger.

Keywords: e-FIR, technology, Blockchain

1   INTRODUCTION

In the current system, people go to police stations to complain against the crimes faced by them. These complaints/E-FIR are registered by the police and they maintain the crime case summary and store it in the database. Whenever there is need of these data, they retrieve it from the database. These data can be accessed by any hacker hence there is no security for these data. There should be a solution adapted to increase the security of crime case summary. Electronic First Information Report (e-FIR) is a basic document filed to the police stations by a victim or someone on his/her behalf when a cognizable offense such as murder, kidnapping, rape, theft, etc. is committed. In the e-FIR database, the offense's record can be compromised due to its centralized nature, and further the intentional registration of false e-FIR can occur. Thus, data integrity and transparency are key concerns in e-FIR database. e-FIR data integrity and false registration appended with police stations in a centralized database are addressed via a consensus-based distributed blockchain solution, as an integral part of a smart city environment. The data owner may adapt attribute-based encryption to encrypt the stored information for attaining access control and keeping data secure, in the cloud. The E-FIR system is proposed to public for indirect interaction with police and to improve the E-governance facility. E-FIR system with E-portal. E-portal is specially designed website that brings information together from diverse sources in a uniform way. Usually each information source gets its dedicated area on the page for displaying information. Generally, many crimes seen by the citizens but they are afraid to complaint in police station due to fear of police department, lack of time and insensibility. Due to this fear

DOI: 10.1201/9781003363781-28

many crimes not reported to the police station. Many cases are registered but due to lack of proofs and evidences and lack of collaboration of public they are not properly investigated. The aim of this study is to develop an online system which is easily accessible to police department, public and administrative department and to achieve e-transparency at various levels like publication, reporting, openness, accountability etc. The main objective of the study is to increase police and citizens interaction without going to nearby police station. It will help to reduce crime percentage and will save the time of people. It will also increase government and citizen interaction and will built an informed society.

2 LITERATURE REVIEW

In the current system, people go to the police stations to complain about the crimes they are facing. These complaints E-FIRs are registered with the police, and they keep a summary of the criminal cases and keep them on the website. Whenever there is a need for this data, they download it from the website. According to the review the technologies used were Object orient language and query language which was very less secure in case to secure the data as like it has a drawback lack of knowledge and lack of less awareness of Technologies [1]. Likewise, the paper which was published in the year 2012 [2] used technologies like Cloud Computing distributed System, Public Key Cryptography, Biometric Features and Radio Frequency Identifications which are improved technologies but still they were struggling in order to give security and privacy to the data stored in the system. We have referred E-FIR using E-Governance [3] which was published in the year 2016 was held a survey based on the data security was not confirmed any particular technology instead they spoke about all the possibilities and came up with the data storage would be difficult which receives using sensors and they also said that there would be chances of Data Loss as well. As the technology Improved in the later days the paper [4] was published in the year 2019 which particularly speaks about Digital Signature Technology to secure the data using Cryptography and faced the data loss kind of issues and the transaction delay as the data got loss there was no recovery option they found in the Digital Signature. Though the Technology grown the data security was left as the greatest issue in the technical field as the well knowledge people concentrated more on the improving technology, hence we had picked a paper [5] 2019 itself which exactly speaks about CPS System where in Prime Security Challenges and Distrust on data accession side occurred. So we had referred very much recent paper [7] which was most relevant to our project research published inthe year 2020, In which the researchers spoke about the trending technology Blockchain Concepts, Though the blockchain technology was very secured and the researchers come up with an Less transaction per second and High risk ofpotential collision kind of Challenges. After all this research we had come to a conclusion that we might have to use Cryptography technology along with the Blockchain conceptin order to keep the data safe and secure.

3 PROPOSED WORK

In a smart city of smart vehicles, smart schools, smart hospitals, smart infrastructure etc., where everything is connected to the Internet (IoE) [6] to share tremendous data volume daily, this city should also provide a smart and secure system for Electronic First Information Report (e-FIR) data management in a police station and Storing Crime case summary as shown in Figure 1. e-FIR is a simple document that has been written out and filed to the police by the victim or someone on his/her behalf when a cognizable offense such as murder, kidnapping, rape, theft, etc shown in the Figure 2. Is committed. Reporting a crime and filing a cognizable offensemanually in a police station consumes a lot of time because thepolice to people ratio in some of the commonwealth countriesare tremendously high as shown in Table 1. Instead, the e-FIR mechanism is usedin some of the common-wealth countries i.e. Pakistan, India, Bangladesh, Malaysia, Japan and Singapore, while the

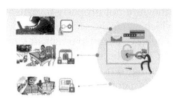

Crime Scenario Infographics

Figure 1. Infographics of data tampering.   Figure 2. Image of the current system with actual. scenarios.

Table 1. Comparison table of complaints registration ratio.

No.	Country	Police and People Ratio
1	Bangladesh	1:1138
2	Malaysia	1:450
3	Singapore	1:614
4	Pakistan	1:625
5	India	1:728

mechanism for filing an offense in Europe and USA is, apparently, different than the aforementioned countries [7]. Non-registration, false registration and integrity of e-FIR and crime case summary data are the main concerned problems connected with it.

These problems are due to police corruption, inefficiency and lack of accountability, data taperers like Hackers as shown in the Figure 1. Initially, e-FIR, related crime case summary data is stored in a central database of police station locally, which is then shared with the head quarter (HQ) of police stations. Here the e-FIR data could easily be manipulated as the control of e-FIR summary database is local within the police station. Therefore, to address this problem, applying blockchain technology can help us to better respond to the security challenges and can endeavor data integrity, as blockchain is a fraud-resilient, distributed ledger, which can record all the transactions in a Peer-to-Peer (P2P) network.

Blockchain has a decentralized architecture, and its popularity in the cryptocurrency world in securing the distributed network communication has been remarkable [8]. In this paper, the major contributions are twofold: Firstly, a blockchain-enabled framework providing efficient integrity to e-FIR data is proposed, which is applicable in, and been an integrated part of, a smart city environment. Secondly, false registration of e-FIR is minimized by resolving it through the concept of blockchain. To the best of our knowledge, this is a first attempt restraining false registration and providing integrity to e-FIR data using blockchain.

The rest of paper is organized as follows. Section II discusses e-FIR and relevant approaches therein. Section III presents the proposed system architecture. The proposed framework implementation and evaluation results are shownin Section IV. Finally, the concluding remarks with future workis given in Section V.

The Police Station register crime case & start investigation on crime case, log crime case summary for further reference to higher officer & court. The crime case summary helps to collect evidence & also future investigation on case, there are chances of data manipulation, hacking data or tamper which would stop the further investigation or mislead the case hence by Using some advance algorithm, block chain concept & with QR Code Image we can secure the crime case summary. We can also develop Public/user provided an option to post complaint on crime activities by their surrounding with respect to police station. These complaints are not disclosed to anyone such that not harmful to public/user. Police station further investigate public complaint & register E-FIR.

Figure 3.   Investigation process.

4   SYSTEM ARCHITECTURE

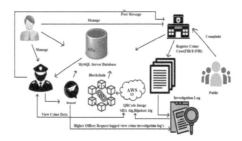

Figure 4.   System design.

Architecture focuses on looking at a system as a combination of many different components, and how they interact with each other to produce the desired result. The focus is on identifying components or subsystems and how they connect. In other words, the focus is on what major components are needed.

The purpose of the design phase is to plan a solution of the problem specified by the requirements document. This phase is the first step in moving from the problem domain to the solution domain. In other words, starting with what is needed; design takes us toward how to satisfy the needs. The design of a system is perhaps the most critical factor affecting the quality of the software; it has a major impact on the later phases particularly testing and maintenance. In high level design identifies the modules that should be built for developing the system and the specifications of these modules. At the end of system design all major data structures, file format, output formats, etc., are also fixed. The focus is on identifying the modules. In other words, the attention is on what modules are needed.

In the detailed design the internal logic of each of the modules is specified. The focus is on designing the logic for each of the modules. In other words, how modules can be implemented in software is the issue. A design methodology is a systematic approach to creating a design by application of a set of techniques and guidelines. Most methodologies focus on high level design.

5   IMPLEMENTATION

As the Figure 4 shows the proposed system we introduce the implementation part step by step with the below example.

The below is an example table which shows an idea to implement the Blockchain concept to secure the crime case summary and e-FIR data in the decentralized system which would be written programmatically to the SQL System. According to the above Table 2. SINO is the Serial Number provided along with the City Police station, PSID is the Police station ID given to the particular Police station area wise individually, CID is the Crime ID given by

195

the Police station according to the Crime occurred and Register in the e-FIR, LDATE is the Log date on which day the Crime occurred, PHV is the Previous Hash Value which can be encrypted using the fields of: SINO, PSID, CID and LDATE by applying AES Algorithm, CHV is the Current Hash value and the FP is File Path. File Path is taken from the AWS S3 Server which means after getting the Current Hash Value that would be written to QRImage Code and then That Image will be Stored in the AWS S3 Server in order to make sure the safety of data which are encrypted and then that file path would be stored in the FP field of the Table for which the Blockchain Concept is implemented programmatically.

Table 2.   Example for Blockchain Concept Implementation.

PSID+CID+LDATE+PHV=CHV

Encrypted Values would be stored in Current Hash Value

Final Encrypted data would be written to QR-Image code and written it to AWS S3 server and that file path would be stored here.

SINO	PSID	CID	LDATE	PHV	CHV	FP
123A1	A09876	A04567	A0451107	0000	A0450X12	FILEPATH0
123A2	A09876	A04567	A0451207	A0450X12	A0450X12YZ	FILEPATH1
123A3	B09877	B05678	B0561307	B0A0450X12YZ	B0A056Y13ZA	FILEPATH2

6   APPROACHED MODEL

The Waterfall Model was first Process Model to be introduced. It is also referred to as a linear-sequential life cycle model. It is very simple to understand and use. In a waterfall model, each phase must be completed fully before the next phase can begin. This type of model is basically used for the project which is small and there are no uncertain requirements. At the end of each phase, a review takes place to determine if the project is on the right path and whether or not to continue or discard the project. In this model the testing starts only after the development is complete. In waterfall model phases do not overlap.

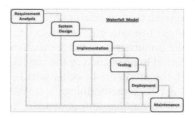

Figure 5.   Approached model.

7   SYSTEM FLOW DIAGRAMS

The below Figure 6 shows How the System data Flow of the 1st actor Application Manager, He would be responsible for managing the Areas and Police stations. The Police station how he can login to the system and log a FIR and Store the Crime case summary using his/her Authenticated Username and the Password. The Higher Officer can access the system, He/she can login to the System and View the details of the E-FiR and Stored Crime case Summary by using his/her Authenticated Username and the password. The Public or the User How he/she

Figure 6.   System flow diagram.

can Register to the System Application and Use the Login Credentials after Registration to post the Crime Occurred Details by selecting the Nearest or Area Police Station.

8   RESULTS AND DISCUSSIONS

The below Screenshot Figure 7 shows the original hash value (not tampered) of the case log which was entered while registering the crime case logs. Where the Blockchain concept was used and encrypted the data and generated Hash Value programmatically.

Figure 7.   Hashed value stored using blockchain concept.

Figure 8. Encrypted data written to QR image Code and AWS S3 Server File path is stored in the Blockchain Concept Table. The above screenshot of Figure 7 shows how the Table looks like with the data written programmatically to the local data base using Blockchain Concept which was Encrypted & written to QR.

Figure 8.   Data tampering occurred.

Figure 9.   The data recovered successfully.

Image Code and stored it into the AWS-S3 Server and the file path written to the last column as shown in the Table 2. The Screenshot of Figure 8 shows the Data Tampering Occurred. This Can be done through the System where we introduced a Third-party Tampering application for the purpose of Showing demo how we can able to find the Data Changes made by the Unauthorized person in the system.

The Above Screenshot of Figure 9. Shows the Recovered data as we implemented the Decentralized Blockchain Concept to store the data in the Local Database System. After knowing that the tampered data can be recovered using the 2nd database node of the System by the Administrator or Higher Officer.

9   CONCLUSION

This paper examines the relatively under-developed area ofrecord management in police stations for the prevention ofdata tampering and false report filing, using the concept of blockchain technology. Research conducted in this paper has presented a consensus-based solution for providing integrity to the offenses data stored in policestation database using blockchain. In the proposed framework, Multiple simulations have been performed to demonstrate the trade between number of transactions occur in asingle block and different hashing security level for e-FIR data. The proposed system will further be investigated in future for improvement of real time implementation as the systemcan show the progress of the cases registered in the portal tothe user and as well as the higher authority so that the case can have track up to the date.

REFERENCES

[1]   *Proposed E-Police System for Enhancement of EGovemment Services of Bangladesh Muhammad Baqer Mollah, Sikder Sunbeam Islam*, Md. Arnan Ullah Dept. of Electrical and Electronics Engineering International Islamic University Chittagong, Chittagong, Bangladesh mbaqer@ieee.org, sikder_-islam@yahoo.co.uk,ullah047@yahoo.com IEEE/OSAIIAPR International Conference on Informatics, Electronics & Vision.

[2]   *Smart Health: A Context-Aware Health Paradigm within Smart Cities Agusti Solanas, Constantinos Patsakis*, Mauro Conti, Ioannis S. Vlachos, Victoria Ramos, Francisco Falcone, Octavian Postolache, Pablo A. Pérez-Martínez, Roberto Di Pietro, Despina N. Perrea, and Antoni Martínez-Ballesté 0163-6804/14/$25.00 © 2014 IEEE

[3]   *E-FIR using E-Governance Kirti Marmat Anand More IJIRST – International Journal for Innovative Research in Science & Technology|* Volume 3 | Issue 02 | July 2016 ISSN (online): 2349-6010

[4] *Framework for Financial Auditing Process Through Blockchain Technology*, using Identity Based Cryptography Shaista Anwar Department of Management and Commerce Amity University, Dubai, U.A.E.

[5] *Realizing the Implementation Platform for Closed Loop Cyber-Physical Systems using Blockchain* Abdullah Bin Masood*†, Hassaan Khaliq Qureshi*, Syed Muhammad Danish*, Marios Lestas† *National University of Sciences and Technology (NUST), Islamabad, Pakistan 978-1-7281-1217-6/19/ $31.00 ©2019 IEEE

[6] Police Complaint Management System using Blockchain Technology Ishwarlal Hingorani, Rushabh Khara, Deepika Pomendkar, Nataasha Raul Department Of Computer Engineering Sardar Patel Institute of Technology Mumbai, India *Proceedings of the Third International Conference on Intelligent Sustainable Systems* [ICISS 2020] IEEE Xplore Part Number: CFP20M19-ART; ISBN:978-1-7281-70893.

[7] P. A. Perez-Martinez *et al.* "Privacy in Smart Cities-A Case Studyof Smart Public Parking," *Proc. 3rd Int'l Conf. Pervasive Embedded Computing and Commun. Sys.*, pp.55–59, 2013.[Online: January, 2019] Urban Population Growth statistics byUN; https://population.un.org/wup/Publications/Files/ WUP2018-Report.pdf

[8] Agusti Solanas *et al.* "Smart Health: A Context-Aware Health Paradigm within Smart Cities", *IEEE Commun. Mag.*, vol, 52, no. 8, 2014.

[9] Jaime Ballesteros *et al.* "*Safe Cities. A Participatory Sensing Approach*", IEEE LCN, 2012.

[10] Paola G. V. *et al.* "FOCAN: A Fog-supported Smart City Network Architecture for Management of Applications in the Internet of Everything Environments", *J. Parallel Distrib. Comput., 2018.[Online: March, 2019]* Website of US department of justice for reportinga crime; https://www.justice.gov/ actioncenter/report-crime.

[11] R. M. Parizi *et al.* "*Empirical vulnerability analysis of automated smartcontracts security testing on blockchains*" CASCON, IBM Corp., 2018.

[12] S. Nakamoto, "*Bitcoin: A Peer-to-peer Electroniccash System*," white paper, 2008.

[13] T. T. A. Dinh *et al.* "Un-tangling Blockchain: A Data Processing Viewof Blockchain Systems," *in IEEE Transactions on Knowledge and Data Engineering*, vol. 30, no.7, pp. 1366–1385, 1 July, 2018.

[14] Jean Bacon *et al.* "*Blockchain Demystified: A Technical and Legal Introduction to Distributed and Centralized Ledgers*", 25 RICH. J.L. and TECH., no. 1, 2018.

[15] Reyna *et al.* "On Blockchain and Its Integration with IoT. Challenges and Opportunities." *Future Generation Computer Systems*, vol 88, 2018.

[16] Antra Gupta *et al.* "*A Method to Secure FIR System using Blockchain*", IJRTE, Vol. 8, Issue-1, 2019.

[17] Maisha A. Tasnim *et al.* "CRAB: Blockchain Based Criminal Record Management System", *SpaCCS, LNCS 11342*, pp.294–303, 2018.

[18] Kirti Marmat *et al.* "E-FIR using E-Governance", *IJIRST*, vol. 3, 2016.

[19] Muhammad Baqer Mollah *et al.* "*Proposed E-Police System for Enhancement of E-Government Services of Bangladesh*", IEEE/OSA/IAPR, 2012.[Online: January, 2017] Personal blockchain for Ethereum development; https://www.trufflesuite.com/docs/ganache/overview. [Online: November, 2019] Josh Cassidy, Article for Online Remix IDE-writing smart contract; https://kauri.io/remix-ide-your-first-smartcontract/124b7db1d0cf4f47b414f8b13c9d66e2/a. [Online: October, 2011] Bangladesh Police's Website, Police to People Ratio; http://www.police.gov.bd/index5.php?category=48.

[20] A. B. Masood *et al.* "Realizing an Implementation Platform for Closed Loop Cyber-Physical System using Blockchain", *IEEE 89th VTC*, 2019.

Recent Trends in Computational Sciences – Gururaj, Pooja & Flammini (Eds)
© 2024 The Author(s), ISBN 978-1-032-42685-3

Disease prediction using machine learning by analysing the symptoms

K.R. Vishwanath
VTU
ORCID ID: 0009-0007-4756-1656

D.C. Vinutha
VVCE
ORCID ID: 0000-0003-3096-2967

A. Amrutha & B.S. Shreya
VTU

V. Rohith & C.P. Vijay
VVCE
ORCID ID: 0000-0002-5206-858X, 0000-0002-0525-2368

ABSTRACT: The incorporation of novel technologies in the biomedical industry is becoming increasingly popular. Key players in this field include Artificial Intelligence (AI), Machine Learning (ML), and Data Science. AI is a term used to describe computer programs that mimic human intelligence. ML is a subset of AI that can automatically extract raw data. With the vast amount of data available today, it is now possible to analyze medical data without human intervention. However, for this to be effective, the available data needs to be of high quality, meaning it should be clean data. ML has the ability to learn from sample data, thereby improving the reliability and efficiency of computational processes while reducing costs.

Keywords: Disease Prediction, Machine learning, Confusion matrix, Naïve Bayes Classifier (Gaussian, Bernoulli), Random Forest and Support Vector (SVM) Classifier

1 INTRODUCTION

The concept of using machine learning to predict human illnesses based on symptoms is a fascinating idea of the decade. This approach uses patient-provided symptoms or user-inputted data to predict diseases. The accuracy of predictions is measured by combining the results of different models and outputting the correct prediction to the user. Gaussian Naïve Bayes Classifier, Bernoulli Naïve Bayes Classifier, Random Forest Classifier, and Support Vector (SVM) Classifier are used to make predictions. As the amount of biomedical and healthcare data increases, accurate analysis of medical data can greatly benefit early disease detection and patient care. The quality of each model is determined using confusion matrices, and combined analysis of the models is performed.

Hospitals generate vast amounts of data that is difficult to analyze and utilize for improving patient outcomes. Machine learning provides a way to leverage this data to predict disease progression for individual patients and provide targeted interventions. By analyzing patient data using Gaussian Naïve Bayes classifier, Bernoulli Naïve Bayes classifier, Random Forest classifier, and Support Vector classifier, predictions can be made and combined to improve accuracy. The K-fold technique is used to determine the best model, and the confusion matrix is used to evaluate the accuracy of each model. In a limited set of data, we were able to achieve 100% accuracy for each prediction model. While the proposed model is a downscaled version of a real-world application, it has significant potential for improving patient outcomes.

DOI: 10.1201/9781003363781-29

2 PROPOSED METHODOLOGY

This project aims to implement an efficient machine learning model which can predict human illness based on the symptoms which the individual provides.

2.1 *Proposed methodology*

Data preparation is the primary step for any machine learning application. We will be using a dataset from Kaggle for this application. This dataset consists of two CSV files one for training and one for testing. There is a total of 133 columns in the dataset out of which 132 columns are symptoms and last column is the prognosis/disease.

2.2 *Cleaning the data*

Cleaning is the most crucial step in this ML project. The quality of the data determines the quality and accuracy of our machine learning model. So, before feeding the data into our model it is necessary to clean the data first.

In our dataset all the columns are numerical and the target column i.e., prognosis is a string type and is encoded to numerical form using a label encoder.

2.3 *Model building*

After gathering and cleaning the data, the data is ready and can be used to train the ML model. We will be using this cleaned data to train the Support vector classifier, Gaussian Naive Bayes classifier, Bernoulli Naïve Bayes Classifier, and Random Forest classifier. Confusion matrix is used to determine the quality of the models.

2.4 *Inference*

After training the four models we will be predicting the disease for the input symptoms by combining the predictions of all the models using mode function. This makes our end prediction much more accurate.

3 EVALUATION AND RESULTS

3.1 *Input data*

An experiment was conducted to detect several diseases based on patient symptoms. The dataset included 120 samples for each of the 40 diseases. Various classifiers were used to obtain precise results for each patient.

3.2 *Implementation*

3.2.1 *Model building*

3.2.1.1 *K-Fold Cross-Validation*
K-Fold cross-validation is a technique used for cross-validation, which involves splitting the entire dataset into k subsets or folds. The model is trained on k-1 subsets and the remaining subset is used to evaluate the model's performance. This process is repeated k times, with each of the k subsets used exactly once for evaluation. K-Fold cross-validation is useful for evaluating the performance of a model when the dataset is limited or when the model is trained on a small dataset.

3.2.1.3 *Gaussian Naïve Bayes Classifier*
It is a probabilistic machine learning algorithm that internally uses Bayes Theorem to classify the data points.

3.2.1.4 *Bernoulli Naïve Bayes Classifier*
It is another useful naïve Bayes model. The assumption in this model is that the features binary (0s and 1s) in nature.

3.2.1.5 Random Forest Classifier

Random Forest is a classification algorithm based on ensemble learning and supervised machine learning. It uses multiple decision trees internally to make classifications. In a random forest classifier, each internal decision tree is a weak learner, and the final prediction is obtained by combining the outputs of all the weak decision trees, typically by taking the mode of all the predictions. In summary, the random forest works as follows:

1. Fix the k side effects from data (clinical data) the sum of m manifestations arbitrarily (here k ≪ m). At that point, it assembles a Decision tree model with the help of side effects of k.
2. Rehashes n number of times with the goal that we have n number of tree model worked from various Random mixes of indications which is denoted as 'k' (or an alternate irregular example of information, known as bootstrap test).
3. Consider every one of the various n-constructed trees and proceeds a variable which is random to foresee the illness. Here it Store the anticipated illness, so that we can have a sum of n illness anticipated from n number of the decision tree model.
4. Computes the decisions in favor of each anticipated illness and consider the mode (which is most continuous illness anticipated) as last expectation from the Random Forest model calculation

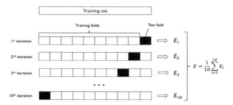

3.2.1.2 Support Vector Classifier

The Support Vector Classifier is a type of machine learning algorithm that is used for classification tasks. It tries to find an optimal hyperplane in hyperspace that can separate the samples into different categories with the maximum possible margin between the decision boundary and the data points. Thehyperplane that is chosen by the algorithm is the one that can best classify the samples into their respective categories. This makes the Support Vector Classifier a powerful tool for solving classification problems, especially when the data is complex or high-dimensional.

3.3 Flow of model

The workflow of the proposed model is sketched below using the flow chart given. The data set is divided into training and testing using the training dataset to train the model and then validating it with the Testing dataset to obtain the expected results.

Figure 1. Flow of the proposed model.

3.4 Performance measure

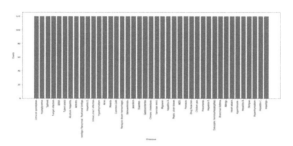

Figure 2. Reading the Train data set to check if the dataset is balanced or not.

3.5 Validating the result

a) Using K-Fold Cross-Validation for model selection to test its accuracy by shuffling the data and providing test scores, mean score for each model:

```
==============================================
SVC

Scores: [1. 1. 1. 1. 1. 1. 1. 1. 1. 1.]
Mean Score: 1.0
==============================================
Gaussian NB

Scores: [1. 1. 1. 1. 1. 1. 1. 1. 1. 1.]
Mean Score: 1.0
==============================================
Random Forest

Scores: [1. 1. 1. 1. 1. 1. 1. 1. 1. 1.]
Mean Score: 1.0
==============================================
Bernoulli NB

Scores: [1. 1. 1. 1. 1. 1. 1. 1. 1. 1.]
Mean Score: 1.0
```

Training all the four models with the train dataset and checking the quality and measuring performance of each model using confusion matrix:

Accuracy of each model for both training80% and testing-20% dataset from Train dataset & Confusion matrix: To get the accuracy of the Model we use a built-in function: **from** sklearn.metrics **import** accuracy score, confusion matrix

3.5.1 Support vector classifier's accuracy & Confusion matrix

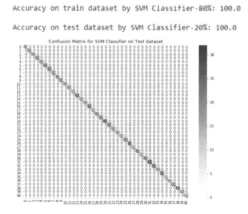

203

3.5.2 *Gaussian Naïve Bayes Classifier's accuracy & Confusion matrix*

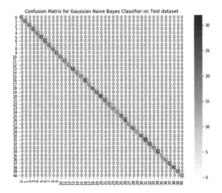

3.5.3 *Bernoulli Naïve Bayes Classifier's accuracy & Confusion matrix*

Accuracy on train dataset by Bernoulli Naive Bayes Classifier-80%: 100.0

Accuracy on test dataset by Bernoulli Naive Bayes Classifier-20%: 100.0

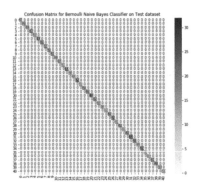

[7] Bernoulli Naive Bayes Algorithm formula

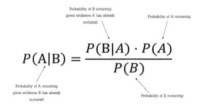

204

3.5.4 Random forest classifier's accuracy & Confusion matrix

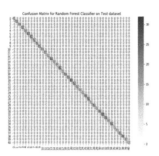

Accuracy on train dataset by Random Forest Classifier-80%: 100.0

Accuracy on test dataset by Random Forest Classifier-20%: 100.0

3.6 Fitting the model on whole data and validating on the Test dataset and the Confusion matrix for final prediction (mode of all three models)

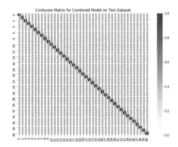

Accuracy on Test dataset by the combined model: 100.0

3.7 Results

3.8 Proposed results

The train data for training (80%) and testing (20%) the model:

Train: (3936, 132), (3936,)
Test: (984, 132), (984,)

3936 is the number of training data and 984 is the number of testing data from the train dataset.

Acquired symptoms must be given as the input for the Model to run. As our model is a downscaled model, the input must be in particular order for the model to understand the input.

The final output of the machine learning model is displayed using a graphical user interface (GUI) implemented in Python using the Tkinter library. The user inputs the symptoms of the illness, and the model predicts the disease based on the input. The prediction is displayed in the GUI, both individually for each model used (such as Gaussian Naïve Bayes, Bernoulli Naïve Bayes, Random Forest, and Support Vector Classifier) and as a combined prediction of all the models.

The GUI provides an easy-to-use interface for the user to input their symptoms and receive an accurate prediction of the disease they may be suffering from. This can be helpful in early disease detection and prompt medical attention can be provided to the patient. With the use of machine learning techniques and graphical user interfaces, medical professionals and patients can benefit from accurate predictions and early detection of diseases.

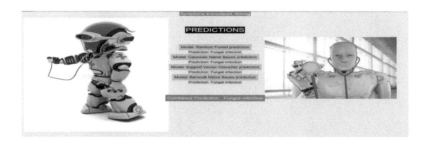

Source:
Train dataset: Disease Prediction Using
Machine Learning | Kaggle
Test dataset: Disease Prediction Using
Machine Learning | Kaggle

4 CONCLUSION AND FUTURE SCOPE

Through the use of advanced machine learning models, we were able to accurately predict diseases based on a limited input of symptoms. Our study utilizes four different models to provide near-perfect predictions, achieving an accuracy score of 100% for the given test dataset. Early detection of diseases using our model can lead to more efficient management of medical resources. We believe that this small change can have a significant impact on the healthcare sector, and we look forward to further implementation and improvement in the future.

REFERENCES

[1] Sebastian Raschka and Vahid Mirjalili, *Python Machine Learning, Machine Learning with Python, Scikit-learn, and TensorFlow.*

[2] *Symptoms Based Multiple Disease Prediction Model using Machine Learning Approach*, Bhanuteja Talasila,Poonati Anudeep and Saipoornachand Kolli.

[3] A Survey on Mathematical, Machine Learning and Deep Learning Models for COVID-19 Transmission and Diagnosis *J. Christopher Clement, Member*, IEEE, VijayaKumar Ponnusamy, Senior Member, IEEE, K.C. Sriharipriya, and R. Nandakumar.

[4] *Machine Learning (ML) in Medicine: Review, Applications, and Challenges* Amir Masoud Rahmani 1, †, Efat Yousefpoor 2, Mohammad Sadegh Yousefpoor 2, Zahid Mehmood 3, Amir Haider 4, †, Mehdi Hosseinzadeh 5, * and Rizwan Ali Naqvi 4, *

[5] *Disease Prediction by Machine Learning Over Big Data from Healthcare Communities MIN CHEN1*, (Senior Member, IEEE), YIXUE HAO1, KAI HWANG2, (Life Fellow, IEEE), LU WANG1, AND LIN WANG3,4

[6] *Integrating Co-Clustering and Interpretable Machine Learning for the Prediction of Intravenous Immunoglobulin Resistance in Kawasaki Disease*, HAOLIN WANG 1, ZHILIN HUANG2, DANFENG ZHANG2, JOHAN ARIEF2, TIEWEI LYU2, AND JIE TIAN2

Recent Trends in Computational Sciences – Gururaj, Pooja & Flammini (Eds)
© 2024 The Author(s), ISBN 978-1-032-42685-3

Blockchain architecture in upcoming digital world

B.M. Promod Kumar & K.S. Manoj Gowda
Department of Computer Science and Engineering, PES College of Engineering, Mandya, India

S.S. Heemashree
Department of Information Science and Engineering, PES College of Engineering, Mandya, India

N.M. Siddesh Kumar
Department of Mechanical Engineering, PES College of Engineering, Mandya, India
ORCID ID: 0000-0002-6905-0634

S. Tejas & Pranadhi Haran
Department of Computer Science and Engineering, PES College of Engineering, Mandya, India

C.M. Vikas
Department of Mechanical Engineering, PES College of Engineering, Mandya, India

ABSTRACT: Blockchain architecture is a multi-layered system with various components that work together to ensure stability, security, and efficiency. The bottom layer, the Data Layer, holds transactions and data, while the Consensus Layer ensures accuracy and validates transactions. The Network Layer manages communication and data transmission, and the Application Layer provides user access. The Contract Layer executes smart contracts and defines rules, and the Security Layer protects the system from attacks. Blockchain architecture is widely used in various industries including finance, healthcare, supply chain, and voting systems, providing secure, transparent, and immutable data. The combination of blockchain and AI technology results in secure, decentralized and automated execution of agreements, transforming industries such as finance, supply chain and healthcare.

Keywords: Blockchain, DigitalCurrency, Blockchain Architecture, Blockchain Layers, Blockchain Layers, SmartContract

1 INTRODUCTION TO BLOCKCHAIN

Blockchain is a decentralized digital currency that was first offered as its foundational technology. Blockchain is a digital ledger technology [1]. Since then, it has gained popularity as a technology for additional uses that need open and secure record-keeping. Blockchains are decentralized databases that keep track of all network transactions. Multiple transactions are recorded in each block of the chain, and once a block is included in the chain, the data it contains becomes unchangeable and permanent. [2] The main characteristic of blockchain technology is that it runs on a decentralized network, which means that no central authority or middleman is in charge of it. As a result, there can be a secure and open system of record-keeping because each network user has a copy of the blockchain and can validate transaction [3]. Blockchain technology is more secure as a result of its decentralisation since data is spread across a larger number of network nodes, making it less vulnerable to manipulation or hacking. Additionally, because only the people participating in the transaction have access to the data, the use of cryptography ensures that transactions are secure and private. [4]

1.1 *An overview of blockchain architecture*

Blockchain technology, a decentralised and secure digital ledger technology, has a form and structure known as blockchain architecture. A blockchain system's architecture is made up of several essential components, such as nodes, blocks, and consensus methods. The machines that take part in the blockchain network and keep a copy of the blockchain are known as nodes. [5] The blockchain can be expanded by each node by adding new blocks and validating transactions. Due to the network's resistance to tampering and hacking and the lack of a single point of failure provided by the decentralised structure, the blockchain is more secure. [6]

The tiny units that make up a blockchain's blocks are where the network's transaction records are kept. Each block has a hash, which serves as a distinctive identifier, and a connection to the block before it is in the chain. The blockchain is built as a result of this blockchain creation. The system that controls how new blocks are added to the blockchain and how transactions are verified is known as the consensus algorithm. [7] Proof-of-Work, Proof-of-Stake, and Delegated Proof-of-Stake are only a few examples of the numerous consensus algorithms. These algorithms govern how network nodes come to a consensus over the truthfulness of transactions and the addition of new blocks to the blockchain. The network's transaction histories are stored in the small units that make up a blockchain's blocks. Every block has a hash, which acts as a unique identity, and a link to the block that came before it in the chain. As a result of this blockchain development, the blockchain is created. [8]

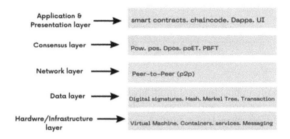

Figure 1.   Blockchain architecture layers.

2   BLOCKCHAIN ARCHITECTURE LAYERS

The application layer, network layer, data layer, and consensus layer are the four fundamental layers that makeup blockchain architecture. [9]

2.1   *Application layer*

The application layer in blockchain refers to the set of software programs, protocols, and services built on top of blockchain technology. This layer is the most visible to end-users and is responsible for providing various applications and services that can interact with the blockchain and use its underlying functionalities. [10] The application layer plays a crucial role in driving the widespread adoption and utilisation of blockchain technology across various domains, including finance, supply chain, real estate, and more. One of the primary functions of the application layer is to provide users with a user-friendly interface that enables them to interact with the blockchain and access its features. This interface can range from simple web-based applications to more complex desktop or mobile applications. [11] The interface provides users with a simple way to initiate transactions, access their digital assets, and monitor the status of their transactions. Another function of the application layer is to provide developers with a platform to build decentralised applications (dApps) that can interact with the blockchain and access its features. [12] The platform allows developers to build and deploy their dApps on the blockchain network, providing users with access to a wide range of decentralised applications that can help them manage and automate various tasks and processes. [13] One of the most popular and widely used applications in the blockchain space is a cryptocurrency exchange. These exchanges provide users with a platform to trade different cryptocurrencies and manage their portfolios. They also provide users with access to a wide range of financial services, such as trading, lending, borrowing, and more. [14].

2.2 *Data layer*

The data layer in a blockchain architecture is responsible for storing the state of the network, including the ledger of transactions and the current state of any smart contracts. It is a critical component of the overall architecture, as it provides the foundation for the network's security, transparency, and decentralisation. [15] In a blockchain network, data is stored in a distributed database, also known as a ledger, that is maintained by multiple nodes. Each node has a copy of the entire ledger, and new transactions are recorded in the ledger by adding blocks to the chain. The blocks in the chain are linked through cryptographic hashes, forming a secure and tamper-evident record of all transactions. [16] The data layer uses a data structure known as a Merkle tree to optimise the storage and retrieval of data. In a Merkle tree, each leaf node represents a transaction, and the root node represents the current state of the blockchain. The tree structure allows for efficient validation of transactions, as only the relevant portions of the tree need to be examined, rather than the entire ledger. In a public blockchain, the data layer must ensure that the ledger is secure and tamper-resistant. [17] To achieve this, the data layer uses cryptographic algorithms such as SHA-256, and consensus algorithms such as Proof of Work (Pow) or Proof of Stake (PoS) to validate transactions and prevent malicious actors from modifying the ledger. The consensus algorithms help to prevent double-spending and ensure that all nodes in the network have the same view of the ledger. [18]

2.3 *Network layer*

The network layer in blockchain refers to the infrastructure that enables communication and data transmission between nodes in a blockchain network. It is responsible for the transmission of transactions and the distribution of blocks among nodes, ensuring the reliability and integrity of the blockchain system. One of the key features of the network layer in the blockchain is the use of peer-to-peer (P2P) communication. [19] P2P communication enables nodes to directly communicate with each other, without the need for a central authority. This eliminates the risk of a single point of failure, as well as the risk of data manipulation by a centralised entity. In P2P networks, nodes are equal participants and work together to maintain the network's stability and security. The network layer also includes consensus protocols, which are used to determine which transactions are valid and how blocks are added to the blockchain. Different blockchain networks have different consensus protocols, but the most common include proof-of-work (PoW), proof-of-stake (PoS), and delegated proof-of-stake (DPoS). Consensus protocols are essential for ensuring that all nodes in the network have the same view of the blockchain and that the blockchain remains secure and tamper-proof. [20]

2.4 *Consensus layer*

The consensus layer in a blockchain network refers to the mechanism that allows all participants to reach an agreement on the state of the network and validate transactions. [21] This layer is critical in ensuring the integrity and reliability of the blockchain, and its design has a significant impact on the security, scalability, and decentralisation of the network. One of the earliest consensus algorithms used in blockchain networks is the Proof-of-Work (PoW) algorithm, which was introduced in the Bitcoin network. PoW requires miners to solve complex mathematical problems to validate transactions and add them to the blockchain. [22] Miners receive rewards for their work in the form of new coins, which incentivizes them to continue validating transactions and maintaining the network's security. Another consensus algorithm is Proof-of-Stake (PoS), which is designed to be more energy-efficient and secure than PoW. In PoS, validators are chosen to validate transactions and add blocks to the blockchain based on the number of tokens they hold, or their "stake". This system provides a more sustainable alternative to PoW and eliminates the need for intensive computational resources.

2.5 *Hardware/infrastructure layer*

The infrastructure layer is one of the critical components of the blockchain architecture and it is responsible for providing the basic building blocks necessary to build and run a decentralised system.This layer consists of the underlying hardware and software systems that provide the technical foundation for the network, and it plays a key role in ensuring the security, scalability, and stability of the blockchain. [23] One of the core elements of the infrastructure layer is the consensus mechanism, which is responsible for validating transactions and maintaining the integrity of the blockchain. This mechanism is critical for ensuring the security of the network and for preventing any malicious actors from compromising the data stored in the blockchain. There are several different consensus algorithms used in blockchain networks, including proof-of-work (PoW), proof-of-stake (PoS), and delegated proof-of-stake (DPoS). Another important element of the infrastructure layer is the network architecture, which determines how nodes communicate with one another and how transactions are propagated throughout the network. [24]

3   APPLICATION OF BLOCKCHAIN ARCHITECTURE

Blockchain is a distributed, decentralised ledger system that enables safe, open, and unchangeable transactions. Due to its incorporation in cryptocurrencies, it has garnered enormous popularity, although its potential uses go beyond finance.

3.1 *Supply chain management*

Blockchain can be used to trace the movement of information and goods in a supply chain, making it possible to efficiently and accurately track products from their point of origin to their final consumer. Participants in the supply chain can safely store, share, and retrieve data on the source, movement, and status of goods and resources using a blockchain-based SCM system. Every member of the supply chain has access to a shared ledger that keeps track of all transactions and movements of materials and goods in real-time. [25]

The following are some advantages of blockchain in SCM:

1. Increased traceability: better ability to trace materials and goods from their point of origin to the final customer is provided by blockchain technology. This serves to lower the possibility of products being counterfeited, promote transparency, and enhance supply chain visibility overall [26].
2. Enhanced security: Blockchain transactions are tamper-proof and safe. Due to the fact that the data is saved on several nodes and is verified via consensus processes, this completely eliminates the possibility of fraud and data tampering [26].
3. Automated processes: Blockchain-based SCM solutions can automate several supply chain operations, including payments, deliveries, and inventory management. This lowers the possibility of errors and improves the supply chain's overall effectiveness. [27]

3.2 *Health care management*

Patient health data can be stored securely and decentralised via a blockchain-based healthcare system. With the patient's consent, authorised healthcare professionals, including doctors and hospitals, can access this data. [28]

The following are some advantages of implementing blockchain in healthcare management:

1. *Enhanced data security*: A blockchain's ability to keep patient health information securely and without human intervention lowers the danger of data breaches while also protecting patients' privacy. [29]

2. *Cost-savings*: Blockchain-based health care systems can automate some tasks, like processing claims, which lowers the administrative expenses of human data entry and reconciliation. [29]

By enhancing data security, accessibility, and interoperability, blockchain technology has the potential to revolutionise the management of health care. Healthcare providers can anticipate better patient outcomes, lower costs, and higher operational efficiency by integrating blockchain-based healthcare solutions.

3.3 *Digital identity*

An individual can own and control their personal information and allow access to it to authorised entities, such as governmental organisations or financial institutions, thanks to a blockchain-based digital identification system. [30]
The following are some advantages of adopting blockchain in digital identity management.

1. Decentralised control: Instead of relying on a centralised authority, people are in charge of their digital identities. This improves privacy and provides people with more power over their personal data. [30]
2. Reduced fraud: By providing safe and reliable identity identification, blockchain-based digital identity systems can lower the risk of fraud. [31]

3.4 *Voting systems*

By storing and managing vote data on a blockchain, a blockchain-based voting system makes voting secure and transparent. [32]
The following advantages of implementing blockchain in voting systems:

1. *Better accessibility:* Voting can be made easier for people who can't physically go to polling places with the use of blockchain-based technologies. [33]
2. *Enhanced effectiveness:* Voting systems built on blockchain technology can automate some voting procedures, saving time and money compared to manual voting methods. [33]
3. *Improved transparency:* A voting procedure that is transparent and auditable is provided by blockchain-based voting systems, which improves the integrity and accuracy of election results. [34]
4. *Decentralised administration:* Blockchain technology offers decentralised administration of voting data, lowering the danger of a single point of failure and boosting the robustness of the voting system. [34]

3.5 *Real estate*

Blockchain-based property registration systems can offer a precise and unchangeable record of property ownership, making the process of transferring property easier and lowering the danger of fraud. [35]
A blockchain-based real estate system enables secure and transparent ownership records and transfer of properties.

1. *Better record-keeping:* Blockchain technology offers accurate and secure tracking of property information such as ownership, transaction history, and property qualities. [35,36]
2. *Decentralised property information management:* Blockchain technology provides decentralised property information management, decreasing the danger of a single point of failure and boosting the robustness of the real estate system. [37]

3. *Transparency is increased:* Blockchain technology offers secure and transparent recording of property ownership and transaction information, lowering the risk of fraud and boosting the efficiency of property transactions. [38]

4 BLOCKCHAIN ARCHITECTURE IN ARTIFICIAL INTELLIGENCE

Although blockchain and artificial intelligence (AI) are two separate technologies with various applications, they can work in conjunction in a number of ways. The development and application of artificial intelligence (AI) could be revolutionised by the usage of blockchain technology [39]. Blockchain technology is the best platform for developing and maintaining transparent, tamper-proof AI algorithms because of its decentralised nature and security. This would lessen the possibility of using unethical or malevolent AI by ensuring that AI applications are created in a responsible and reliable manner [39]. The development of decentralised AI markets where data owners may make money off of their data and AI developers can obtain high-quality data to train their algorithms is one example of how blockchain is being used in AI. This might improve the amount of data that is available for AI applications, resulting in more precise and efficient AI solutions [40]. Blockchain in AI is the use of smart contracts for the automation of AI decision-making. Smart contracts can be used to ensure that AI algorithms are only executed when certain conditions are met, providing a level of control and transparency over AI systems. [41]

5 BLOCKCHAIN ARCHITECTURE IN SMART CONTRACTS

Blockchain architecture is a decentralised system of records (blocks) linked together through cryptography. It is the backbone of cryptocurrencies such as Bitcoin and Ethereum, and it can be used to build other decentralised applications, such as smart contracts [42]. A smart contract is a self-executing agreement with the terms of the agreement between buyer and seller being directly written into lines of code. This code is stored, verified, and executed on the blockchain network, eliminating the need for intermediaries. In the blockchain, each node in the network maintains a copy of the ledger and executes the same code. When a contract is executed, all nodes validate the transaction, ensuring its integrity and security [43]. This eliminates the risk of tampering and provides a tamper-proof, secure platform for the execution of smart contracts. [44]

The architecture of a blockchain network that supports smart contracts can be broken down into several components:

1. *Consensus algorithm:* This determines how the network reaches consensus on the state of the ledger and is crucial to the security of the network. The most commonly used consensus algorithms are proof-of-work (PoW) and proof-of-stake (PoS). [45]
2. *Blockchain nodes:* These are the individual computers that make up the network and maintain a copy of the ledger. [47]
3. *P2P Network:* This is the communication network that connects the nodes and enables them to communicate with each other. [46]
4. *Cryptographic functions:* These are used to secure the network and protect the privacy of transactions. The most commonly used cryptographic functions in blockchain are hashing and digital signatures. [47]
5. *Virtual Machine:* This is a component of the blockchain network that executes the smart contract code. It ensures that the code is executed as intended and that the contract is fulfilled. [48]
6. *Ledger:* This is the database that maintains a record of all transactions on the network. The ledger is stored on every node and is updated when a new block is added to the chain. [49]
7. *API:* This allows applications to interact with the blockchain network, making it possible to build decentralised applications that run on the network. [50]

6 CONCLUSION OF BLOCKCHAIN ARCHITECTURE

Blockchain is a decentralised, distributed database that maintains a continuously growing list of records, called blocks. Each block contains a cryptographic hash of the previous block, a time-stamp, and transaction data. The combination of these features allows blockchains to be secure, transparent, and tamper-proof. The most well-known use of blockchain technology is in crypto-currencies, but its potential extends far beyond that. Blockchains can be used to build decen-tralised applications, track supply chains, manage digital identities, and much more. The core components of a blockchain architecture are nodes, blocks, and consensus algorithms. Nodes are computers that maintain a copy of the blockchain, validate transactions, and compete to create new blocks. The blocks contain transaction data, a timestamp, and the cryptographic hash of the previous block. The consensus algorithm is the mechanism by which nodes agree on the state of the blockchain, ensuring its integrity and consistency. One of the most important aspects of blockchain architecture is its security. The cryptographic hash functions and consensus algorithms used in blockchains make it very difficult for attackers to modify the data contained in blocks. Additionally, the decentralised nature of blockchains means that there is no single point of failure, making them highly resistant to attack. Another key advantage of blockchain architecture is its transparency. All transactions are publicly available for anyone to view, providing an unprece-dented level of accountability and transparency.

REFERENCES

[1] Nofer M., Gomber P., Hinz O., and Schiereck D., "Blockchain," *Business and Information Systems Engineering*, vol. 59, no. 3, pp. 183–187, Jun. 2017, doi: 10.1007/s12599-017-0467-3.
[2] "10.1109@COMST.2019.2928178".
[3] Yaga D., Mell P., Roby N., and Scarfone K., *"Blockchain Technology Overview,"* Jun. 2019, doi: 10.6028/NIST.IR.8202.
[4] Li X., Jiang P., T. Chen, X. Luo, and Q. Wen, "A Survey on the Security of Blockchain Systems," *Future Generation Computer Systems*, vol. 107, pp. 841–853, Jun. 2020, doi: 10.1016/j.future.2017.08.020.
[5] Iansiti M. and R K. Lakhani, *"The Truth About Blockchain."*
[6] Swan M., *Blockchain: Blueprint for a New Economy*. O'Reilly Media, 2015. [Online]. Available: https://books.google.co.in/books?id=RHJmBgAAQBAJ
[7] Chetan Kumar V., Shiva Prakash S. P. and Balandin S., "Performance Enhancement Of Optimized Link State Routing Protocol For Health Care Applications In Wireless Body Area Networks," *2018 23rd Conference of Open Innovations Association (FRUCT)*, Bologna, Italy, 2018, p195 p.-203, doi: 10.23919/FRUCT.2018.8588070.
[8] Nada Ratkovic, "Improving Home Security Using Blockchain," *International Journal of Computations, Information and Manufacturing (IJCIM)*, vol. 2, no. 1, May 2022, doi: 10.54489/ijcim.v2i1.72.
[9] Zheng Z., Xie S., Dai H., Chen X., and Wang H., "An Overview of Blockchain Technology: Architecture, Consensus, and Future Trends," in *Proceedings – 2017 IEEE 6th International Congress on Big Data, BigData Congress 2017*, Sep. 2017, pp. 557–564. doi: 10.1109/BigDataCongress.2017.85.
[10] Marianna Belotti, Nikola Bozic, Guy Pujolle, and Stefano Secci, "A Vademecum on Blockchain Technologies: When, Which and How," *IEEE*.
[11] Cachin C., Schubert S., and Vukolić M., "Non-determinism in Byzantine Fault-tolerant replication," in *Leibniz International Proceedings in Informatics, LIPIcs*, Apr. 2017, vol. 70, pp. 24.1–24.16. doi: 10.4230/LIPIcs.OPODIS.2016.24.
[12] Wang S., Yuan Y., Wang X., Li J., Qin R., and Wang F.-Y, *An Overview of Smart Contract: Architecture, Applications, and Future Trends*. 2018. doi: 10.0/Linux-x86_64.
[13] W.G. Peters and R.G. Vishnia, *"Overview of Emerging Blockchain Architectures and Platforms for Electronic Trading Exchanges."* [Online]. Available: https://ssrn.com/abstract=2867344
[14] O. and A. A. Jaradat Ashraf and Ali, "Blockchain Technology: A Fundamental Overview," in *Blockchain Technologies for Sustainability*, S. S. Muthu, Ed. Singapore: Springer Singapore, 2022, pp. 1–24. doi: 10.1007/978-981-16-6301-7_1.

[15]   M.L. de Rossi, Salviotti G., and Abbatemarco N., *Towards a Comprehensive Blockchain Architecture Continuum*. [Online]. Available: https://hdl.handle.net/10125/59898

[16]   Kan L., Wei Y., Hafiz A. Muhammad, Siyuan W., C.L. Gao, and Kai H., "A Multiple Blockchains Architecture on Inter-Blockchain Communication," in *2018 IEEE International Conference on Software Quality, Reliability and Security Companion (QRS-C)*, 2018, pp. 139–145. doi: 10.1109/QRS-C.2018.00037.

[17]   Neudecker T. and Hartenstein H., "Network Layer Aspects of Permissionless Blockchains," *IEEE Communications Surveys & Tutorials*, vol. 21, no. 1, pp. 838–857, 2019, doi: 10.1109/COMST.2018.2852480.

[18]   Shahid Latif, Zeba Idrees, Zil e Huma, and Jawad Ahmad, "Blockchain Technology for the Industrial Internet of Things: A Comprehensive Survey on Security Challenges, Architectures, Applications, and Future Research Directions," *Emerging Telecommunications Technology*, 2021.

[19]   Zhao H., Zhang M., Wang S., Li E., Guo Z., and Sun D., "Security Risk and Response Analysis of Typical Application Architecture of Information and Communication Blockchain," *Neural Comput Appl*, vol. 33, no. 13, pp. 7661–7671, 2021, doi: 10.1007/s00521-020-05508-z.

[20]   Chen J., Ma X., Du M., and Wang Z., "*A Blockchain Application for Medical Information Sharing*."

[21]   Zhai S., Yang Y., Li J., Qiu C., and Zhao J., "Research on the Application of Cryptography on the Blockchain," *J Phys Conf Ser*, vol. 1168, no. 3, p. 032077, 2019, doi: 10.1088/1742-6596/1168/3/032077.

[22]   F.A. Zorzo, Nunes H. C., Lunardi R. C., Michelin R. A., and Kanhere S. S., "Dependable IoT Using Blockchain-Based Technology," in *2018 Eighth Latin-American Symposium on Dependable Computing (LADC)*, 2018, pp. 1–9. doi: 10.1109/LADC.2018.00010.

[23]   Gupta S., Sinha S., and Bhushan B., "*International Conference on Innovative Computing and Communication (ICICC 2020) Emergence of Blockchain Technology: Fundamentals, Working and its Various Implementations.*" [Online]. Available: https://ssrn.com/abstract=3569577

[24]   Dey K and Shekhawat U., "Blockchain for Sustainable E-agriculture: Literature Review, Architecture for Data Management, and Implications," *J Clean Prod*, vol. 316, Sep. 2021, doi: 10.1016/j.jclepro.2021.128254.

[25]   Dedeoglu V., Jurdak R., Putra G. D., Dorri A., and Kanhere S. S., "A Trust Architecture for Blockchain in IoT," in *Proceedings of the 16th EAI International Conference on Mobile and Ubiquitous Systems: Computing, Networking and Services*, 2020, pp. 190–199. doi: 10.1145/3360774.3360822.

[26]   Paik H. Y., Xu X., Bandara H. M. N. D., Lee S. U., and Lo S. K., "Analysis of Data Management in Blockchain-based Systems: From Architecture to Governance," *IEEE Access*, vol. 7, pp. 186091–186107, 2019, doi: 10.1109/ACCESS.2019.2961404.

[27]   Li W., He M., and Haiquan S., "An Overview of Blockchain Technology: Applications, Challenges and Future Trends," in *2021 IEEE 11th International Conference on Electronics Information and Emergency Communication (ICEIEC)2021 IEEE 11th International Conference on Electronics Information and Emergency Communication (ICEIEC)*, 2021, pp. 31–39. doi: 10.1109/ICEIEC51955.2021.9463842.

[28]   Malamas V., Kotzanikolaou P., Dasaklis T. K., and Burmester M., "A Hierarchical Multi Blockchain for Fine Grained Access to Medical Data," *IEEE Access*, vol. 8, pp. 134393–134412, 2020, doi: 10.1109/ACCESS.2020.3011201.

[29]   Applications M., "*United States US 20080051989A1 (12) Patent Application Publication.*"

[30]   Liang X., Shetty S., Tosh D., Kamhoua C., Kwiat K., and Njilla L., "ProvChain: A Blockchain-Based Data Provenance Architecture in Cloud Environment with Enhanced Privacy and Availability," in *2017 17th IEEE/ACM International Symposium on Cluster, Cloud and Grid Computing (CCGRID)*, 2017, pp. 468–477. doi: 10.1109/CCGRID.2017.8.

[31]   Liang X., Shetty S., Tosh D., Kamhoua C., Kwiat K., and Njilla L., "ProvChain: A Blockchain-Based Data Provenance Architecture in Cloud Environment with Enhanced Privacy and Availability," in *2017 17th IEEE/ACM International Symposium on Cluster, Cloud and Grid Computing (CCGRID)*, 2017, pp. 468–477. doi: 10.1109/CCGRID.2017.8.

[32]   Xu X. *et al.*, "The Blockchain as a Software Connector," in *2016 13th Working IEEE/IFIP Conference on Software Architecture (WICSA)*, 2016, pp. 182–191. doi: 10.1109/WICSA.2016.21.

[33]   Zhang X., Li R., and Cui B., "A Security Architecture of VANET Based on Blockchain and Mobile Edge Computing," in *2018 1st IEEE International Conference on Hot Information-Centric Networking (HotICN)*, 2018, pp. 258–259. doi: 10.1109/HOTICN.2018.8605952.

[34]   Zaabar B., Cheikhrouhou O., Jamil F., Ammi M., and Abid M., "HealthBlock: A Secure Blockchain-based Healthcare Data Management System," *Computer Networks*, vol. 200, p. 108500, 2021, doi: https://doi.org/10.1016/j.comnet.2021.108500.

[35] el Kafhali S., Chahir C., Hanini M., and Salah K., "Architecture to Manage Internet of Things Data Using Blockchain and Fog Computing," in *Proceedings of the 4th International Conference on Big Data and Internet of Things*, 2020. doi: 10.1145/3372938.3372970.

[36] Zhai S., Yang Y., Li J., Qiu C., and Zhao J., "Research on the Application of Cryptography on the Blockchain," *J Phys Conf Ser*, vol. 1168, no. 3, p. 032077, 2019, doi: 10.1088/1742-6596/1168/3/032077.

[37] Pahlajani S., Kshirsagar A., and Pachghare V., "Survey on Private Blockchain Consensus Algorithms," in *2019 1st International Conference on Innovations in Information and Communication Technology (ICIICT)*, 2019, pp. 1–6. doi: 10.1109/ICIICT1.2019.8741353.

[38] Kaur M., Khan M. Z., Gupta S., Noorwali A., Chakraborty C., and Pani S. K., "MBCP: Performance Analysis of Large Scale Mainstream Blockchain Consensus Protocols," *IEEE Access*, vol. 9, pp. 80931–80944, 2021, doi: 10.1109/ACCESS.2021.3085187.

[39] Sankar L. S., Sindhu M., and Sethumadhavan M., "Survey of Consensus Protocols on Blockchain Applications," in *2017 4th International Conference on Advanced Computing and Communication Systems (ICACCS)*, 2017, pp. 1–5. doi: 10.1109/ICACCS.2017.8014672.

[40] Milutinovic M., He W., Wu H., and Kanwal M., "Proof of Luck: An Efficient Blockchain Consensus Protocol," in *SysTEX 2016 – 1st Workshop on System Software for Trusted Execution, colocated with ACM/IFIP/USENIX Middleware 2016*, Dec. 2016. doi: 10.1145/3007788.3007790.

[41] Chalaemwongwan N. and Kurutach W., "Notice of Violation of IEEE Publication Principles: State of the Art and Challenges Facing Consensus Protocols on Blockchain," in *2018 International Conference on Information Networking (ICOIN)*, 2018, pp. 957–962. doi: 10.1109/ICOIN.2018.8343266.

[42] S. and S. P. Sriman B. and Ganesh Kumar, "Blockchain Technology: Consensus Protocol Proof of Work and Proof of Stake," in *Intelligent Computing and Applications*, 2021, pp. 395–406.

[43] Cañete A., Amor M., and Fuentes L., "Supporting IoT Applications Deployment on Edge-based Infrastructures Using Multi-layer Feature Models," *Journal of Systems and Software*, vol. 183, p. 111086, 2022, doi: https://doi.org/10.1016/j.jss.2021.111086.

[44] Queralta J. P., Qingqing L., Zou Z., and Westerlund T., "Enhancing Autonomy with Blockchain and Multi-Access Edge Computing in Distributed Robotic Systems," in *2020 Fifth International Conference on Fog and Mobile Edge Computing (FMEC)*, 2020, pp. 180–187. doi: 10.1109/FMEC49853.2020.9144809.

[45] Nagarathna, alli S., Manjunath D., "Using an Evolution Model for Efficient Estimation and Tracking of Dynamic Boundaries". TENCON 2012 – 2012 *IEEE Region 10 Conference*, CEBU, pp. 1–6, Publisher: IEEE, INSPEC Accession Number: 13249789, ISSN: 2159-3442,

[46] Li Z., Barenji A. V., and Huang G. Q., "Toward a Blockchain Ccloud Manufacturing System as a Peer to Peer Distributed Network Platform," *Robot Comput Integr Manuf*, vol. 54, pp. 133–144, 2018, doi: https://doi.org/10.1016/j.rcim.2018.05.011.

[47] Liang G., Weller S. R., Luo F., Zhao J., and Dong Z. Y., "Distributed Blockchain-Based Data Protection Framework for Modern Power Systems Against Cyber Attacks," *IEEE Trans Smart Grid*, vol. 10, no. 3, pp. 3162–3173, 2019, doi: 10.1109/TSG.2018.2819663.

[48] Ghiasi M., Dehghani M., Niknam T., Kavousi A.-Fard, Siano P., and Alhelou H. H., "Cyber-Attack Detection and Cyber-Security Enhancement in Smart DC-Microgrid Based on Blockchain Technology and Hilbert Huang Transform," *IEEE Access*, vol. 9, pp. 29429–29440, 2021, doi: 10.1109/ACCESS.2021.3059042.

[49] Ali T. Syed, Alzahrani A., Jan S., Siddiqui M. S., Nadeem A., and Alghamdi T., "A Comparative Analysis of Blockchain Architecture and its Applications: Problems and Recommendations," *IEEE Access*, vol. 7, pp. 176838–176869, 2019, doi: 10.1109/ACCESS.2019.2957660.

[50] Saberi S., Kouhizadeh M., Sarkis J., and Shen L., "Blockchain Technology and Its Relationships to Sustainable Supply Chain Management," *Int J Prod Res*, vol. 57, no. 7, pp. 2117–2135, Apr. 2019, doi: 10.1080/00207543.2018.1533261.

Recent Trends in Computational Sciences – Gururaj, Pooja & Flammini (Eds)
© 2024 The Author(s), ISBN 978-1-032-42685-3

Need for Artificial Neural Networks in today's world

N.M. Siddesh Kumar
Department of Mechanical Engineering, PES College of Engineering, Mandya, India
ORCID ID: 0000-0002-6905-0634

B.M. Promod Kumar
Department of Computer Science and Engineering, PES College of Engineering, Mandya, India

Bheem Kumar Haloor
Department of Electrical and Engineering

Pranadhi Haran & K.S. Manoj Gowda
Department of Computer Science and Engineering, PES College of Engineering, Mandya, India

S.S. Hemashree
Department of Information Science and Engineering, PES College of Engineering, Mandya, India

ABSTRACT: Artificial Neural Networks (ANNs) are widely used in various fields due to their versatile applications. They are capable of handling complex functions and can adapt to new conditions. Their applications range from data interpretation, drug design, pattern recognition, text translation, credit card fraud detection, medical diagnosis and more. ANNs have become an essential tool in today's world due to their ability to interpret large amounts of data, making them crucial in fields such as medical diagnosis and credit card fraud detection. The versatility of ANNs makes them applicable in many different industries and continues to contribute to their growing popularity. The current work focuses on exploring the wide range of applications of ANNs and their significance in our world today.

Keywords: ANN models, Human brain, neural network, Nicosia, DNA, CGM

1 INTRODUCTION

Human brain consists of many neurons, which perform various functions. Similarly Artificial neural networks (ANN's) work like a human brain.

Artificial neural networks (ANN's) are also biologically similar neuron types as they are the programs used to gather specific data/information like a human brain processes the given data. The Input data is received by the input layer and transmitted to the hidden layer and then transmitted to the Output layer. ANN is being used in all fields of research. It is used to solve the programs by following a sequence of rules. Each connection in the neuron is associated with a numeric number, which can be called as weight [1].

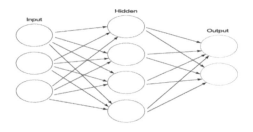

Figure 1. Neural network.

DOI: 10.1201/9781003363781-31

As the biological neurons connect the brain nerves, similarly neural networks also connect the data/information by collecting the relations or connections in the data.

Its advantage is also that It has the ability to work with insufficient knowledge like if we know input, the output can be easily predicted by using ANN. Sometimes ANN can be forced to make some predictions in many unsuitable conditions.

ANN can also be easy to analyse multi-dimensional input into two dimensional input, so it is used as an alternative for PCA(principal component analysis). Neural networks may take hours or may even take months to train, but time has a very small scope when compared to ANN's vast scope. All visionaries have reached the neural network due to the question "To what extent can the human brain be replicated? If we talk in terms of calculation/ computation, the computer network will be more efficient than human brain which takes some time to calculate that the work of normal computer is that it converts input data into output data by means of algorithm or some programming languages. Neural networks learn and train and make improvements in the output data. Finally Neural networks are those which can accept any number of inputs given by the user and will predict two to three outputs which can have 0.005% error in the actual output value. Neural network applications are used in various fields like Engineering, Psychology, Medical, Research field, etc. [1].

Even ANN has found its usage in pharmaceutical research fields like interpretation of data, drug delivery, etc. [1]. Now let's see the applications of ANN in the medical field one by one.

2 APPLICATIONS OF ANN IN MEDICAL SECTOR

In the medical field, ANN finds a wide range of applications, it is used to segment some of the infected parts. CT scans are performed to differentiate widespread and well known COVID-19 from other non-covid pneumonia infections [2]. It is even used for making urine tests by recognizing urine particles by using some digital and microscopic pictures of urine sediments [3]. Neural networks are also used in IoT technologies. ANN are also used in fruit processing and fish processing and also ANN finds its applications in intelligent food processing by checking its quality, ensuring the safety of food and avoiding cases of food poisoning [4]. ANN is also developed in the medical field to treat cancer patients, it is used to predict the visiting of emergency departments visiting patients having cancer [5]. Even ANN is also used in the predictions of some heart disease and many algorithms are made and also used for heart disease predictions [6]. In Surgical research also, artificial neural networks are used [7].

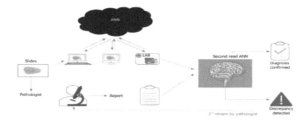

Figure 2. Process flow of ANN in medical field.

Diabetic patients have to track their food intake to maintain their glucose level in their body. CGM (continuous glucose monitoring) is used to predict the amount of glucose in diabetic patients. ANN also finds its development in prediction of diabetes in the initial stage [8]. ANN can also be used in the early detection of cardiac arrest. Till now, the number of people who can have cardiac arrest is not clear so it can be prevented by using ANN [9]. ANN has also established its fame in dentistry. It also plays a very important role in assisting dental professionals. It performs the tasks within no time. But it has some drawbacks also, like complex mechanisms, high cost, and requires an adequate period of training [10]. ANN can also be used to detect kidney diseases. At first, kidney images are classified/identified as normal/abnormal. Abnormal images of kidney are kidney diseases like kidney stones, tumours, etc.. And then ANN sufficient technique is used to solve the abnormal kidney problems by the segmentation of abnormal images of the kidney [11]. Even ANN can be used for the estimation of sex i.e. to identify sex as male or female. The sex of the skeleton of some person as male or female can be identified/Predicted using ANN. But it also has some drawbacks like it's expensive, difficult to access and also it requires qualified and experienced personnel and by using ANN, DNA can't be obtained [12]. To Diagnose corneal Diseases also, a neural network is used [13]. Using neural networks, X-ray images can be taken and it is used to diagnose covid-19 symptoms for a covid-19 patient. It is even used to classify the X-ray images as covid-19 symptoms or pneumonia type symptoms. Similarly X-rays are used to diagnose many chest-related problems also [14].

3 APPLICATIONS OF ANN IN ENGINEERING SECTOR

Using ANN, wastewater treatment can be conducted and even desalination of water can be undertaken. Like this, the modern technique(ANN) is used to save water [15]. Even ANN applications are also used to compute the signal strength, that computation will be based on some weather factors like temperature, pressure, speed of the air and also relative humidity, these factors will be used as input for ANN [16]. To study micromechanical behaviour of some polycrystalline metals also, ANN is used [17]. Even ANN's are used to solve some issues of solid wastes by knowing its causes and by using ANN's applications to solve those issues [18]. By using ANN applications, engine behaviours can be studied and it has accuracy more than 95%. It is used to study many uses of biodiesel in many engines without conducting any practical lab experiments, it's applications are used for predicting the engine performance, it's COP(coefficient of performance) and emissions of engines which are fuelled with a diesel which is produced from many waste materials i.e. biodiesel [19]. ANN's has the ability to accurately predict upcoming future droughts. The data will be provided as inputs for ANN for predicting droughts. By knowing the upcoming droughts, some methods of its prevention or some control measures could be taken so that it would reduce the harmfulness caused on the population [20]. Applications of ANN can even be used for predicting the concentration of acids(of only two types of acids). ANN models can be used for predicting some percentage concentrations of some number of acid samples, maybe of 14 acid samples [21]. ANN model can also be used to predict metakaolin-based concrete's compressive strength [22]. Artificial neural network(ANN) model can also be used as a data mining tool, it is also used to predict Biochemical Oxygen Demand(BOD) and also to predict the concentration of effluents during waste-water treatment. ANN is also used as a powerful tool for modelling the removal of pollutants [23]. ANN can also be used to predict fuel consumption in vehicles. Fuel consumption of the vehicles depends on many factors like vehicle weight, engine displacement, number of cylinders and valves, driving pressure, tire

Figure 3.

pressure, etc. ANN is successful in measuring fuel consumption by considering the above factors [24]. Air pollution is one of the major pollution affecting almost all the living organisms. ANN is used to predict the concentration of air pollution and some measures can be taken to reduce air pollution and protect the environment. The same was done in a place called Nicosia. The performance of ANN in predicting the concentration of pollutants was satisfactory in Nicosia [25]. Artificial Neural Networks is also used for the prediction of quality of water. It can also be used for predicting urban sustainability and contamination of water [26]. ANN is also used to predict the heat transfer in soil temperature [27]. Artificial Neural Network(ANN) and its applications are also used to detect fraud and its applications can also be used in direct marketing [28]. ANN applications are used for wireless communication [29]. ANN model was employed to know polycrystalline material behaviour. Even for studying the micromechanical behaviours of polycrystalline metal, ANN is used [30].

4 APPLICATIONS OF ANN IN RESEARCH SECTOR

The waste management was very concerned because of growing environmental consciousness. ANN is also useful in the minimization of solid waste effluents [31]. Using Artificial Neural Network(ANN), we can even forecast road surface conditions especially during the rainy season. By using ANN, the drivers would get the information of unsafe road conditions [32]. Even for predicting minimum boiling temperature for some experiments or studies using artificial neural networks [33]. Even Artificial Neural Networks(ANN's) are used for optimization and studying COVID-19 [34]. ANN is also used for the performance analysis of logical gates [35]. Even Artificial Neural Networks can be used to recognize some environmental sounds or a little far away noises [36]. By using ANN, even many cybercrimes can be detected. One such example is the detection of credit card fraud by using ANN. Credit card fraud is very common nowadays because most people use credit cards to do online payments very frequently. Even credit card fraud can be classified into 10 types. ANN is 100% best suited for detecting any type of credit card fraud [37]. Even ANN finds its use in forecasting air compressor load. To reduce the electricity consumption in air compressor systems, therefore ANN can be used as a replacement for electricity and can be used in air compression systems [38]. For the management of environmental odour also, ANN can be used. ANN is used to measure odour intensity prediction and its concentration and by knowing all this, management of odour can be done using ANN [39]. ANN applications can be used for the prediction of Chemical Oxygen Demand(COD) from wastewater treatment plants [40]. Even to manage the heating system of solar water, ANN is used for the optimization purpose. Also, ANN can be used for the design of hot water storage tanks [41]. Artificial Neural Networks(ANN's) nowadays are used for the prediction of price of Crude Oil. The advantage of using ANN here is that to know the variation of Crude Oil Prices [42]. In rivers, for the quality prediction of water, ANN can be used. Like this, the performance of ANN models could be improved [43]. Even ANN shows its impact on crop productivity and human health. For crops, the quality of nutrients and chemicals are applied based on the suggestion of ANN method [44]. Biogas can even be produced from the waste food, wastes of fruits and vegetables using artificial neural networks(ANN) [45].

Artificial Neural Networks(ANN's) also play a major role in the minimization of solid waste effluents. ANN if used in papermaking processing steps, it would be a measure to control environmental pollution caused by paper mills [46]. Even to predict Vehicular traffic noise, ANN can be used [47]. ANN can be used for predicting the relative liking of food even in the presence of background noise [48]. Even for the mapping of landslide susceptibility, ANN is used [49]. Even in the educational field, ANN applications are widely used. ANN is used to predict the academic performance of students, it could be useful to train academically poor students [50].

5 CONCLUSION

Artificial neural networks (ANN's) are similar to biological neurons in that they are programs used to gather specific data/information like a human brain processes the given data. Each connection in the neuron is associated with a numeric number, which can be called a weight. ANN is being used in all fields of research and is used to solve programs by following a sequence of rules. Its advantage is that it has the ability to work with insufficient knowledge and can be forced to make predictions in unsuitable conditions.

Neural networks (ANNs) are used to analyse multi-dimensional input into two-dimensional input, and are an alternative to PCA (principal component analysis). They can take hours or months to train, but time has a very small scope when compared to ANN's vast scope. Neural networks can learn and train and make improvements in the output data, and can accept any number of inputs given by the user and predict two to three outputs which can have 0.005% error in the actual output value. ANN applications are used in various fields like Engineering, Psychology, Medical, Research field, etc., and even in pharmaceutical research fields like interpretation of data, drug delivery, etc. In the medical field, ANN is used to segment some of the infected parts, and CT scans are performed to differentiate widespread and rare diseases.

REFERENCES

[1] Agatonovic-Kustrin, S., & Beresford, R. (2000). Basic Concepts of Artificial Neural Network (ANN) Modelling and Its Application in Pharmaceutical Research. In *Journal of Pharmaceutical and Biomedical Analysis* (Vol. 22). www.elsevier.com/locate/jpba

[2] Khanday, N. Y., & Sofi, S. A. (2021). Deep Insight: Convolutional Neural Network and Its Applications for COVID-19 Prognosis. In *Biomedical Signal Processing and Control* (Vol. 69). Elsevier Ltd. https://doi.org/10.1016/j.bspc.2021.102814

[3] Suhail, K., & Brindha, D. (2021). A Review on Various Methods for Recognition of Urine Particles Using Digital Microscopic Images of Urine Sediments. In *Biomedical Signal Processing and Control* (Vol. 68). Elsevier Ltd. https://doi.org/10.1016/j.bspc.2021.102806

[4] Nayak, J., Vakula, K., Dinesh, P., Naik, B., & Pelusi, D. (2020). Intelligent Food Processing: Journey From Artificial Neural Network to Deep Learning. In *Computer Science Review* (Vol. 38). Elsevier Ireland Ltd. https://doi.org/10.1016/j.cosrev.2020.100297

[5] Sutradhar, R., & Barbera, L. (2020). Comparing an Artificial Neural Network to Logistic Regression for Predicting ED Visit Risk Among Patients With Cancer: A Population-Based Cohort Study. *Journal of Pain and Symptom Management, 60*(1), 1–9. https://doi.org/10.1016/j.jpainsymman.2020.02.010

[6] Mienye, I. D., Sun, Y., & Wang, Z. (2020). Improved Sparse Autoencoder Based Artificial Neural Network Approach for Prediction of Heart Disease. *Informatics in Medicine Unlocked, 18*. https://doi.org/10.1016/j.imu.2020.100307

[7] Velanovich, V., & Walczak, S. (2020). Artificial Neural Networks in Surgical Research. In *American Journal of Surgery* (Vol. 220, Issue 6, pp. 1532–1533). Elsevier Inc. https://doi.org/10.1016/j.amjsurg.2020.06.074

[8] Cichosz, S. L., Jensen, M. H., & Hejlesen, O. (2021). Short-term Prediction of Future Continuous Glucose Monitoring Readings in Type 1 Diabetes: Development and Validation of a Neural Network Regression Model. *International Journal of Medical Informatics, 151*. https://doi.org/10.1016/j.ijmedinf.2021.104472

[9] Jang, D. H., Kim, J., Jo, Y. H., Lee, J. H., Hwang, J. E., Park, S. M., Lee, D. K., Park, I., Kim, D., & Chang, H. (2020). Developing Neural Network Models for Early Detection of Cardiac Arrest in the Emergency Department. *American Journal of Emergency Medicine, 38*(1), 43–49. https://doi.org/10.1016/j.ajem.2019.04.006

[10] Tandon, D., & Rajawat, J. (2020). Present and Future of Artificial Intelligence in Dentistry. In *Journal of Oral Biology and Craniofacial Research* (Vol. 10, Issue 4, pp. 391–396). Elsevier B.V. https://doi.org/10.1016/j.jobcr.2020.07.015

[11] Nithya, A., Appathurai, A., Venkatadri, N., Ramji, D. R., & Anna Palagan, C. (2020). Kidney Disease Detection and Segmentation Using Artificial Neural Network and Multi-kernel k-means Clustering for Ultrasound Images. *Measurement: Journal of the International Measurement Confederation, 149*. https://doi.org/10.1016/j.measurement.2019.106952

[12] Oner, Z., Turan, M. K., Oner, S., Secgin, Y., & Sahin, B. (2019). Sex Estimation Using Sternum Part Lengths by Means of Artificial Neural Networks. *Forensic Science International, 301*, 6–11. https://doi.org/10.1016/j.forsciint.2019.05.011

[13] Elsawy, A., Eleiwa, T., Chase, C., Ozcan, E., Tolba, M., Feuer, W., Abdel-Mottaleb, M., & Abou Shousha, M. (2021). Multi Disease Deep Learning Neural Network for the Diagnosis of Corneal Diseases. *American Journal of Ophthalmology, 226*, 252–261. https://doi.org/10.1016/j.ajo.2021.01.018

[14] Ozturk, T., Talo, M., Yildirim, E. A., Baloglu, U. B., Yildirim, O., & Rajendra Acharya, U. (2020). Automated Detection of COVID-19 Cases Using Deep Neural Networks With X-ray Images. *Computers in Biology and Medicine, 121*. https://doi.org/10.1016/j.compbiomed.2020.103792

[15] Jawad, J., Hawari, A. H., & Javaid Zaidi, S. (2021). Artificial Neural Network Modelling of Wastewater Treatment and Desalination Using Membrane Processes: A Review. In *Chemical Engineering Journal* (Vol. 419). Elsevier B.V. https://doi.org/10.1016/j.cej.2021.129540

[16] Igwe, K. C., Oyedum, O. D., Aibinu, A. M., Ajewole, M. O., & Moses, A. S. (2021). Application of Artificial Neural Network Modelling Techniques to Signal Strength Computation. *Heliyon, 7*(3). https://doi.org/10.1016/j.heliyon.2021.e06047

[17] Dai, W., Wang, H., Guan, Q., Li, D., Peng, Y., & Tomé, C. N. (2021). Studying the Micromechanical Behaviours of a Polycrystalline Metal by Artificial Neural Networks. *Acta Materialia, 214*. https://doi.org/10.1016/j.actamat.2021.117006

[18] Xu, A., Chang, H., Xu, Y., Li, R., Li, X., & Zhao, Y. (2021). Applying Artificial Neural Networks (ANNs) to Solve Solid Waste-related Issues: A Critical Review. In *Waste Management* (Vol. 124, pp. 385–402). Elsevier Ltd. https://doi.org/10.1016/j.wasman.2021.02.029

[19] Tuan Hoang, A., Nižetić, S., Chyuan Ong, H., Tarelko, W., Viet Pham, V., Hieu Le, T., Quang Chau, M., & Phuong Nguyen, X. (2021). A Review on Application of Artificial Neural Network (ANN) for Performance and Emission Characteristics of Diesel Engines Fueled With Biodiesel-based Fuels. *Sustainable Energy Technologies and Assessments, 47*. https://doi.org/10.1016/j.seta.2021.101416

[20] Khan, M. M. H., Muhammad, N. S., & El-Shafie, A. (2020). Wavelet Based Hybrid ANN-ARIMA Models for Meteorological Drought Forecasting. *Journal of Hydrology, 590*. https://doi.org/10.1016/j.jhydrol.2020.125380

[21] Sang, T. T., An, D. H., Chuong, H. D., Hang, N. T., Nhat, L. D., Kim Anh, N. T., My Duyen, T. T., & Tam, H. D. (2021). ANN Coupled with Monte Carlo Simulation for Predicting the Concentration of Acids. *Applied Radiation and Isotopes, 169*. https://doi.org/10.1016/j.apradiso.2020.109563

[22] Moradi, M. J., Khaleghi, M., Salimi, J., Farhangi, V., & Ramezanianpour, A. M. (2021). Predicting the Compressive Strength of Concrete Containing Metakaolin With Different Properties Using ANN. *Measurement: Journal of the International Measurement Confederation, 183*. https://doi.org/10.1016/j.measurement.2021.109790

[23] Kiiza, C., Pan, S. qi, Bockelmann-Evans, B., & Babatunde, A. (2020). Predicting Pollutant Removal in Constructed Wetlands Using Artificial Neural Networks (ANNs). *Water Science and Engineering, 13*(1), 14–23. https://doi.org/10.1016/j.wse.2020.03.005

[24] Zargari Nejad, S., Dashti, R., & Ahmadi, R. (2019). Predicting Vehicle Fuel Consumption in Energy Distribution Companies Using ANNs. *Transportation Research Part D: Transport and Environment, 74*, 174–188. https://doi.org/10.1016/j.trd.2019.07.020

[25] Cakir, S., & Sita, M. (2020). Evaluating the Performance of ANN in Predicting the Concentrations of Ambient air Pollutants in Nicosia. *Atmospheric Pollution Research, 11*(12), 2327–2334. https://doi.org/10.1016/j.apr.2020.06.011

[26] Dawood, T., Elwakil, E., Novoa, H. M., & Gárate Delgado, J. F. (2021). Toward Urban Sustainability and Clean Potable Water: Prediction of Water Quality Via Artificial Neural Networks. *Journal of Cleaner Production, 291*. https://doi.org/10.1016/j.jclepro.2020.125266

[27] Jebamalar, S., Christopher, J. J., & Ajisha, M. A. T. (2021). Random Input-based Prediction and Transfer of Heat in Soil Temperature Using Artificial Neural Networks. *Materials Today: Proceedings, 45*, 1540–1546. https://doi.org/10.1016/j.matpr.2020.08.091

[28] Zakaryazad, A., & Duman, E. (2016). A Profit-driven Artificial Neural Network (ANN) with Applications to Fraud Detection and Direct Marketing. *Neurocomputing, 175*(PartA), 121–131. https://doi.org/10.1016/j.neucom.2015.10.042

[29] Alapuranen, P., & Schroeder, J. (2021). Complex Artificial Neural Network With Applications to Wireless Communications. *Digital Signal Processing: A Review Journal, 119*. https://doi.org/10.1016/j.dsp.2021.103194

[30] Dai, W., Wang, H., Guan, Q., Li, D., Peng, Y., & Tomé, C. N. (2021). Studying the Micromechanical Behaviours of a Polycrystalline Metal by Artificial Neural Networks. *Acta Materialia, 214*. https://doi.org/10.1016/j.actamat.2021.117006

[31] Almonti, D., Baiocco, G., & Ucciardello, N. (2021). Pulp and Paper Characterization by Means of Artificial Neural Networks for Effluent Solid Waste Minimization—A Case Study. *Journal of Process Control, 105*, 283–291. https://doi.org/10.1016/j.jprocont.2021.08.012

[32] Kim, S., Lee, J., & Yoon, T. (2021). Road Surface Conditions Forecasting in Rainy Weather Using Artificial Neural Networks. *Safety Science, 140*. https://doi.org/10.1016/j.ssci.2021.105302

[33] Bahman, A. M., & Ebrahim, S. A. (2020). Prediction of the Minimum Film Boiling Temperature Using an Artificial Neural Network. *International Journal of Heat and Mass Transfer, 155*. https://doi.org/10.1016/j.ijheatmasstransfer.2020.119834

[34] Elhag, A. A., Aloafi, T. A., Jawa, T. M., Sayed-Ahmed, N., Bayones, F. S., & Bouslimi, J. (2021). Artificial Neural Networks and Statistical Models for Optimization Studying COVID-19. *Results in Physics, 25*. https://doi.org/10.1016/j.rinp.2021.104274

[35] Hamedi, S., & Dehdashti Jahromi, H. (2021). Performance Analysis of an All-optical Logic Gate Using an Artificial Neural Network. *Expert Systems with Applications, 178*. https://doi.org/10.1016/j.eswa.2021.115029

[36] Simonović, M., Kovandžić, M., Ćirić, I., & Nikolić, V. (2021). Acoustic Recognition of Noise-like Environmental Sounds by Using Artificial Neural Networks. *Expert Systems with Applications, 184*. https://doi.org/10.1016/j.eswa.2021.115484

[37] RB, A., & KR, S. K. (2021). Credit Card Fraud Detection Using Artificial Neural Networks. *Global Transitions Proceedings, 2*(1), 35–41. https://doi.org/10.1016/j.gltp.2021.01.006

[38] Wu, D. C., Bahrami Asl, B., Razban, A., & Chen, J. (2021). Air Compressor Load Forecasting Using Artificial Neural Networks. *Expert Systems with Applications, 168*. https://doi.org/10.1016/j.eswa.2020.114209

[39] Zarra, T., Galang, M. G., Ballesteros, F., Belgiorno, V., & Naddeo, V. (2019). Environmental Odour Management by Artificial Neural Network – A Review. In *Environment International* (Vol. 133). Elsevier Ltd. https://doi.org/10.1016/j.envint.2019.105189

[40] Abba, S. I., & Elkiran, G. (2017). Effluent Prediction of Chemical Oxygen Demand from the Wastewater Treatment Plant Using Artificial Neural Network Application. *Procedia Computer Science, 120*, 156–163. https://doi.org/10.1016/j.procs.2017.11.223

[41] Kulkarni, M.V., Deshmukh, D. S., & Shekhawat, S. P. (2020). An Innovative Design Approach of Hot Water Storage Tank for Solar Water Heating System Using Artificial Neural Network. *Materials Today: Proceedings, 46*, 5400–5405. https://doi.org/10.1016/j.matpr.2020.09.058

[42] Gupta, N., & Nigam, S. (2020). Crude Oil Price Prediction using Artificial Neural Network. *Procedia Computer Science, 170*, 642–647. https://doi.org/10.1016/j.procs.2020.03.136

[43] Kim, S. E., & Seo, I. W. (2015). Artificial Neural Network Ensemble Modelling with Conjunctive Data Clustering for Water Quality Prediction in Rivers. *Journal of Hydro-Environment Research, 9*(3), 325–339. https://doi.org/10.1016/j.jher.2014.09.006

[44] Elahi, E., Weijun, C., Zhang, H., & Abid, M. (2019). Use of Artificial Neural Networks to Rescue Agrochemical-based Health Hazards: A Resource Optimisation Method for Cleaner Crop Production. *Journal of Cleaner Production, 238*. https://doi.org/10.1016/j.jclepro.2019.117900

[45] Gonçalves Neto, J., Vidal Ozório, L., Campos de Abreu, T. C., Ferreira dos Santos, B., & Pradelle, F. (2021). Modelling of Biogas Production From Food, Fruits and Vegetables Wastes Using Artificial Neural Network (ANN). *Fuel, 285*. https://doi.org/10.1016/j.fuel.2020.119081

[46] Almonti, D., Baiocco, G., & Ucciardello, N. (2021). Pulp and Paper Characterization by Means of Artificial Neural Networks for Effluent Solid Waste Minimization—A Case Study. *Journal of Process Control, 105*, 283–291. https://doi.org/10.1016/j.jprocont.2021.08.012

[47] Nourani, V., Gökçekuş, H., Umar, I. K., & Najafi, H. (2020). An Emotional Artificial Neural Network for Prediction of Vehicular Traffic Noise. *Science of the Total Environment, 707*. https://doi.org/10.1016/j.scitotenv.2019.136134

[48] Alamir, M. A. (2021). An Enhanced Artificial Neural Network Model Using the Harris Hawks Optimiser for Predicting Food Liking in the Presence of Background Noise. *Applied Acoustics, 178*. https://doi.org/10.1016/j.apacoust.2021.108022

[49] Bragagnolo, L., Silva, R. V. da, & Grzybowski, J. M. V. (2020). Artificial Neural Network Ensembles Applied to the Mapping of Landslide Susceptibility. *Catena, 184*. https://doi.org/10.1016/j.catena.2019.104240

[50] Rodríguez-Hernández, C. F., Musso, M., Kyndt, E., & Cascallar, E. (2021). Artificial Neural Networks in Academic Performance Prediction: Systematic Implementation and Predictor Evaluation. *Computers and Education: Artificial Intelligence, 2*, 100018. https://doi.org/10.1016/j.caeai.2021.10001

Current trends in cyber security

Recent Trends in Computational Sciences – Gururaj, Pooja & Flammini (Eds)
© 2024 The Author(s), ISBN 978-1-032-42685-3

Location recommender system

Deepali Suraj Attavar, R. Deekshitha, S. Nikhitha & B. Deepthi
Department of Computer Science, PES University, India
ORCID ID: 0009-0007-2998-4954, 0009-0006-8024-5084

P. Evlin Vidyu Latha
PES University, India

ABSTRACT: Location Recommender System is a Web Application that recommends the best places to visit in and around a location. Once the User logs in, his/her location coordinates are obtained and the nearby places are listed down based on the Location-Aware Recommendation model built; the user can also do the same for a custom location. The system displays the recommendations along with ratings. The system interface displays these suggestions based upon Machine Learning and Data Analysis tools: Recommender system and optimal search algorithm to obtain the finest nearby places with minimal buffering and time consumption. User-Item-based collaborative filtering algorithm is also used to give a personalised recommendation.

Keywords: Location Aware Recommender System, Location Coordinates, User Preference, Content Filtering, Hybrid model, User's Context

1 INTRODUCTION

In the current world, users have been successfully provided with objects of interest (such as movies, music, books, news, and photos) using recommendation algorithms. On the other hand, the traditional recommendation algorithms, mostly do not consider the location to be an important factor when making these suggestions. Our aim is to design a web application that recommends the best places to visit in and around a location. We are taking Bangalore as the experimental case in this phase of our research. The web application assists in attracting and being a great help to the tourists or the people new to a place in saving time while exploring the places in that location or to visit places of their interest. By providing the ratings, it aids the user to decide better as to which places to choose for instance, historic places, a place for leisure, eatery, shopping, park etc. The user's profile caters in giving a preferable and a personalized recommendation.

2 RELATED WORK

In [1], a cloud-based hybrid recommendation framework is used. Multi-level collaborative filtering and scalar optimization are included in the recommender system. It generates Top K recommendations. However, it does not employ multi-agent systems to investigate a user's journey pattern in order to develop a preference-aware location recommender system. Concerns about privacy are not properly addressed. [2] talks about a context aware tourist app that includes location, time, weather, social media sentiment, and penalisation. The perceptron model using Fuzzy logic and ANN is used; also employs PCA. Although they state that the number of factors included increases the complexity. The work that is done in [3] is to provide a survey on location aware recommendation systems in mobile computing. Various concepts like traditional recommendation system, location aware recommendation system (LARS) are discussed here. They [4] created a personalised recommendation system for the web and mobile that will provide recommendations based on relevant location based data collected from users (GPS routes and photos). Recommending routes and scoring criteria introduced to the recommendation system require refinement, posing additional obstacles. In [5], the focus is on personalised sightseeing tour recommendations using adaptive recommendation techniques and profile modelling integration. The method allows for the inheritance of features by collecting information on tourist profiles, specialised hobbies, and so on. The "tour box" stores tourists' trip information. In [6] the primary goal is to propose a framework for a Personalized Location-based Traveller Recommender System for a mobile application. It aims to provide an appropriate recommendation mechanism for use in an Individualized Location-based Traveler Recommender System (PLTRS) and location-based services (LBS). This software discussed in [7] is for tourists who want to visit destinations using their Android devices. This is useful for those who want to visit nearby locations or work on similar projects. The study in [8] is based on a system for recommending tourism destinations. Collaborative Filtering is employed, and Material Based Filtering is used to propose places based on similar content. The machine learning model is trained and pickled before being utilised in the front-end, and the data is saved in a MySQL database. Content Based Filtering, Web Scraping, and MySQL are all the terms used in the index.

3 PROPOSED METHODOLOGY

- The Approach to the project would begin by first clustering the areas in the target location based on the latitude and longitude coordinates.
- Using the Unsupervised Machine Learning technique: K-Means Clustering and Elbow Curve method, the least error optimal number of clusters is found and clustered accordingly. The K-Means algorithm identifies k number of centroids and then allocates every data point to the nearest cluster, while keeping the centroids as small as possible.
- User's location is accessed via the Geo-location API through which co-ordinates are obtained of the host device.
- Based on the latitude and longitude coordinates obtained, the user is identified to which cluster he/she belongs to by taking the distance as the metric in order to make the best classification of the cluster for it's representation.
- The Bangalore venue data is cleaned, pre-processed and stored after handling the missing, duplicate and inconsistent values.
- On the basis of the location aware recommendation system built, the user is provided with recommendations. For this Hybrid Recommender system: Top K recommendations and Collaborative Filtering is used as the tool.
- API's are used to provide the user details with regard to the brief overview, the timings of the place, reviews etc. For this, Google PlaceID API and Google Place API are used.
- In addition, Weather API – RapidX API fetches real-time weather details of the place based on it's location coordinates.

4 DESIGN AND IMPLEMENTATION

4.1 *Design*

4.1.1 *Models*

- Sign-up, Authorization: Registration of a new user, giving necessary credentials.
- User login: Login of a user using credentials.
- Authentication: Following sign up and login, user data is verified in database.

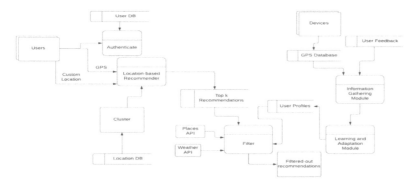

Figure 1.   Workflow diagram.

- Cluster finder: It is used to find the cluster, the location coordinate belongs to.
- Main: Main is the Heart of processing modules.
- Top K recommendations: It recommends the Top K nearby places and far off places for a location.
- Collaborative Filtering: It recommends places to users according to their previously visited places and ratings using user-item-based collaborative filtering.

4.1.2   *APIs*

- Geolocation API: Geolocation API is used for obtaining user's current location.
- Weather API: Weather API helps in getting weather details for coordinates from RapidX API.
- Place ID API: Place ID API is used for getting place ID for the coordinates.
- Place Details API: Place Details API helps in getting details such as opening hours, delivery, dine-in from Google API.

4.1.3   *Database*

- User DB: User's Login details are stored.
- User GPS DB: Stores current position of the users.
- User Feedback DB: User's ratings for the visited places are stored.
- Location Cluster DB: Stores the clustered area data.

4.2   *Implementation*

4.2.1   *Top K recommender*
For Top K recommender, K-means Clustering is done. Various clusters are formed. The cluster to which the user's current location belongs to is found; to which recommendations are given accordingly. Two types of recommendations are made: 1) Nearby Top K recommendations 2) Slightly Far Top K recommendations. In Far away and Nearby Top K recommender, the number of clusters considered is different from each other as the radius of the distance taken into consideration varies consequently. Based on Elbow curve method, for the Nearby Top K recommender the number of clusters considered is 5 whereas for Slightly Far Top K recommender the number of clusters considered is 3.

4.2.2   *Collaborative filtering*
Collaborative filtering is used to give a personalised recommendation based on the user's history of visited places and their ratings. User-Item-based is used for the same. The

database consisting of the previously visited places and ratings by users are stored. A particular user's previously visited places are considered and ratings given are checked. Then a similarity matrix is obtained regarding the similar category of places a particular user has visited and rated and recommendations are given accordingly. The collaborative filtering helps in finding places similar to what the user has already visited and rated, and gives top "n" recommendations personalised to that user.

5 RESULTS

K-means clustering is an unsupervised machine learning algorithm that finds a rule to group similar data together.

$$J(V) = \sum_{i=1}^{c} \sum_{j=1}^{c_i} (||x_i - v_j||)^2$$

where, $[x_i - y_{jic}]$ is Euclidean distance between x_i and y_j, c_i is the number of data points in i^{th} cluster, 'c' is the number of cluster centers.

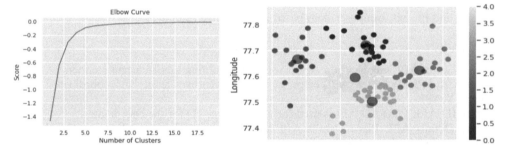

Figure 2.   Elbow curve method.   Figure 3.   K-Means clustering visualized of areas.

Here, it's used to organize large location coordinates of unlabelled data (latitiude and longitude) to generate clusters of Bangalore areas.
The Elbow Method is one of the most popular methods to determine this optimal value of k which was found to be 5 for our dataset.

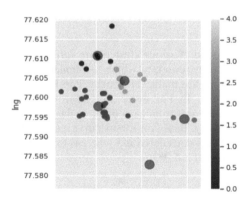

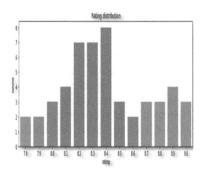

Figure 4.   K-means clustering visualized of venues.

Figure 5.   Ratings distribution.

228

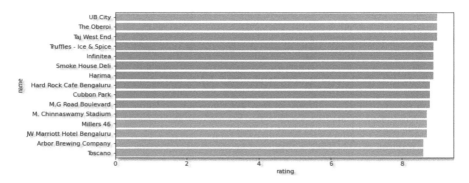

Figure 6.   Top 15 venues based on ratings.

6   DISCUSSIONS

The recommendation system with time, the personalized recommendations gets better as the model gathers more data. The data is also limited as the range taken into consideration is just Bangalore alone. Recommendations for city center will always provide more preferable suggestions as the places are largely concentrated in the central part of the city. This also gives many cafeterias, pizzerias and bars as suggestions as they are popular and highly rated and have been taken from the database that are large in numbers. In addition, the system also provides additional recommendations such as gyms, shops and parks.

7   CONCLUSION AND FUTURE WORK

We have designed a Location-Aware Recommendation Web Application system for the tourists as well as commoners who are busy in their day-to-day life. Typically, one thinks to visit some place that is a well-known touristic locations or places they have already been before and rarely choose to try something new. Knowingly or unknowingly people tend to not recall or ignore the great places to visit that aren't well known. Many approaches for tourist guides have been proposed and have mainly been focused on the well-known tourist spots. With the aid of our system, a user's decision making process is simplified and is enhanced to discover new places without much effort from their side. The developed solution consists of an easy-to-use interface with usability being an important factor. This application contributes to the tourism industry as well as to an everyday localites as it would be promoting touristic locations and lesser known places keeping proximity of the user's location a vital factor while making recommendations.

REFERENCES

[1]   Logesh Ravi, Malathi Devarajan, Vijayakumar V, Arun Kumar Sangaiah, Lipo Wang, Sasikumar A & V Subramaniyaswamy (2021) "An Intelligent Location Recommender System Utilising Multi-agent Induced Cognitive Behavioural Model", *Enterprise Information Systems*, 15:10, 1376–1394, DOI: 10.1080/17517575.2020.1812003

[2]   K. Meehan, T. Lunney, K. Curran and A. McCaughey, "Context-aware Intelligent Recommendation System for Tourism," *2013 IEEE International Conference on Pervasive Computing and Communications Workshops (PERCOM Workshops)*, San Diego, CA, USA, 2013, pp. 328–331, doi: 10.1109/PerComW.2013.6529508.

[3] Rodríguez Hernández, María del Carmen & Ilarri, S. & Trillo, Raquel & Hermoso, Ramon. (2015). Location-aware Recommendation Systems: Where We are and Where We Recommend to go. *CEUR Workshop Proceedings*. 1405. 1–8.

[4] K. Waga, A. Tabarcea and P. Fränti, "Recommendation of Points of Interest from User Generated Data Collection," *8th International Conference on Collaborative Computing: Networking, Applications and Worksharing (CollaborateCom)*, Pittsburgh, PA, USA, 2012, pp. 550–555, doi: 10.4108/icst. collaboratecom.2012.250451.

[5] Anacleto, Ricardo & Figueiredo, Lino & Almeida, Ana & Novais, Paulo. (2013). Mobile Application to Provide Personalized Sightseeing Tours. *Journal of Network and Computer Applications*. 41. 10.1016/j. jnca.2013.10.005.

[6] Husain, Wahidah & Dih, Lam. (2012). A Framework of a Personalized Location-based Traveler Recommendation System in Mobile Application. *International Journal of Multimedia and Ubiquitous Engineering*. 7.

[7] Umanets, Artem & Ferreira, Artur & Leite, Nuno. (2014). GuideMe – A Tourist Guide with a Recommender System and Social Interaction. *Procedia Technology*. 17. 10.1016/j.protcy.2014.10.248.

[8] G. Badaro, H. Hajj, W. El-Hajj and L. Nachman, "A hybrid Approach with Collaborative Filtering for Recommender Systems," *2013 9th International Wireless Communications and Mobile Computing Conference (IWCMC)*, Sardinia, Italy, 2013, pp. 349–354, doi: 10.1109/IWCMC.2013.6583584.

Recent Trends in Computational Sciences – Gururaj, Pooja & Flammini (Eds)
© 2024 The Author(s), ISBN 978-1-032-42685-3

Novel technique for malicious node detection in wireless sensor networks

Inzimam Ul Hassan, Hilal Ahmad Shah, Abdul Hafiz & Inam Ul Haq
Chandigarh University, India
ORCID ID: 0000-0001-7475-7039, 0000-0001-7408-3409, 0000-0001-8868-2293, 0000-0001-8186-0632

ABSTRACT: Wireless sensor networks are the self-configuring varieties of networks whereby the sensor nodes are put in such a way that they can join or leave the network whenever they wish. Due to the decentralized nature of this type of network, it is possible for many malicious nodes to join the network. Attacks of both the active and passive variety may occur as a result of the presence of such malicious nodes. The term "DDoS form of attack" refers to an attack where a flood of raw packets is sent to the victim node. It's a much more active type of attack. The DDoS attack reduces the network's lifespan and increases its energy consumption when it occurs inside the network. Therefore, a novel technique is going to be suggested in this research effort to find the malicious nodes from the community that are responsible for the DDoS attack.

Keywords: WSN, Threshold, Delay, DoS, DDoS

1 INTRODUCTION

A wireless sensor network (WSN) has one base station and several sensor nodes placed inside it. The sensor nodes are small, low-cost, low-power devices with limited memory, processing power, and communication resources. The network contains a large number of geographically dispersed autonomous sensors that collect data from their backgrounds and transmit it to the base station. The nodes that are used in these networks gather data from their surroundings. The network's base station, which serves as a gateway between the sensor networks and the outside world, receives all the data obtained and transmits it to it. Base stations have enormous amounts of storage space, as well as multiple data processing tools, making them useful to the network [1]. Receiving the data that is sent by the sensor nodes is the base station's main responsibility. The end user can access this information and use it in accordance with their needs as shown in Figure 1.

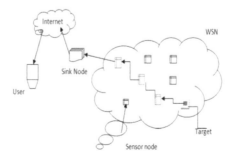

Figure 1.   Traditional wireless sensor network.

DOI: 10.1201/9781003363781-33

The sensor nodes are placed around the base station and can be organized into groups depending on the needs of the application. The sensor nodes are smaller in size thus results their battery size dimension. Due to their one-time deployment in very vast areas, the sensors batteries run out very quickly and are difficult to recharge. As a result, the network's lifetime shortens, which is a serious problem. A crucial requirement for wireless sensor networks is that security be complete. These constraints not only ensure the security of sensitive data but also enable each sensor node to have restricted resources, keeping the sensor network operational. Power is a basic requirement for carrying out a variety of tasks in wireless sensor networks [5]. Data collection, processing, and communication are all tasks that demand energy. The components of nodes require a tremendous quantity of energy even when they are not engaged in any tasks. These constraints not only ensure the security of sensitive data but also enable each sensor node to have restricted resources, keeping the sensor network operational. Power is a basic requirement for carrying out a variety of tasks in wireless sensor networks [5]. Data collection, processing, and communication are all tasks that demand energy. When certain specialized processes need to be carried out, even when the nodes are not in use, a significant quantity of energy is required in the components of the nodes. The nodes' batteries need to be replaced or recharged once all their energy has been used up. The replacement of these batteries is essentially difficult because the nodes are placed in vast locations where humans are unable to travel. This issue is therefore particularly important for the design and development of wireless sensor networks. Proposed hardware and software protocols are used in these networks because of their energy efficiency. The upkeep of these networks' security poses another difficult problem in terms of WSN architecture. The WSNs are used not just in military applications but also in large buildings to provide alarms and monitor the environment. Two variables create the potential for an attacker to have an impact on wireless sensor networks: attacker motivation and vulnerabilities, as well as opportunity [2].

2 LITERATURE REVIEW

In the most recent networking scenario based on cloud computing, SDN, and NFV settings, Aljuhani *et al.* [18] (2021) the author discussed DDoS mitigation options based on ML/DL techniques. Additionally, it offers ML defenses against DDoS assaults in IoT contexts. The author of Nassif *et al.* [19] (2021) conducts a systematic study of machine learning methods for cloud security. In this study, KDD, and KDD CUP'99 are mostly used. In this study, eleven cloud security domains are identified. SVM employs 30 ML techniques in standalone and hybrid models. The application of deep learning methods to cloud security has received little attention. Recent datasets for intrusion detection, such as CICIDS2017, CSE-CIC-IDS2018, and Kyoto 2006+, are absent from the paper.

In the context of edge computing, Liu *et al.* [10] Cyber-Physical-Social-System's CPSS LR DDoS Scenario was proposed in 2020. The suggested algorithm can identify unusual traffic in networking devices. The principles of surface learning neural networks, K-means, and support vector machines are employed in this paper.

MTD-based defense strategy was proposed by Debroy *et al.* [16] (2020) in which GENI Cloud testbed for news and video feeds were considered. In this the 40% reduction of DDoS attack was observed and 30% resource utilization was increased

Low-level DDoS attacks and existing defense strategies are categorized by Zhijun *et al.* [17] (2020) based on the time and frequency domains from which detection and defense are carried out. There was discussion on the inventiveness and aggression of L-DDoS attacks.

Shivam Dhuria & Monika Sachdeva 2018 [6]. In this study, two solutions are suggested, and one of them uses a lightweight two-way authentication method to block most WSN attacks. Another method, traffic analysis, which is based on data filtering method, is used to detect and block DDoS attacks from WSN. The Network Simulator 2 (NS2) is used to verify several characteristics, including throughput, latency, packet loss, energy usage, and PDR. On the basis of authentication and data filtering techniques, most DDoS attempts are recognized and avoided. This evaluation demonstrates how easy-to-use the proposed technique is and how it has been implemented at every node. DDoS attacks completely drain the power source, which can be avoided by making tiny calculations to monitor the data rates from nearby nodes.

M. Hassan *et al.* 2020, [2], suggested that apply sources of pandemic modelling to WSN-based IoT networks. Researchers create a suggested framework to identify unusual defense activity. IoT-specific characteristics, such as inadequate processing capacity, power restrictions, and node density, have a substantial impact on the development of a botnet. They employ common datasets for two active well-known assaults, like Mirai. To identify unusual activity like DDOS features, we also used a variety of machine learning and data mining methods, including LSVM, Neural Network, and Decision Tree. According to the experimental findings, the merging of decision trees with random forests produced a high level of accuracy in identifying attacks.

Jiangtao Pei *et al.* 2019, [7] used a DDoS attack tool to carry out local attacks. The packet capture tool analyses the capture attack in comparison to regular packets, finds the data attack laws, and then transforms it into data attack traits. To identify the DDoS attack, machine learning used the random forest approach. To conduct the characteristics on a large scale, it first extracts the feature and converts the format. The random forest algorithm, which recognizes the DDoS attack, receives these collected features.

Chunnu Lal's presented that the denial-of-service attack is regarded as the biggest attack among the many threats that affect the functionality of wireless sensor networks [8]. The development of efficient and lightweight security mechanisms that reduce and avoid various attacks on WSN, such as Denial-of-Service (DoS) attacks, becomes a significant issue for many academics as a result. To identify the presence of a DoS attack and lower the power consumption in the wireless sensor network, the author in this paper only considers efficient detection algorithms. There are numerous detection mechanisms in the network, however sensor nodes have limited power and processing power, making it important to create an energy-preserving DoS detection mechanism in WSNs to reduce the power consumption.

3 RESEARCH METHODOLOGY

Sensor nodes can join or leave the wireless sensor network at any time because it is a decentralized sort of network. Because of the network's dynamic nature, hostile nodes can infiltrate the system and start a variety of active and passive attacks. Active attacks are ones that have an impact on specific network performance metrics. The active form of attack known as a denial of service involves a malicious node bombarding legitimate nodes with erroneous packets to slow down the network. The most sophisticated type of DOS attack is known as distributed denial of service (DDoS), in which a malicious node picks a slave, and the slaves overwhelm the legal node with erroneous packets, lowering network performance. The key servers are created in the network according to the suggested technique, and each network node registers with the key server node with their data rate and bandwidth consumption. Malicious node identification procedure begins when all nodes in the network begin transmitting data, when a DDOS attack is launched against the network, and when network throughput reaches a threshold value. In the process of detecting malicious nodes, nodes that transfer data above the threshold value are regarded as malicious nodes, and the watch dog technique is used to determine if these nodes are sending control packets or data packets. The nodes are regarded as slave nodes when they are transmitting data packets. The slave nodes that can analyse network traffic are given the monitor mode approach to use. The node that sent the control packets is identified as a malicious node in the network when the slave nodes receive them from the other node. The suggested method is used in a simulated environment to identify malicious network nodes that are the source of DDoS attacks on the network. The proposed technique is applied under the simulated environment so that presence of malicious nodes can be determined easily which is responsible of causing DDOS attack in the network. The threshold is depicted using the formula shown below:

set timer = rt(msg*data) node "" - (tt(msg*data) node ""+ size of(msg*data nsnode "")
check node_()*address_alloc[];
select timer;

Where, data = represents data rate that how much data is transmitted in given amount of time.

size = represents total byte of data

node_()*address_alloc[] = represents node number and their IP address which can prevent from the spoofing.

timer = represents number of data transmitted by particular node in the given amount of time.

3.1 *Formula explanation*

The timer is set which can check the data rate which is define with the variable name "data" that how much data is transmitted in given amount of time. The total byte of data is defined by the function called size. The node number and their IP address are defined which we prevent from the spoofing. The timer is returned which define that this number of data is transmitted by particular node in the given amount of time.

3.2 *Proposed algorithm*

Input: Sensor nodes
Output: Detected malicious nodes
1.   Deploy wireless sensor network with finite number of sensor nodes
2.   Define key server in the network and each node register with the key server with their data rate
3.   If (Throughput reduces to threshold value) Check node which is sending data above threshold value
        If (Node send data with high data rate== true)
        Check traffic type of node
if (traffic == data traffic)
        Define node= slave node
        Apply monitor mode technique and check with node is sending control packets
        Node which is sending control packets to slave node is detected as malicious nodes
4.   Else
5.   Communication continue in the network

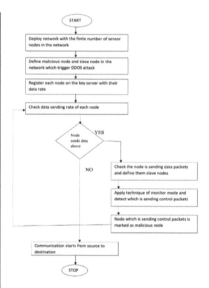

Figure 2.   Flowchart.

4   EXPERIMENTAL RESULTS

The planned approach is applied in NS2, and the results are evaluated by making comparisons amongst proposed and existing methods in terms of packet loss, throughput, and energy consumption in the networks.

234

Figure 3.   Network deoloyment of sensor nodes.

Figure 4.   Discovery of topology using flat grid.

Figure 5.   Checking Data rates of Sensor nodes.

Figure 6.   Applying Monitor Mode Technique.

Figure 7.   Attacker Nodes Identification.

Figure 8.   Communication starts in the network through maximum reliable nodes.

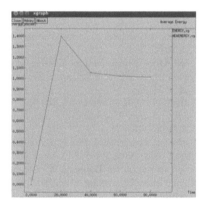

Figure 9.   Energy consumption.

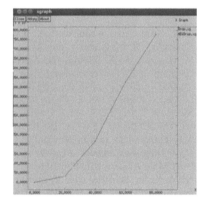

Figure 10.   Packet loss.

As shown in Figure 9, the sensor nodes are self-configuring and have a very compact size. Therefore, it is necessary to lower energy usage in order to lengthen network lifetime. This graph shows the graph of energy use.

As shown in Figure 10, the technique of neural networks is applied in this research work. The proposed improvement leads detection of malicious nodes due to which packet loss get reduced in the network.

As shown in Figure 11, the proposed scenario's throughput is contrasted with the current scenario. Throughput increases steadily when an attack is isolated from the network, according to analysis.

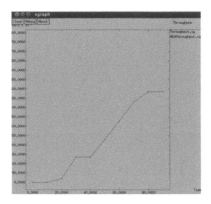

Figure 11.   Throughput comparison.

5   CONCLUSIONS

It has been determined via this research that wireless sensor networks are self-configuring networks because of which some malicious nodes enter the network and cause active and passive attacks to occur. The malicious nodes overwhelm the victim with raw packets during a distributed denial of service attack, or DDoS. The threshold technique, which isolates malicious nodes from the network and detects them, are suggested. The suggested technique enhancement results in an increase in throughput, network longevity, and network delay reduction.

REFERENCES

[1]   M.H. Anisi, A.H. Abdullah, S.A. Razak, "Energy-Efficient Data Collection in Wireless Sensor Networks", *Wireless Sensor Networks*, vol. 3, pp. 329–333, 2011.
[2]   M. H. Aysa, A. A. Ibrahim and A. H. Mohammed, "IoT Ddos Attack Detection Using Machine Learning," *2020 4th International Symposium on Multidisciplinary Studies and Innovative Technologies (ISMSIT)*, 2020, pp. 1–7, doi: 10.1109/ISMSIT50672.2020.9254703.
[3]   L. Xie, X. Xiao, Y. Shi, C. Zhang and J. Jiang, "An Activatable DDoS Defense for Wireless Sensor Networks," *2021 IEEE 9th International Conference on Information, Communication and Networks (ICICN)*, 2021, pp. 154–158, doi: 10.1109/ICICN52636.2021.9673926.
[4]   A.K. Pathan, "Security in Wireless Sensor Networks: Issues and Challenges", *Proc. 8th International Conf. Advanced Communication Technology*, vol. 2, pp. 1043–1048, 2006.
[5]   Patel MM, Aggarwal A, "Two Phase Wormhole Detection Approach for Dynamic Wireless Sensor Networks in Wireless Communications Signal Processing and Networking (WiSPNET)", *2016 International Conference on IEEE*, vol. 5, pp. 2109–2112, 2016.
[6]   Shivam Dhuria and Monika Sachdeva, *"Detectionand Prevention of DDoS Attacks in Wireless Sensor Networks"* Springer Nature Singapore Pte Ltd. 2018
[7]   Jiangtao Pei, Yunli Chen, Wei Ji. *"A DDoS Attack Detection Method Based on Machine Learning"* published in ICSP 2019.
[8]   Chunnu Lal, "A Survey on Denial-of-Service Attacks Detection and Prevention Mechanisms in Wireless Sensor Networks", 2017, *International Journal of Current Engineering and Scientific Research (IJCESR)*, volume-4, Issue-10
[9]   Inzimam Ul Hassan and A. Kaur, *"Prevention and Detection of DDoS Attack on WSN,"* pp. 245–249, 2018.
[10]   Liu, Z., Yin, X. and Hu, Y., 2020. CPSS LR-DDoS Detection and Defense in Edge Computing Utilizing DCNN Q-learning. *IEEE Access*, 8, pp.42120–42130
[11]   Inzimam Ul Hassan and A. Kaur, "Literature Review of Prevention and Detection of DDoS Attack on WSN," Volume XII, Issue IV, April 18, pp. 260–266, 2018.

[12]   D. Kim and S. An PKC-Based DoS Attacks-Resistant Scheme in Wireless Sensor Networks, in *IEEE Sensors Journal*, 16(8) (2016) 2217–2218. doi: 10.1109/JSEN.2016.2519539

[13]   J. Wu, K. Ota, M. Dong, and C. Li, A Hierarchical Security Framework for Defending Against Sophisticated Attacks on Wireless Sensor Networks in Smart Cities, in *IEEE Access*, 4 (2016) 416–424. doi: 10.1109/ACCESS.2016.2517321.

[14]   L. Shi, Q. Liu, J. Shao, and Y. Cheng, Distributed Localization in Wireless Sensor Networks Under Denial-of-Service Attacks, in *IEEE Control Systems Letters*, 5(2) (2021) 493–498. doi: 10.1109/ LCSYS.2020.3003789.

[15]   Ghildiyal S, Mishra A K, Gupta A, *et al.* Analysis of Denial of Service (dos) Attacks in Wireless Sensor Networks[J]. *IJRET: International Journal of Research in Engineering and Technology*, 3 (2014) 2319–1163.

[16]   Debroy, S., Calyam, P., Nguyen, M., Neupane, R.L., Mukherjee, B., Eeralla, A.K. and Salah, K., 2020. Frequency-minimal Utilitymaximal Moving Target Defense Against DDoS in SDN-based Systems. *IEEE Transactions on Network and Service Management*, 17(2), pp.890–903

[17]   Zhijun, W., Wenjing, L., Liang, L. and Meng, Y., 2020. Low-rate DoS Attacks, Detection, Defense, and Challenges: A Survey. *IEEE Access*, 8, pp.43920–43943

[18]   Aljuhani, A., 2021. Machine Learning Approaches for Combating Distributed Denial of Service Attacks in Modern Networking Environments. *IEEE Access*, 9, pp.42236–42264.

[19]   Nassif, A.B., Talib, M.A., Nassir, Q., Albadani, H., and Albab, F.D., 2021. Machine Learning for Cloud Security: A Systematic Review. IEEE Access.

Recent Trends in Computational Sciences – Gururaj, Pooja & Flammini (Eds)

A novel method to improve the performance of mobile nodes using a proactive approach in military MANET's

Nayanashree

Department of Information and Communication Technology, Manipal Institute of Technology, Manipal Academy of Higher Education, Manipal, India

Padmanabhan Ramaswamy

EISHAA Communications System Private Limited

S. Raghavendra & Ramyashree

Department of Information and Communication Technology, Manipal Institute of Technology, Manipal Academy of Higher Education, Manipal, India
ORCID ID: 0000-0003-2733-3916, 0000-0002-0237-2444

ABSTRACT: A Mobile Ad hoc network (MANET's) is a collection of self-configuring, multi-hop wireless networks. These networks are self-forming and self-healing, which makes deployment easier and reduces the need for centralized networks. Network performance is one of the most important research issues that MANET implemented in the military domain encounters. Maintaining a tactical network's low power consumption, high throughput, minimal latency, little routing overhead, and higher packet delivery ratios is the key difficulty. When compared to the prior work, the suggested algorithm can improve some of the following factors: throughput, energy consumption, latency, packet delivery ratio, routing overhead, and packet loss. In this paper, we introduce a novel approach to improve the performance of mobile ad hoc networks in areas like military. We propose a new routing algorithm named Actor-critic reinforcement algorithm for this purpose. Simulation results (based on NS3 version 2.6) showed that our protocol is very efficient. The simulation results show that our proposed protocol works effectively to improve throughput, reduce bandwidth, save energy as well as improve packet delivery ratio at major nodes.

Keywords: MANET's, Actor-critic reinforcement algorithm, mobile ad hoc, major nodes and minor nodes

1   INTRODUCTION

Over the past few decades, there has been a progressive shift from wired networks to wireless networks. Among the wireless networks currently in use, MANET is the most distinctive and significant application. The limited transmission range of MANET nodes makes it impossible for source and destination to directly communicate when they are outside of their transmission zones.

A mobile ad hoc network (MANET's) is an Internet made up of mobile nodes that is not supported by any infrastructure or base stations. routing policy is another name for a routing protocol. There are three different categories for routing protocols. There are several different proactive, Reactive and hybrid routing methods that can be used to create a safe and effective path from source to destination. In proactive routing systems, the source node establishes a path to the destination using predetermined network information. Routing protocols that are proactive that are those based on table-driven routing. The sensor nodes update the routing path on a periodic basis, while routing protocols that are reactive are those whose routing paths are kept on demand by the network. As

DOI: 10.1201/9781003363781-34

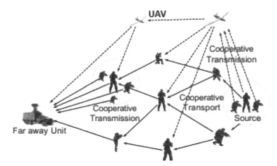

Figure 1.   Battlefield communication scenarios using tactical MANET'S.

opposed to proactive and reactive routing protocols, hybrid routing protocols combine their features. In a hybrid routing protocol, the source collects network data while simultaneously using preset data to construct a path to the destination. Routing, security, and quality of service (QoS) are the three main challenges in the MANET's that can be rectified in the future to increase communication's effectiveness and dependability. Zones are defined in hybrid routing protocols. A table-driven routing path is followed within each zone, and a demand-driven routing path is followed outside the zone.

There are numerous protocols developed to enhance MANET's network performance, however the current protocol still needs to be improved. To enhance the performance at military nodes, Special Routing Protocol for Military MANET's was developed. It was able to demonstrate improvement in significant nodes and is based on the reactive protocol. The actor-critic algorithm, which is based on Deep Reinforcement Learning, is used to pick the cluster heads. Path traversal is done using the Special Routing Protocol for Military MANET method. Better bandwidth usage between the nodes is the major goal.

Figure 1: Individual or company units must communicate with one another on missions and battlefield status because they are dispersed over the battlefield. The main strategy and concept are what we want to concentrate on describing. We only distinguish between two categories of network nodes: big nodes, such as commanders or operational equipment with a crucial function, are referred to as Super-Peers (SP) or major nodes, and normal nodes, such as soldiers or military equipment, are referred to as Peers (P) or minor nodes. High throughput, interference avoidance, high packet delivery ratios, and node scalability are the major goals. Reliability is maintained using proactive routing technology.

The primary goals are to prevent flooding, which causes data loss, to offer high throughput, scalability, which allows any number of nodes to be added or deleted, and to assure reliability. There is a possibility of lowering energy consumption and reducing delay between nodes.

2   LITERATURE SURVEY

Abida Sharif *et al.* [1] has demonstrated the viability of applying DRL approaches for cluster and data management via edge computing in IoV networks. The key goal is to choose the CH that can best allocate resources inside a network to meet user service and SLA standards while considering the environment's dynamic nature. Additionally, the suggested technique requires less convergence time to train the network to behave optimally. In the IoV network's dynamic and loud environment.

Shruthi *et al.* [2], investigated various proactive routing protocols to enhance MANET's power management and effectiveness. These protocols include source tree adaptive routing, global state routing, destination sequenced distance vector routing, wireless routing, etc. Each of these procedures' key traits and operational specifics are explored. The outcomes of several protocols are contrasted and compiled.

Quy Vu Khanh *et al.* [3], Proposed a revolutionary technique to enhance mobile adhoc network performance. The author distinguishes between major nodes and regular nodes.

Major nodes are those nodes that are given a higher priority than other nodes, such as commander nodes. The suggested protocol works well to increase throughput, decrease latency, conserve energy, and increase the packet delivery ratio at significant nodes, according to simulation findings based on NS2 ver3.26

Lee *et al.* [4] A brand-new technique was put forward. To improve tactical MANET's unmanned vehicle system dependability and survivability. To support QoS for the tactical MANET's and maximize energy efficiency, the authors used a central TDMA slot and four power scheduling algorithms. The findings indicated that, when compared to existing protocols for tactical MANET's scenarios, the suggested protocol had enhanced aspects in QoS and energy efficiency.

For two important routing protocols, such as the proactive and reactive routing protocols, Yuxia Bai *et al.* [5]. Provided a performance evaluation and analyses. This study evaluates the throughput, end-to-end delay MANET's to present the simulation results. Finally, the outcomes of the performance justification of stimulation are displayed.

To meet the QOS demands by decreasing end-to-end delay, Gwangjin Wi *et al.* [6]. Presented a centralized TDMA slot scheduling based on deep reinforcement learning (DRL). The key obstacle is timely traffic delivery while meeting QOS standards for mission success and survivability. The results of the simulation showed that the suggested technique ensures the tactical systems' need for QOS.

In order to compare the Packet Delivery Ratio (PDR) and Average End-To-End Delay of four alternative routing protocols, Russell Skaggs-Schellenberg *et al.* [7]. Presented a simulation in which the protocols were stimulated in NS-3 at various movement speeds and area sizes. Simulations were performed with the following node speeds: 0, 5, 10, 15, 20, 25, and 30 using protocols such AODV, DSDV, DSR, and OLSR. For each set of simulations, 500 m2, 750 m2, and 1000 m2 of space were used.

The study by Ying Song *et al.* [8], explains the fundamentals of the graph kernel and the theory behind optimizing single- and multi-graph kernels. In this research, we provide a clustering algorithm for MANET's based on graph kernels (GKCA). The GKCA algorithm introduces the fundamental idea of a graph kernel, covers the idea of optimizing a graph kernel and a multi-graph kernel, and suggests the fundamental idea based on a d-hop graph kernel. GKCA algorithm connects nodes using the shortest path (SP).

For packet transfer, various cluster head nodes. Using network simulation (NS2) software, the performance of the GKCA algorithm is empirically assessed in terms of the control packets ratio, packets loss ratio, and average end-to-end delay. Interoperability for wireless ad hoc routing technologies was the main emphasis of Marco Antonio To *et al.* [9]. Interoperability is the ability of two or more networks to exchange data even though they are using different protocols. The proactive strategy for strip interoperability used in this paper has been shown to successfully connect diverse networks only using layer 3 protocols.

A novel approach to route discovery in MANET's was put out by Vignesh *et al.* [10]. In this research, two routing methods for mobile ad hoc networks are taken into consideration. They are: Ad hoc On-Demand Distance Vector Routing and Destination Sequenced Distance Vector (DSDV) (AODV). The suggested strategy is a cutting-edge technique built on an ant colony optimization and a multi-agent system. In order to maximize node connectivity, reduce route discovery latency, and end-to-end delay, the suggested algorithm has a route-finding scheme.

3 PROPOSED SYSTEM

To The homogeneous network assumption is the foundation of general-purpose ad-hoc network technologies. All nodes are considered under this assumption as having equivalent transmission, computing, and storage capacities.

The intricacy of a vulnerable environment brought on by unforeseen physical and cyber-attacks from the enemy would have a significant negative impact on the effectiveness and viability of these network routing methods in realistically severe conditions, such as a battlefield. Over time, industry and academia have become interested in the military MANET's communication field to find a solution to this issue. In order to overcome the difficulties of choosing the shortest path in MANET's networks, network clustering algorithms have been proposed. It has been suggested to use traditional optimization techniques to effectively manage the network's resources. With reinforcement learning, an agent learns from its surroundings to determine the best policies. This method is model-free. We suggest the Actor-Critic Reinforcement algorithm, a strategy based on experience, for effectively choosing a cluster head (CH) to manage the network's resources.

The suggested approach evaluates nodes' structural data to determine how similar they are, then uses these similarities to forecast linkages and cluster-heads.The efficient route is selected using SRPMM (Special Routing Protocol for Military MANET's) protocol and finally according to the Packets send/receive the major nodes will follow the shortest route and minor nodes will be restricted to travel through the major nodes unless it is the only path. Here to evaluate the results simulation tool NS3ver 2.6 is used, Source: [https://www.nsnam.org/releases/ns-3-26/] Utilizing the Spectrum module is a new class called Spectrum WIFI Phy. Because it reuses the same Interference Helper and Error Model classes for this version, its functionality and API are currently very identical to those of the YansWifiPhy model.

The FQ-CoDel, PIE, and Byte Queue Limits models have been added to the traffic control module. Additionally, modifications were performed to integrate the traffic control sublayer with the Wi-Fi module.

Better IEEE 802.11e features, such as TXOP limitations, are supported by the Wi-Fi module, and the API for defining the Wi-Fi channel number, center frequency, and standard has been improved.

3.1 Selection of routing protocol

Most mobile ad hoc network (MANET's) routing protocols fall into one of two categories: proactive or reactive. When there is an immediate need for routing information, such as when one of the nodes needs to send a packet, reactive or on-demand routing protocols update it (and there is no working route to the destination). i.e., a change in the network topology prevents the packet from being sent any longer. Therefore, as a proactive measure, we are employing cluster gateway switch routing protocol to avoid this. Better bandwidth usage between the nodes is the major goal.

3.2 Cluster gateway switch routing protocol

The cluster-based hierarchical routing method known as CGSR is common. The entire network is divided into clusters using the reliable clustering algorithm Least Cluster head Change (LCC), and a cluster head is chosen in each cluster. A mobile node that is a part of two clusters or more acts as a gateway between the clusters.

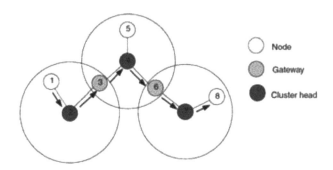

Figure 2. Cluster gateway switch routing protocol.

3.3 *Cluster head selection using game theory*

MANET's working in the military will have sensor nodes with low battery life. Working with sensor nodes in an energy-efficient manner is challenging since they are dispersed throughout the world and are challenging to replace when their batteries run out. In this study, we suggest a Game Theory Based Energy Efficient Cluster-Head Selection Approach to address this difficulty.

3.4 *Actor-critic algorithm for energy calculation*

The actor-critic approach is a final subset of RL algorithms that results from the fusion of policy-based and value-based approaches. Two distinct learning agents are defined in an actor-critic algorithm, as seen in Figure 3:

- The actor chooses which actions to perform.
- The critic comments on the actor's performance and suggests improvements.

The actor critic model combines methodologies that are value- and policy-based. Both of these techniques are used to choose the cluster head and to reconstruct the cluster head.

The critic generates a signal that modifies the preferences for action selection, favoring actions associated with higher-value functions more frequently. Due to the use of value function information to direct the evolution of the policy, this type of algorithm provides the gradient estimation with the least amount of distance, and the highest amount of bandwidth and energy. Additionally, choosing an action only requires a minimal amount of processing because there is no need to compare various state-action values and an explicitly stochastic policy may be learned.

3.5 *Path selection*

The quantity of links and the dependability of each link that makes up a path determine its availability. Numerous routing metrics have been developed that take into account the quantity of links, including the Special Routing Protocol (SRPMM) routing protocol. To forward the data to the destination node, SRPMM chooses the path with the lowest cost. The choice of the SRPMM routing protocol is based on the direct flow of traffic from source to destination, maximizing network performance and lowering costs. Shortest path routing can improve network performance, but it also depends on the routing protocol's functioning and the settings chosen for shortest path routing. It will be prohibited for packets to send and receive from minor nodes unless there is no other option.

4 RESULTS AND DISCUSSION

In today's environment, when the majority of gadgets, including laptops and mobile phones, are portable, MANETs are absolutely necessary. The numerous features of a MANET, includingScalability, flexibility, and self-healing have all shown to be beneficial in the current situation. The metrics used to measure the performance of a MANET include the packet delivery ratio, throughput, average end-to-end delay, path optimality, and normalised routing overhead.

This section uses a graph to compare the proposed routing protocol to earlier work. Gnu plot is a tool used to create a graph from the NS3 data. ACRLSRP(Actor critic Reinforcement Learning Special Routing Protocol), the proposed routing method, is contrasted with AODV and MANET's Graph kernel-based clustering approach (GKCA).

Figure 4 Shows the proposed algorithm works over the number of nodes in x-axis, which shows delay which is in meter per second is calculated by varying number of nodes. Proposed

242

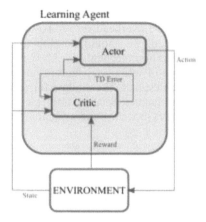

Figure 3.   The structure of an actor-critic algorithm.

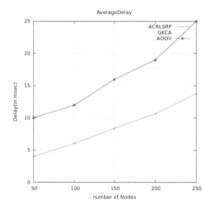

Figure 4.   Average delay is calculated for proposed algorithm.

Figure 5.   Average delay is calculated for node speed.

algorithm shows 43% delay on average but less than other routing protocols. Figure 5 shows average delay which is in meter per second is calculated by varying the speed of nodes. Proposed algorithm shows 11.5% delay on average but 24.8% less than GKCA and 33.9% AODV.

5   CONCLUSION AND FUTURE WORK

Network experts face a few problems when setting up and maintaining WSNs across a range of applications, including dealing with battery limitations and node lifespans. Wireless sensor networks are one of the examples is where energy awareness can greatly increase network life. An algorithm based on game theories is presented in this research for routing and cluster selection. In this routing and CH selection protocol, information is sent according to data priorities, while the CH selection is based on game theories. Neither authors nor groups have analyzed all parameters involved in a particular solution.

This work involved the detection of energy, bandwidth, and distance on CHs in shortest paths, which can be extended to multi-class detection and identification of specific types of random node failure in the future and the future work can also be detected using network encoding and decoding to reduce the overhead delay. To improve sniffing approaches by reducing time and packet overhead, we should develop a better way to calculate the threshold and present effective detection mechanisms. The performance of a routing protocol is affected by a great number of additional factors, including power control, traffic control, and security. To accurately estimate the total performance of each routing protocol, further research is needed on these factors. Security implementations in these protocols are extremely difficult because the nodes move around and interact wirelessly.

REFERENCES

[1]   Abida Sharif, Jian Ping Li, Muhammad Asim Saleem, Gunasekaran Manogran, Seifedine Kadry, Abdul Basit, Muhammad Attique Khan, "A Dynamic Clustering Technique Based on Deep Reinforcement Learning for Internet of Vehicles", *Journal of Intelligent Manufacturing (2021)* 32:757–768, January 2021

[2]   Shruthi S. "Proactive Routing Protocols For a MANET'S- A Review", *International Conference on I-SMAC (IoT in Social, Mobile, Analytics and Cloud) (I-SMAC 2017)*, IEEE

[3]   Quy Vu Khanh, Han Nguyen Dinh, Linh Manh Dao, "A Novel Method to Improve Performance of Major Nodes in Military MANET'S", *August 2021 IAENG International Journal of Computer Science* 48(3):776–781

[4]   Jae Seang Lee, Yoon-Sik Yoo, Taejoon Kim "Energy-Efficient TDMA Scheduling for UVS Tactical MANET'S", *August 2019IEEE Communications Letters* PP(99):1-1, researchgate, DOI:10.1109/LCOMM.2019.2936472

[5]   Yuxia Bai, Yefa Mai, Nan Wang, "Performance Comparison and Evaluation of the Proactive and Reactive Routing Protocols for MANET'Ss", April 2017, DOI:10.1109/WTS.2017.7943538, *Conference: 2017 Wireless Telecommunications Symposium (WTS)*

[6]   Gwangjin Wi, Sunghwa Son, Kyung-Joon Park, "Delay-aware TDMA Scheduling with Deep Reinforcement Learning in Tactical MANET'S", Department of Information and Communication Engeenering, *2020 IEEE*

[7]   Russell Skaggs-Schellenberg, Dr. Nan Wang, Daniel Wright, "Performance Evaluation and Analysis of MANET'S Protocols at Varied Speeds", *IEEE Conference*, Las Vegas, NV, USA

[8]   yingsong, hongwei luo, shangchao Pi, Chao Gui, and Baolin Sun, "Graph Kernel Based Clustering Algorithm in MANET'Ss", 2020, *IEEE Transactions*, Volume 8

[9]   Floriano De Rango, Marco Fotino, Salvatore Marano, *"EE-OLSR: Energy Efficient OLSR Routing Protocol for Mobile Ad-hoc Networks"*, DOI:10.1109/MILCOM.2008.4753611, IEEE Xplore

[10] Vignesh Ramamoorthy, "A New Proposal for Route Finding in Mobile AdHoc Networks", *June 2013 International Journal of Computer Network and Information Security* 5(7), DOI:10.5815/ijc-nis.2013.07.01,researchgate

Recent Trends in Computational Sciences – Gururaj, Pooja & Flammini (Eds)
© 2024 The Author(s), ISBN 978-1-032-42685-3

Implementation of reverse geocoding using Python and C# tools

R. Jyothi
Global Academy of Technology

K. Janani
Koneru Lakshmaiah Educational Foundation

R. Varshinee
Global Academy of Technology
ORCID ID: 0000-0002-4050-8322

ABSTRACT: Reverse geocoding is a process in which a location is accessed using latitude and longitude coordinates. Reverse Geocoding helps lot of people in locating addresses. It helps industries, postal department, transport and communication. Traffic police personnel often have great difficulties in locating places where traffic rules are violated. It takes time for them to arrive at the place of violation. Traffic rules violators use this time lapse to escape from the scene of crime. Reverse Geocoding plays a paramount role in locating the place quickly so that the traffic police can catch the culprits without any hassles or delay. We propose to develop an application wherein by providing the latitude and longitude coordinates corresponding to a place, a map consisting of the location of the place is displayed using python modules. We implemented Reverse Geocoding using Python and C# tools. Implementation using C# tools, places can be detected even without internet connection. This notably lessens connectivity issues during the usage of the system in remote areas.

Keywords: Basemap, Reverse Geocoding, matplotlib, geopy, Coordinates based service

1   INTRODUCTION

The world has become a global Village due to the digital revolution. The internet provides great service in connecting people all over the world. People all over the world speak different languages. The rapid technological development in the field of digital communication led to connection of people all over the world. It also paved way for industrial development. It plays a vital role in e-commerce. It has become a paramount power in economic growth of both individual and countries.

Reverse Geocoding acts as a very useful tool in detection of places just by providing coordinates of the place. It is a unifying mechanism that helps people of different places speaking different languages to locate places easily.

Reverse Geocoding can be implemented using various tools for effortless detection of places using coordinates. Certain tools may rely on internet infrastructure and some may not. The use of tools may depend upon varieties of applications. The advantage of such a system that works both online and offline is that using the application does not require any prerequisites such as knowledge about software. Another benefit of such a system is rapid yielding of results. A drawback of the application obtained by using Python tools could be internet. This means that the application requires Internet connection for its successful implementation. Without internet it does not yield the output. This may be a disadvantage for places lacking internet connection. To overcome that one can use the application obtained by using C# tools. It mainly uses cache memory for feeding the map data. The map data becomes the database. For sharing the cache with multiple users secondary cache database with SQL server can be used. The GPS coordinates can be converted into human readable Street addresses easily and swiftly. It also helps people to find the nearby location of their places.

DOI: 10.1201/9781003363781-35

Reverse geocoding is highly useful in remote place for detection of districts and other areas. [3] It is also used in governmental bodies for security reasons.

Reverse Geocoding is of immense help to army, navy, police department. It paved way for more exploration and research on unknown places. It acts a bridge between human knowledge and landscape of the earth. [2] There is a possibility of unravelling many mysteries to reduce crime rate.The role played by reverse Geocoding is laudable in the times of natural disasters viz floods, earthquakes, cyclone, etc. Reverse Geocoding tools help in rescuing stranded people swiftly.

Python tools used:

1. The Python module geopy: This tool helps to locate latitude and longitudes of addresses, cities, countries across the globe. It uses geocoders and other data sources. It is also used to calculate the distance between different places.
2. Nominatim: This tool helps to search open street map data. The process involves usage of address or location. Nominatim is included in the geopy Python library.mpl toolkits is a python package for plotting the multi vectors of geometric algebra using matplotlib.
3. Matplotlib is a library in Python. [8] It is numerical-mathematical extension for NumPy library which is used for working with arrays, domains of linear algebra and matrices. [1] Pyplot is a state-based interface to a matplotlib module. It can be used in line plot, histogram, 3-D plot, etc.
4. Basemap is an amazing tool for creating maps using Python without any hassles. It is a matplotlib extension.It has got all features to create data visualizations used to plot coastlines,countries, etc.

C# tools used:

We used GMap.NET Windows Forms & Presentation which is a free and powerful open source cross-platform.NET control. We downloaded the package via nuget as we used nuget as package manager. GMap.Net can be used to place a map control on a form and it can also be used to initialize the map to show the desired coordinates. It also offers the functionality of adding markers and polygons to the map. We created a new C# Windows Forms project in Visual Studio 2019. In GMap.NET download, DLLs named GMap.NET.Core.dll and GMap.NET.WindowsForms.dll can be found. They have to be placed in a subfolder of the project, and a reference must be added to both. It allows maps from Google, Yahoo!, Bing, OpenStreetMap, ArcGIS, Pergo, SigPac, Yendux, Mapy.cz, Maps.lt, iKarte.lv, NearMap, HereMap, CloudMade, etc. It is also used for routing, geocoding,etc.

2 WORKING METHODOLOGY

With Python implementation,using Basemap,different maps can be used for the display of location. We used a Miller cylindrical projection map is used. The size and resolution of the map can be altered. Basemap includes coastline dataset as well as datasets for rivers, state and country boundaries from GMT. These datasets can be used to draw coastlines, rivers and political boundaries on maps at several different resolutions.

The address on the map is printed using annotate() method in which the address obtained on the screen, latitude and longitude coordinates and the font size specifications are passed as arguments.

To display the address on the screen, an object of the class Nominatim is created which when used along with the reverse() function obtains the address.The reverse() function requires the latitude and longitude coordinates to be passed as string with a comma separated between them. The user is prompted whether he/she wants to use the application again. If yes,the user is prompted to press 1 else press 0.

With C# implementation,using GMap.Net different map data providers can be selected as per choice of the programmer. Some of the data map providers include BingMapProvider, GoogleMapProvider, YahooMapProvider. [9] We used GoogleMap-Provider for the display of location of the specified coordinates.

Markers and polygons can also be shown on the map using GMap.Net. Placing the DLL files in a subfolder of the project and adding a reference to both of them allows the written code to access

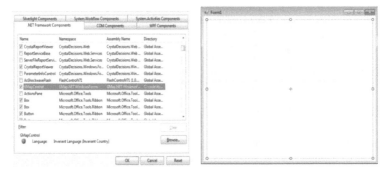

Figure 1.   Choosing toolbox items and adding a GMap to a windows form.

classes of GMap.Net.GMap.Net can be added to the toolbox since GMap.Net is a user control. Control can be added to the Toolbox by simply right-clicking the Toolbox and selecting'Choose Items'. The newly created C# Windows Application consists of a new form and for the control to display as an empty rectangle that can be resized,GMapControl has to be dragged to it.

The offline working of the application requires cache memory. Initially the Access-Mode should be ServerAndCache mode. The application has to be first run in the presence of internet to store the location of entered coordinates. It also stores the location of nearby places that show on the map tiles. For offline usage of the application the desired places need to be first run in the presence of internet whose map location is stored in the memory and it can be accessed offline. For offline usage, the application works when the AccessMode is either ServerAndCache mode or CacheOnly mode. It must be noted that during the initial run, the AccessMode must be ServerAndCache mode only. For large requirements of map data, larger cache memory may be required. Larger memory efficient systems with higher capacity Central Processing Units could be used.

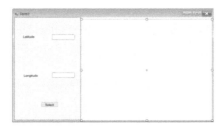

Figure 2.   Adding input taking options to the windows form.

3   IMPLEMENTATION

We implemented and tested both these applications as part of our internship programme in a private organization located in Bangalore, Karnataka, India which supplies digital tools to help traffic police of Karnataka.

This helped the police personnel to track the culprits who violated traffic rules and help the public. Our application benefited the department immensely to get a comprehensive coverage of address point data with the help of coordinates. This allows users to take advantage of a number of location based benefits.

Any organization which controls the fleet of vehicles needs a system that can keep track of where each unit is at any given time. The geographic coordinates convert data into a human readable address on the map on the exact location. This is the greatest advantage of Reverse Geocoding.

Table 1. Table structure of reverse geocoding.

Sl.No	Latitude	Longitude	Address
1	12.9507	77.5848	Lalbagh Botanical Gardens, South Zone, Bangalore, Karnataka, India
2	43.092461	−79.047150	542, Portage Road, Niagara Falls, New York, United States

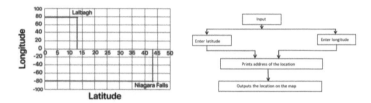

Figure 3. Graphical representation of reverse geocoding and the flow of working methodology.

4 RESULT

The proposed Python application uses Basemap to display the address and the location on the map. Based on the latitude and longitude coordinates given as input by the user, the corresponding address is displayed on the screen.

Locate & Get the address on the map with Latitude & Longitude!!

Please enter Latitude 12.9507
Please enter Longitude 77.5848

Address:
Lalbagh Botanical Gardens, 2 Cross Road, Basavanagudi, Jayanagar Ward, South Zone, Bengaluru
, Bangalore North, Bangalore Urban, Karnataka, 560004, India

Figure 4. Coordinates and address display.

The user can view a world map which displays the address along with the location on the map. Then the user is prompted whether he/she wants to continue using the application.

Do you want to use the application again?(Enter 1 for Yes and Enter 0 for No)1

Please enter Latitude 43.092461
Please enter Longitude -79.047150

Figure 5. User prompt for choosing option and Reuse of application after pressing 1.

This gives an opportunity to the user to explore more places on the globe. This option is useful because the user can the application multiple number of times without exiting from the program which saves ample of time. The users can keep on using the application until they wish to exit from the application.

When the user wants to continue the application he/she enters another set of coordinates. After entering the coordinates, the address of the corresponding coordinates are displayed on the screen.

If the user wants to continue then it is required to press 1 or else press 0 to exit. The proposed C# application uses GMap.Net and Cache memory for the display of map data offline. After the user enters the input in ServerAndCache mode in the presence of internet, the corresponding map consisting of the desired map data is displayed. It gets stored in the cache memory which can be used offline. The user can use the cache memory as database for offline use of the application.

Address:
542, Portage Road, Niagara Falls, Niagara County, New York, 14301, United States

Figure 6.   Display of address for the entered coordinates and Display of address on map for the specified coordinates.

Figure 7.   Input coordinates and map display.

5   CONCLUSION

Reverse Geocoding is a fantastic tool to locate address with the help of coordinates. In this paper we proposed an efficient Reverse Geocoding method using Python modules and C# tools. Experiments show that it is highly useful in all fields of data collection based on coordinates. However, internet connection is required for the successful implementation of the proposed Python application. Further research and analysis can be done to implement the proposed Python application offline which serves as a vital tool in detection of locations in remote areas such as forest areas, hilly terrains, villages and other areas devoid of internet. Improvements can be done in increasing the size of the cache memory during the usage of the proposed C# application. Further enhancement can be done to directly download the map data and use it for Reverse Geocoding.

REFERENCES

[1]   *Mastering Matplotlib* by Duncan M. McGreggor
[2]   *Reverse Geocoding in Python [Online]* Available: https://towardsdatascience.com/reverse-geocoding-in-python-a915acf29eb6
[3]   *Python Geospatial Development* by Erik Westra
[4]   *Matplotlib:Visualization in Python [Online]* Available: https://matplotlib.org/
[5]   *NumPy [Online]* Available: https://numpy.org/
[6]   *What is Numpy in Python – Everything You Need to Know About [Online]* Available: https://www.mygreatlearning.com/blog/python-numpy-tutorial/
[7]   *Python and Matplotlib Essentials for Scientists and Engineers* by M A Wood
[8]   *Customize your Maps in Python using Matplotlib [Online]* Available: https://www.earthdatascience.org/courses/intro-to-earth-analytics/spatial-data-vector-shapefiles/python-customize-map-legends-geopandas/
[9]   GMap.NET – *Maps For Windows – GitHub [Omline]* Available: https://github.com/judero01col/GMap.NET
[10]   *Python for Everybody* by Charles Severance.

Recent Trends in Computational Sciences – Gururaj, Pooja & Flammini (Eds)
© 2024 The Author(s), ISBN 978-1-032-42685-3

Role of social engineering in cyber security

V. Ambika, M. Spoorthi & A.D. Radhika
Vidyavardhaka College of Engineering, Mysuru, India

ABSTRACT: Social engineering refers to wider ways of tactics, where an attacker sets up a social situation in which a potential victim is encouraged to let down his or her guard. Social engineering schemes usually play some kind of mind games with their target. Major social engineering techniques used by an attacker are "instant messages", "fake antivirus", "emails", "phone calls" and many more. Actions by victims includes revealing details, downloading and installing malware, engaging in an illegal act in transaction. Social Engineering is most frequently used as a part of an information security attack and easier to pretend an employee into performing an act than it is to circumvent and control such SE Attacks. This paper highlights the social engineering tactics used by an attackers, classifications of SE attacks, recent social engineering attacks took place globally and different approaches to detect SE attacks, mitigations and preventions to avoid social engineering tactics.

Keywords: social engineering attacks, cyber-crimes, psychology, security

1   INTRODUCTION

Social engineering entails the use of fraud and deception to gain access to private or sensitive information. It is one of the current greatest dangers to information security. The subject is complex, and it is becoming more popular among individuals and businesses. The social engineer is concerned with the human aspect of organisations. In general, cyber-attacks are increasing at an alarming rate, with one occurring every 39 seconds. [1]. Current COVID-19 pandemic conditions have moved a wide range of everyday activities onto internet. This has led to an increase in the number of users on these networks as well as a boost the amount of time the user already engage in online activities. The rise in web usage has almost never been accompanied by education about Infosec and the different hazards that a typical Web user can encounter. According to the recent report, around 98% of cyber-attacks worldwide are somehow in the form of social engineering attacks. Every year, 700 plus social engineering threats are presented to the typical organization [2,3]. Cyber criminals persuade their victims to violate security protocols, exposing confidential information that could be used in a more targeted attack. Many distinguishing characteristics make SE a popular attack in the tech community and a serious, universal, and persistent threat to cyber-security. There are numerous of attacks because, phishing attacks have been conducted to take advantage of not only technical but also social vulnerabilities. Facebook now (Meta), Instagram, Twitter etc., have billions of users. A

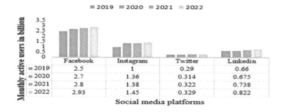

Figure 1.   Social media active users in a month.

Table 1.   Average active annual users.

SL NO	Phishing Platforms	In Billions
	(AAAU)	
1	Facebook	32.79
2	Twitter	15.57
3	Instagram	8.52
4	LinkedIn	3.76

DOI: 10.1201/9781003363781-36

large portion of the world's population uses one or more social media platforms, and as digital platforms gain popularity, they become a more attractive target for cybercrime [1].

$$AAAU = \frac{\sum Total\ number\ of\ users\ for\ past\ four\ years}{number\ of\ years}$$

Depending on the privacy settings, an attacker might be able to access people's contact, location, and areas of interest. Attackers can use the same channel to gather data, customise a strategy, and launch it. As there is a huge growth in social media users, phishers are the most fan followers for these platforms [4]. Due to the enormous possibility this creates for spear phishing cyberattacks specific to the victim, the number of attacks is rising. (Table 1) By analysis it is the Facebook, which takes top place in number of active users and more number of cyber-attacks occurs in this platform when compare to rest of the SMP.

This study gives an outline to social engineering attacks, as well as how user responses to social engineering attacks. These attacks are increasingly becoming a vector for phishing, malware, cyber stalking, email jacking, web jacking etc., and are wreaking havoc all over the world. As a result, it is critical to take all necessary precautions to keep your social media accounts, websites, email accounts, phone numbers secure. Social engineering attacks can be reduced through education and training, and following proper mitigation steps also this paper highlights challenges facing from social engineering tactics. In section (II) gives the review about the literature. In section (III) phases or life cycle of social engineering attacks. In section (IV) classification of SE attacks and its types. In section (V) describes the attack framework (VI) current challenges and also different approaches for detection of SE attacks.

2 LITERATURE REVIEW

Review of literature is what this portion is made up of. In this literature, according to study, the best defence against cyber security issues involving social engineering attempts is a knowledgeable computer user. The attacker is especially looking for personally identifiable information (PII), making new employees within a business the most vulnerable group to assault. The psychological factors that affect user vulnerability are also confirmed by this research. This study comes to the conclusion that education can still have a positive influence on people's behaviour, illogical tendencies, and psychological behavioural patterns even while technology can help decrease the consequences of social engineering attacks. Investment in organisational education programmes gives hope that social engineering attacks can be curbed, although no definitive answer to these cyber security issues has yet been proposed [4].

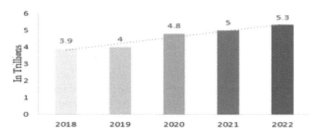

Figure 2. Intrusion attempts (2018–2022).

The research focuses on user experiments, structures, assessment, ideas, architectures, modelling techniques, and techniques for avoiding social engineering. Prevention interventions, human as safety sensor frameworks, user-centric guidelines, and user vulnerability models are examples of existing social attack preventative measures, models, and frameworks. The framework for human as a security sensor requires direction that will start investigating cyber security as extremely, potentially policing a secure system. The purpose of this paper is to conduct a critical and rigorous review of previous research on SE technology attack preventative measures, models, and frameworks [1]. This review describes existing social manipulation models and frameworks and proposes a new social engineering framework encompassing the session and dialogue concept. A social manipulation session is a full-fledged social engineering assault (SES) [12]. (Figure 3) According to the Sonic-Wall report, intrusion attempts will increase linearly between 2018 and 2022.

3   SOCIAL ENGINEERING ATTACK SCENARIO

Social engineering may entail researching the victim's interests, a routine, a related organisation such as a place of work, place of study, or even close family. After studying this relevant data, the attacker could use it to trick the target into sending them information such as bank account information and passwords in exchange for friendly requests and favours [5].

Perspectives of SE: A perpetrator who assumes they need the victim's personal data to finish an important task is generally the one who launches the con. By pretending to be a co-worker, police officer, bank or tax official, or any other person with the power to know anything, the attacker often starts by winning the victim's trust.

Attacker perspective: The term "attacker perspective" refers to "a worldview in which every piece of information can be used for personal gain or leveraged to achieve a desired goal." They can identify a target, set a goal, and devise a strategy to achieve it. They have the ability to look at a building, identify its flaws, and devise various ways to attack it. This same viewpoint is employed in social engineering attacks. The goal is to obtain information from the target while speaking with them on the phone or making direct eye contact. Sophisticated attackers can examine available information about you or your company and determine how to best use it against you and checks whether it is successful, understands the tactics and emotional behaviour of victims [15] example for phishing shown below:

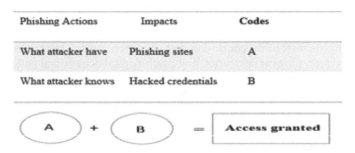

Phishing Actions	Impacts	Codes
What attacker have	Phishing sites	A
What attacker knows	Hacked credentials	B

A + B = Access granted

Figure 3.   Impact of phishing: an attacker perspective.

Organization perspective: Social engineering has a huge effect on business. Every year, thousands of businesses are victimised by some form of social engineering mechanism every year. As a result, more and more hackers are focusing on the human component of a company's security rather than the systems themselves. This creates a great impacts that leads to loss of money, credential loss, hits ones reputation without the knowledge of victim [6,15].

4   CLASSIFICATIONS OF SE ATTACKS

SE Attacks are classified into following types, based on activity performed by the attacker.

(A) *Human based attacks:* The hacker performs the threat in person by direct interaction with the target in order to gather the desired information in this type of attack. As a result, they can only have a limited impact on a restricted number of victims. It is the most common human-based social engineering technique in which the attacker poses as a legitimate or authorised person, may impersonate a legitimate or authorised person in person or through a communication medium such as phone, email. Impersonating a legitimate end user helps attackers trick a target into providing critical information. [7].

(B) *Computer based attacks:* To obtain information from targets, computer-based attacks use software tools. Hoax Letters: These are bogus emails that send out warnings about malware, viruses, and worms that can harm computers. Chain letters: requesting that people forward emails or messages in exchange for money. Spam Messages: These are unsolicited emails that attempt to gather information about users.

(C) *Mobile based attacks:* What exactly is mobile social engineering? Social engineers in the online and mobile generations try to trick unsuspecting users into clicking on malicious links and/or disclosing sensitive information by posing as an acquaintance, trusted authorities, or even a recognisable app. SMS-based communication: Sending a bogus SMS claiming that the user has won a bounty and urging him or her to register with sensitive information or attempt to collect other important details [21].

5 SE ATTACK FRAMEWORK

[26] A single point of contact between the attacker and the target in a larger SE attack. A comprehensive social engineering attack method is created by the sequential combining of many SE mechanism.

(A) *Attack Preparation:* In this SE framework Four types of stages are required in the preparation state that is P = {target, participants, physical objects (Tools and Techniques), attack vectors}. Based on a SE attack preparation step, the stage of the attack goal is added.

(B) *Attack Implementation:* A broad SE attack framework is a methodical approach to using trust and reaching phase objectives. For a SE to achieve its end goals, the sequencing of the SE is crucial. The attack goal will probably not be accomplished if the order is improperly set up. Initially an attack has to be started that includes medium, techniques, compliance principles, relationship exploitation.

(C) *Attack Goal:* This stage deals about the attack satisfaction and failure. The result of an attack depends upon the information collected, order of stage, physical objectives, and trust build for the next stage attack.

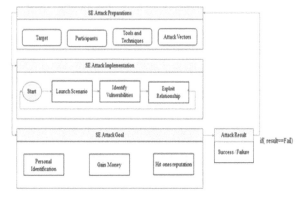

Figure 4.   SE attack framework.

Table 2. SE attack description.

SL NO		Description
1	Attack path	Attacker → Target 1, 2, … .,Target n → Attacker
2	Condition	Attacker{Target 1, … .,Target n} ∪ Information = {Username, Password and sensitive details}
3	Attack medium	Physical objects = {USB} ∪ Malicious Attachments{files, scripts} ∪ Trust{Target 1, … , Target n}
4	Attack goal	Goal{Personal Identification, Money, Hit ones reputation}
5	Attack Results	Success or Failure

The practice of "human hacking" through social engineering is used to commit fraud and identity theft. Hackers manipulate people's emotions subtly by invoking fear, excitement, or urgency. They'll use your heightened emotional state against you to skew your judgment once you're in it. One human error is all it takes for someone to fall victim to a SE. And because of this weakness, criminals are employing social engineering strategies more frequently [6].

6   LIMITATIONS AND CHALLENGES

Organizations are investing significant funds and resources in developing effective pro engineering strategies. Methods for detecting survivors have limitations and defensive measures are ineffectual when dealing with the increase in number of SE vectors. The arbitrary judgements of people limit human-based methods [14]. Technology faults can be exploited, computer-based and mobile techniques may also be constrained. These attacks evolve on a daily basis, with attackers becoming smarter and more powerful. As a result, more impactful diagnosis and actions to prevent techniques for detecting and reducing the consequences of these attack vectors are critical. Training students at a young age can help in to reduce the number of victims in the forthcoming. Countries must also invest in cyber defence education [15].

6.1   *Approaches in detection of SE attacks*

1. Natural language processing: Cyber attackers target the weakest link in a security system, and people are frequently more vulnerable than even the most secure computer system. If a user reveals a password or other critical information. Social engineering is a modern version of the confidence trick that con artists have always used. Phishing emails, which fraudulently request private information, are a common form of the attack, but social engineering takes many forms designed to exploit the target's psychological weaknesses [16], if a statement requests secure information or the performance of a secure operation, it is considered inappropriate [16].
2. Machine Learning: Machine learning algorithms which are commonly used in power grid functions for control and monitoring. From the standpoint of social engineers, ML models is used to train AI and its algorithms to specifically target types of files in order to zero in on their metadata. Attackers can use type-specific ML classifiers to train their models [17]. Machine Learning algorithms are able to train and detect cyber-attacks. As soon as the attack is detected, a confirmation email can be sent to cyber security professionals or users [18].
3. Internet of things: Attacks as well as anomalies in IoT infrastructure are becoming a growing concern inside the domain. As IoT infrastructure is used more often across all domains, there are more attacks and threats against it. Denial of service attacks, data type probing, malicious control, malicious operation, scanning, espionage, and wrong configuration are just a few of the attacks and anomalies that can make an IoT system malfunction [18].

4. Artificial Intelligence: Using sophisticated algorithms, AI systems are being trained to identify malware, run predictive modelling, and detect even the smallest behaviour patterns of malware as well as ransomware attacks before they enter the system. The system tries to imitate human intelligence. It holds tremendous promise in the security field. When used correctly, artificial intelligence (AI) systems can be given training to generate risk alerts, discover potential kinds of malware, and safeguard sensitive information for organisations [17].

6.2 *Mitigation techniques*

In earlier sections of this paper, we discussed various methods for detecting social engineering (SE) attacks, how tactics work by exploiting psychological threats or human characteristics, and the reasons why social engineering attacks are carried out, such as financial gain, politics, or personal interest and to hit one's reputation [25]. Below table shows the comparison of different mitigation techniques.

Table 3.   Mitigation techniques.

SL No.	Mitigation Techniques			
	Technique	Description	Advantages	Limitations
1	Education and training	Providng training to staffs and employees	Proper education required to avoid emotional blackmails and manipulation of one's mind set.	Trust factor, easily influenced
2	Software tools	Anti-virus, anti-phishing tools, anti-scams	Alerts users when there is an intrusion in the system	Expensive, ignorance towards alerts
3	Biometric	Verify users based on biological aspects	Identifies fake profiles,	Can imitate or mimic
4	Sensors	Senses unauthorized access	Identifies and alerts unauthorized access with alarms and capturing of thefts	Expensive
5	Artificial Intelligence and Machine learning based	Train models, learning based approach	Adaptive, high accuracy	Expensive and complex
6	IDS	Intrusion Detection System	Detects intrusion attempts in system	False alarm rates is high

7   CONCLUSION

This study is a survey on social engineering tactics, current detection methods, and its countermeasures. Social engineers exploits human behaviour and natural reactionary tendencies. SE employs human psychology to try and fool individuals into making security errors and disclosing sensitive information. Attackers do this by making proposals that are too good to be true. Since it is clear that social engineers can wrongly affect victims, one must be very vigilant when detecting fake offers. Unfortunately, these attacks are unstoppable by technical means alone, and Social engineering without cybersecurity knowledge can easily circumvent even the most robust security systems. SE strikes are becoming increasingly violent and common, causing financial and emotional harm to individuals and businesses. As a result, there is a high demand for new detection and mitigation methods and strategies, also the worker skills training. Countries also need to invest in cyber security learning to build a skilled and knowledgeable workforce.

REFERENCES

[1]   Syafitri W., Shukur Z., Mokhtar U. A., Sulaiman R. and Ibrahim M. A., "Social Engineering Attacks Prevention: A Systematic Literature Review," in *IEEE Access*, vol. 10, pp. 39325–39343, 2022, doi: 10.1109/ACCESS.2022.3162594.

[2]   Venkatesha, Sushruth, K. Rahul Reddy, and B. R. Chandavarkar. "Social Engineering Attacks During the COVID-19 Pandemic." *SN Computer Science* 2.2 (2021): 1–9.

[3]   Salahdine, Fatima, and Naima Kaabouch. "Social Engineering Attacks: A Survey." *Future Internet* 11.4 (2019): 89.

[4]   Koyun, Arif, and Ehssan Al Janabi. "Social Engineering Attacks." *Journal of Multidisciplinary Engineering Science and Technology (JMEST)* 4.6 (2017): 7533–7538.

[5]   Yasin, Affan, et al. "Understanding and Deciphering of Social Engineering Attack Scenarios." *Security and Privacy* 4.4 (2021): e161.

[6]   Frumento, Enrico, et al. *"The Role of Social Engineering in Evolution of Attacks."* (2016).

[7]   Dorr, Bonnie, et al. "Detecting Asks in Social Engineering Attacks: Impact of Linguistic and Structural Knowledge." *Proceedings of the AAAI Conference on Artificial Intelligence.* Vol. 34. No. 05. 2020.

[8]   Hijji, Mohammad, and Gulzar Alam. "A Multivocal Literature Review on Growing Social Engineering Based Cyber-attacks/threats During the COVID-19 Pandemic: Challenges and Prospective Solutions." *Ieee Access* 9 (2021): 7152–7169.

[9]   Arabia-Obedoza, Maha Rita, et al. "Social Engineering Attacks a Reconnaissance Synthesis Analysis." *2020 11th IEEE Annual Ubiquitous Computing, Electronics & Mobile Communication Conference (UEMCON).* IEEE, 2020.

[10] Sawa, Yuki, et al. "Detection of Social Engineering Attacks Through Natural Language Processing of Conversations." *2016 IEEE Tenth International Conference on Semantic Computing (ICSC).* IEEE, 2016.

[11] *COVID-19 Exploited by Malicious Cyber Actors* | CISA. Accessed: Oct. 17, 2020. [Online]. Available: https://us-cert.cisa.gov/ncas/alerts/aa20-099a

[12] Sayghe, Ali, et al. "Survey of Machine Learning Methods for Detecting False Data Injection Attacks in Power Systems." *IET Smart Grid* 3.5 (2020): 581–595.

[13] Haji, Saad Hikmat, and Siddeeq Y. Ameen. "Attack and Anomaly Detection in IOT Networks Using Machine Learning Techniques: A Review." *Asian Journal of Research in Computer Science* 9.2 (2021): 30–46.

[14] Salahdine, F.; Kaabouch, N. Social Engineering Attacks: A Survey. *Future Internet* 2019, 11, 89. https://doi.org/10.3390/fi11040089

[15] Zulkurnain, Ahmad Uways, et al. "Social Engineering Attack Mitigation." *International Journal of Mathematics and Computational Science* 1.4 (2015): 188–198.

[16] Khlobystova, Anastasiia, Maxim Abramov, and Alexander Tulupyev. "An Approach to Estimating of Criticality of Social Engineering Attacks Traces." *International Conference on Information Technologies.* Springer, Cham, 2019.

[17] Fan, Wenjun, Lwakatare Kevin, and Rong Rong. "Social Engineering: IE Based Model of Human Weakness for Attack and Defense Investigations." *IJ Computer Network and Information Security* 9.1 (2017): 1–11.

[18] AL-Otaibi, Abeer F., and Emad S. Alsuwat. "A Study on Social Engineering Attacks: Phishing Attack." *International Journal of Recent Advances in Multidisciplinary Research* 7.11 (2020): 6374–6380.

[19] Alzahrani, Ahmed. "Coronavirus Social Engineering Attacks: Issues and Recommendations." *International Journal of Advanced Computer Science and Applications* 11.5 (2020).

[20] Zheng, Kangfeng, et al. "A Session and Dialogue-based Social Engineering Framework." *IEEE Access* 7 (2019): 67781–67794.

[21] Beckers, Kristian, Leanid Krautsevich, and Artsiom Yautsiukhin. "Analysis of Social Engineering Threats with Attack Graphs." *Data Privacy Management, Autonomous Spontaneous Security, and Security Assurance.* Springer, Cham, 2014. 216–232.

[22] Wang, Zuoguang, et al. "Social Engineering in Cybersecurity: A Domain Ontology and Knowledge Graph Application Examples." *Cybersecurity* 4.1 (2021): 1–21.

Recent Trends in Computational Sciences – Gururaj, Pooja & Flammini (Eds)
© 2024 The Author(s), ISBN 978-1-032-42685-3

Leveraging channel capacity of wireless system with multibranch selection combiner impacted by κ − μ fading and co-channel interference for quantum machine learning QoS level prediction

Dragana Krstic
Faculty of Electronic Engineering, University of Nis, Nis, Serbia

Suad Suljovic
The Academy of Technical Professional Studies Belgrade, Belgrade, Serbia

Nenad Petrovic
Faculty of Electronic Engineering, University of Nis, Nis, Serbia

Dalibor Dobrilovic
Technical Faculty "Mihajlo Pupin", University of Novi Sad, Zrenjanin, Serbia

Devendra S. Gurjar
Department of ECE, National Institute of Technology Silchar, Assam, India

Suneel Yadav
Department of ECE, Indian Institute of Information Technology Allahabad Prayagraj, India

ABSTRACT: This paper considers the general κ − μ fading distribution, for which fading model this distribution is proposed. First, an expression for the channel capacity (CC) of multi-branch selection combiner influenced by κ − μ small-term fading and co-channel interference (CCI) with κ − μ distribution, is derived. κ − μ distribution is matched for line-of-sight (LOS) conditions in wireless systems. Then, a few figures are plotted to stand out the impact of small-term fading and CCI on the CC. Finally, the CC expression is leveraged for quantum machine learning-based Quality of Service (QoS) level and consumer number predictions inside network modeling, planning and simulation environment.

Keywords: Channel capacity, SC combiner, κ − μ Fading, κ − μ co-channel interference (CCI), machine learning

1 INTRODUCTION

The propagation of signals in wireless radio communication environment is distinguished by waves which interact with different non-ideal surfaces in processes of diffraction, reflection, absorption, and scattering. The interaction of waves with objects in the environments leads to varying both, amplitude and phase of those waves according the surfaces' characteristics. Because of that, the signal from transmitter arrives at receiver via multiple paths, and consequently the resulting combined signal fades rapidly causing the small term fading [1].

Different distributions are used for describing small term fading. Earlier, the Nakagami-m distribution received the greatest attention since has broad application and simple mathematical form for mathematical manipulation [2]. By further investigations, it was discovered that the ends of the Nakagami-m distribution did not fit the experimental data well. Thus, the κ − μ distribution was arrived at by measurements and calculations [3]. Due to the good agreement with the measured data from the environment, it is often used in an analysis of some wireless system. The κ − μ distribution is good suited for line-of-sight (LOS) scenarios

[4]. This is general fading distribution which contains the Rician, Rayleigh, One-sided Gaussian, and Nakagami-m distributions. The log-normal distribution, responsible for describing shadowing, may also be good approximated by the $\kappa - \mu$ distribution.

In order to diminish the impact of multipath fading on the system features, different diversity techniques for combining received signals were investigated. Between them are: switch and stay combining (SSC), selection combining (SC), equal gain combining (EGC), maximal ratio combining (MRC), etc [5]. While the MRC diversity technique gives the best results of fading mitigation, the SC diversity method is simple and less complex than others. This type of diversity receiver is of space diversity type, where there are multiple antennas at the receiver, but only the signal from one of branches (antennas) is led to the receiver. After receiving, the branch with highest signal or signal to interference ratio (SIR) is chosen [6]. That fact encouraged us to study SC diversity technique in our work.

The fading and CCI with $\kappa - \mu$ distribution were investigated in [7] and [8], whereby in [7] LCR of the SIR at the output of SC combiner with several branches was obtained. Afterwards, the GPU accelerated simulation and also linear optimization for the purpose of optimal network planning in a smart city was done. The average bit error probability (ABEP) of the SIR at the output of the multi-branch SC combiner under the of $\kappa - \mu$ fading and CCI was derived at the beginning of [8]. Later in [8], an approach to determining the level of QoS based on classification was suggested, whereby one of inputs was the previously calculated value of ABEP.

This paper is composed of four sections, with the first section being the introduction. The derivation and presentation of the channel capacity is made in the second section. The adoption of quantum machine learning for QoS prediction is presented in the third section. Fourth section is conclusion.

2 DERIVATION AND PRESENTATION OF THE CHANNEL CAPACITY

In this part of the paper, we analyze SC combiner with multiple branches in a wireless radio communication system disturbed by $\kappa - \mu$ fading and $\kappa - \mu$ CCI. The block-scheme of this type of combiner is presented in [8, Figure 1]. Its input signals are x_1, x_2, ..., x_L. The input CCI envelopes are $y_1, y_2, \ldots, y_L$. The output values are x and y. The SC receiver selects the antenna from the branch where the SIR value is the highest.

2.1 *Distribution of the SIR at the output of SC combiner*

The PDF of the envelope at the input branch has the κ-μ distribution [9]:

$$p_{x_i}(x_i) = \frac{2e^{-\frac{\mu(1+\kappa)}{\Omega_i}x_i^2}}{e^{\mu\kappa}}\sum_{i_1=0}^{+\infty}\frac{\mu^{2i_1+\mu}\kappa^{i_1}x_i^{2i_1+2\mu-1}}{\Gamma(i_1+\mu)i_1!}\left(\frac{1+\kappa}{\Omega_i}\right)^{i_1+\mu}, \tag{1}$$

where parameter κ is called Rician factor. It is actually the ratio of the dominant component and scattered components. The parameter μ denotes a number of clusters in the propagation area. $\Gamma(t)$ is Gamma function, and mean signals powers are marked with Ω_i, $i = 1, 2, \ldots, L$. The CCI is described by κ-μ distribution also:

$$p_{y_i}(y_i) = \frac{2e^{-\frac{\mu(1+\kappa)}{s_i}y_i^2}}{e^{\mu\kappa}}\sum_{i_2=0}^{+\infty}\frac{\mu^{2i_2+\mu}\kappa^{i_2}y_i^{2i_2+2\mu-1}}{\Gamma(i_2+\mu)i_2!}\left(\frac{1+\kappa}{s_i}\right)^{i_2+\mu}, \tag{2}$$

with positive y_i. The mean values of the CCI powers are s_i, $i = 1, 2, \ldots, L$.

The ratios of signals and CCIs at each SC receiver input are $z_i = x_i/y_i$, where their PDFs are [10]:

$$p_{z_i}(z_i) = \int_0^{+\infty} y_i p_{x_i}(y_i z_i)p_{y_i}(y_i)dy_i = \frac{2}{e^{2\mu\kappa}}\sum_{i_1=0}^{\infty}\sum_{i_2=0}^{\infty}\frac{\mu^{i_1+i_2}}{i_1! i_2!}\cdot$$
$$\cdot\frac{\kappa^{i_1+i_2}z_i^{2i_1+2\mu-1}s_i^{i_1+\mu}\Omega_i^{i_2+\mu}\Gamma(i_1+i_2+2\mu)}{(\Omega_i+s_iz_i^2)^{i_1+i_2+2\mu}\Gamma(i_1+\mu)\Gamma(i_2+\mu)}. \tag{3}$$

The CDF of z_i are [11]:

$$F_{z_i}(z_i) = \int_0^{z_i} p_{z_i}(t)dt = \frac{1}{e^{2\mu K}} \sum_{i_1=0}^{\infty} \sum_{i_2=0}^{\infty} \frac{(\mu K)^{i_1+i_2}}{i_1!i_2!} \cdot$$
$$\cdot \frac{\Gamma(i_1+i_2+2\mu)}{\Gamma(i_1+\mu)\Gamma(i_2+\mu)} B_{\frac{s_i z_i^2}{\Omega_i + s_i z_i^2}}(i_1+\mu, i_2+\mu). \tag{4}$$

If the incomplete Beta function $B_x(p,q)$, appearing in previous expression, is presented in the shape of sum from [8, eq. (6)], based on [12, eq.8.39], the CDF is obtained in the next form:

$$F_{z_i}(z_i) = \frac{1}{e^{2\mu\kappa}} \sum_{i_1=0}^{\infty}\sum_{i_2=0}^{\infty}\sum_{i_3=0}^{\infty} \frac{(i_1+\mu)_{i_3}(1-i_2-\mu)_{i_3}}{i_1!i_2!i_3!(i_1+\mu)} \cdot$$
$$\cdot \frac{(\mu\kappa)^{i_1+i_2}\Gamma(i_1+i_2+2\mu)}{(i_1+\mu+1)_{i_3}\Gamma(i_1+\mu)\Gamma(i_2+\mu)} \left(\frac{s_i z_i^2}{\Omega_i + s_i z_i^2}\right)^{i_1+i_3+\mu}. \tag{5}$$

The SC receiver output SIR z is maximal value of all SIRs z_i, $z = max\ (z_1, z_2, \dots, z_L)$. Accordingly, the output SIR z will have now the PDF [13]:

$$p_{z_i}(z) = L p_{z_i}(z_i)(F_{z_i}(z_i))^{L-1} =$$
$$= \frac{2L}{e^{2\mu\kappa}} \sum_{i_1=0}^{\infty}\sum_{i_2=0}^{\infty} \frac{(\mu\kappa)^{i_1+i_2}}{i_1!i_2!} \cdot \frac{\Omega_i^{i_2+\mu} s_i^{i_1+\mu} z_i^{2i_1+2\mu-1}\Gamma(i_1+i_2+2\mu)}{\Gamma(i_1+\mu)\Gamma(i_2+\mu)(\Omega_i + s_i z_i^2)^{i_1+i_2+2\mu}} \cdot$$
$$\cdot \left(\frac{1}{e^{2\mu\kappa}} \sum_{i_3=0}^{\infty}\sum_{i_4=0}^{\infty}\sum_{i_5=0}^{\infty} \frac{(i_3+\mu)_{i_5}(1-i_4-\mu)_{i_5}}{i_3!i_4!i_5!(i_3+\mu)(i_3+\mu+1)_{i_5}} \cdot \frac{(\mu\kappa)^{i_3+i_4}\Gamma(i_3+i_4+2\mu)}{\Gamma(i_3+\mu)\Gamma(i_4+\mu)}\right) \tag{6}$$
$$\cdot \left(\frac{s_i z_i^2}{\Omega_i + s_i z_i^2}\right)^{i_3+i_5+\mu}\Bigg)^{L-1}.$$

2.2   Derivation of the channel capacity

Channel capacity, as one of the most important performance measure of wireless system, is defined as maximal speed at which data can be transferred through the channel [14]. The normalized CC is:

$$\frac{CC}{B} = \frac{1}{\ln(2)} \int_0^{\infty} \ln(1+z)p_{z_i}(z)dr, \tag{7}$$

where CC is Shannon capacity expressed in bits/s, and by B is marked transmission bandwidth in Hz. When the logarithmic function introduced in the shape [15]:

$$\ln(1+x) = \sum_{i=0}^{\infty} (-1)^i \frac{x^{i+1}}{(i+1)!} \tag{8}$$

is used, normalized CC will be:

$$\frac{CC}{B} = \frac{1}{\ln(2)} \sum_{i_1=0}^{\infty} \frac{2L}{e^{2\mu\kappa}} \sum_{i_2=0}^{\infty}\sum_{i_3=0}^{\infty} \frac{(\mu\kappa)^{i_2+i_3}(-1)^{i_1}}{(i_1+1)\Gamma(i_2+\mu)} \cdot \frac{\Gamma(i_2+i_3+2\mu)}{i_2!i_3!\Gamma(i_3+\mu)} \cdot$$
$$\cdot \left(\frac{1}{e^{2\mu\kappa}} \sum_{i_4=0}^{\infty}\sum_{i_5=0}^{\infty}\sum_{i_6=0}^{\infty} \frac{(i_4+\mu)_{i_5}}{i_4!i_5!i_6!} \cdot \frac{(1-i_5-\mu)_{i_6}(\mu\kappa)^{i_4+i_5}\Gamma(i_4+i_5+2\mu)}{(i_4+\mu)(i_4+\mu+1)_{i_6}\Gamma(i_4+\mu)\Gamma(i_5+\mu)}\right)^{L-1} \cdot \tag{9}$$
$$\int_0^{\infty} \frac{z_i^{i_1+2i_2+2Li_4-2i_4+2Li_6-2i_6+2L\mu}(\Omega_i/s_i)^{i_2+Li_4-i_4+Li_6-i_6+L\mu}}{(1+(\Omega_i/s_i)z_i^2)^{i_2+i_3+Li_4-i_4+Li_6-i_6+L\mu+\mu}} dr.$$

If the formula with β function from [12; 3.251] is involved in (9), the normalized CC is:

$$\frac{CC}{B} = \frac{L}{\ln(2)e^{2\mu\kappa}} \sum_{i_1=0}^{\infty}\sum_{i_2=0}^{\infty}\sum_{i_3=0}^{\infty} \frac{(\mu\kappa)^{i_2+i_3}(-1)^{i_1}\Gamma(i_2+i_3+2\mu)}{i_2!i_3!(i_1+1)!\Gamma(i_2+\mu)\Gamma(i_3+\mu)} \cdot$$
$$\cdot \left(\frac{1}{e^{2\mu\kappa}} \sum_{i_4=0}^{\infty}\sum_{i_5=0}^{\infty}\sum_{i_6=0}^{\infty} \frac{(i_4+\mu)_{i_5}(1-i_5-\mu)_{i_6}(\mu\kappa)^{i_4+i_5}}{i_4!i_5!i_6!(i_4+\mu)(i_4+\mu+1)_{i_6}} \cdot \frac{\Gamma(i_4+i_5+2\mu)}{\Gamma(i_4+\mu)\Gamma(i_5+\mu)}\right)^{L-1} \tag{10}$$
$$\cdot \left(\frac{\Omega_i}{s_i}\right)^{\frac{i_1+4i_2+4(L-1)(i_4+i_6)+4L\mu+1}{2}} \cdot B\left(\frac{i_1+2i_2+2(L-1)(i_4+i_6)+2L\mu+1}{2}, \frac{2i_3-i_1+2\mu-1}{2}\right).$$

259

2.3 *Graphical presentation of the channel capacity*

Graphical representation of the normalized CC at output of SC receiver with multiple branches is in Figure 1 versus $w = \Omega_i/s_i$.

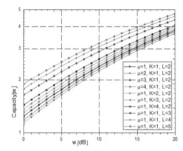

Figure 1.   The normalized capacity, CC/B, at the SC receiver output for variable parameters κ, μ, and L.

When the μ parameter increases, the CC increases and the system has better performances. When the κ parameter increases, the CC versus the SIR decreases, and the system performance get worse. The system has the best characteristics with the increase in the number of input branches L, as can be seen from the picture.

If we compare the obtained values, we can notice that the biggest improvement is achieved for enlarging the number of input branches from 2 to 3. Further increase makes the system more expensive without much economic justification in terms of improved performance.

3   ADOPTING QUANTUM MACHINE LEARNING FOR QOS PREDICTION

Predictions enabled thanks to machine learning models belong among key enablers for many innovative usage scenarios like proactive adaptation and infrastructure maintenance with aim to enhance Quality of Service (QoS) perceived by the end users, when it comes to latest generation wireless and mobile networks [16]. In our paper, a method aiming classification-based estimation of QoS is proposed, relying on the previously calculated value of channel capacity along with other factors as input. Furthermore, the emergence of so-called quantum machine learning has brought various improvements with respect to traditional techniques – including both the processing time reduction due to faster quantum-based linear algebra operations and also greater expressiveness of predictive models making use of higher-dimensional feature spaces that results with increase of prediction accuracy [17]. In this paper, it is leveraged for QoS level estimation making use of service consumer count prediction.

In Figure 2, the underlying workflow for wireless network planning and simulation adopting quantum machine learning is illustrated. First, users draw smart city mobile network diagram within graphical user interface run in web browser, while its structure is defined by metamodel from [16]. Additionally, value of CC is calculated relying on consumer-grade GPU hardware running NVIDIA CUDA kernels [18]. About 67 times speed-up is achieved this way compared against Mathematica CPU-only program thanks to general purpose GPU (GPGPU) programming.

After that, the values of CC values are used as one of independent variables for quantum machine learning predictive model. In the last step, base station modules based on software-defined radio (SDR) are either turned off (existence of anomalies) or on (increasing requests from consumers).

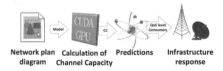

Figure 2.   Network planning and simulation workflow based on quantum machine learning predictions.

Table 1. Dataset headers.

Problem	Input 1	Input 2	Input 3	Input 4	Output
Consumer num QoS level	Location	Avg temp. Cons. num	Season Base station	Special occasion Channel capacity	Consumer num QoS level [0-3]

Table 2. Experimental results.

Problem	Training time [s]	Performance
Consumer number (QSVR regression)	94	MRE – 4.4 %
QoS determination (QSVC classification)	81	Accuracy – 94.8 %

The use of quantum predictable components adopts Qiskit [19] for Python, a publicly available open source quantum computing library provided by IBM Research Group. It offers a wide set of quantum kernels, while we make use of the following two: Quantum Support Vector Regressor (QSVR) – service consumer count prediction at given location within smart city, treated in form of regression-alike problem; Classifier (QSVC) – in case of QoS level determination, treated as classification.

In the regression problem, we have the following inputs: location identifier, average daily temperature, ordinal number of current year's season, special occasion flag (lockdown, major holiday). In case of the second one, the previously calculated value of CC at given locations where base stations are placed are leveraged in synergy with predicted number of service consumers.

There are four possible categorical outcomes: 0 – defect/anomaly presence; 1 – low QoS; 2 – medium QoS; 3- high QoS. In Table 1, the layout of dataset headers is presented. Table 2 summarizes experimental results in case of both quantum predictive models, considering two different aspects of evaluation: time necessary for training and prediction quality (for regression – mean relative error; for classification – accuracy). The free quantum computing simulation surrounding of IBM Quantum Lab was used to execute the experiment.

The datasets were constructed by exporting 50 000 records from the network simulation and planning framework (described in [16]) and further divided on two subsets: test and training (75%). It can be noticed that quantum-based machine learning enhances prediction performance, compared against traditional methods from our past works [16,20]. However, more time was spent for training, considering that we did not leverage real quantum computer in our experiment.

4 CONCLUSION

In our work, the multi-branch SC combiner under the presence of $\kappa - \mu$ fading and CCI was evaluated. We derived the formula for CC, presented it graphically and analyzed the impact of two fading parameters, κ and μ, and also the number of input branches in SC receiver. Then, the formula for CC is used for quantum machine learning-based Quality of Service (QoS) level and consumer number predictions inside network modeling, planning and simulation environment. Adoption of quantum machine learning leads to significant betterments against the known methods, but training took more time as real quantum computing hardware was not leveraged.

ACKNOWLEDGMENT

This paper is made under project of Serbian Ministry of Science, Technological Development and Innovation, and project between Republics of Serbia and India "Development of Secure and Spectral Efficient Simultaneous Wireless Information and Power Transfer Systems for Large-Scale Wireless Networks" (Science and Technological Cooperation Projects 2022–2024).

REFERENCES

[1] Simon M.K. and Alouini M.S, *Digital Communication Over Fading Channels*, John Wiley and Sons, 2005.

[2] Nakagami N., "The m-distribution – A General Formula of Intensity Distribution of Rapid Fading", *Statistical Methods in Radio Wave Propagation*, W. C. Hoffman, Ed. Elmsford, NY: Pergamon, pp. 3–36, 1960.

[3] Yacoub M.D., "The κ–μ Distribution: A General Fading Distribution", *IEEE Atlantic City Fall Vehicular Technology Conf.*, pp. 1427–1432, October 2001.

[4] Yacoub M.D., "The κ-μ Distribution and the η-μ Distribution", *IEEE Antennas and Propagation Magazine*, 49(1), pp. 68–81, 2007.

[5] Ko Y.C., Alouini M.S., Simon M.K., "Analysis and Optimization of Switched Diversity Systems", *IEEE Trans. Veh. Technol.*, vol. 49, no. 5, pp. 1813–1831, Sep. 2000.

[6] Ansari I.S., Al-Ahmadi S., Yilmaz F., Alouini M.S., Yanikomeroglu H., "A New Formula for the BER of Binary Modulations with Dual-branch Selection Over Generalized-K Composite Fading Channels", *IEEE Transactions on Communications*, 59(10), pp. 2654–2658, 2011.

[7] Krstic D., Suljovic S., Petrovic N., Popovic Z., Stefanovic M., "Level Crossing Rate of Next Generation Wireless Systems with Selection Combining in the Presence of k-μ Fading and Interference: Derivation and Simulation", *16th International Conference on Telecommunications – ConTEL 2021*, Zagreb, Croatia, pp. 4–9, June 30–July 2, 2021.

[8] Krstic D., Suljovic S., Petrovic N., Minic S., Popovic Z., "Determining the ABEP Under the Influence of K-μ Fading and CCI with SC Combining at L-branch Receiver Using Moment Generating Function", *International Conference on Software, Telecommunications and Computer Networks (SoftCOM)*, Split, Croatia, 22–24. September 2022.

[9] Suljović S. and Milić D., "Performance of Relay Signal Transmission by AF Technique Influenced by k-μ Fading", *27th Telecommunications forum TELFOR*, Belgrade, 26–27 November 2019.

[10] Milenkovic V, Sekulovic N., Stefanović M., Petrovic M., "Effect of Microdiversity and Macrodiversity on Average Bit Error Probability in Gamma Shadowed Rician Fading Channels", *ETRI Journal*, 32, no. 3, pp. 464–467, 2010.

[11] Krstić D., Suljović S., Milić D., Panić S., Stefanović M., "Outage Probability of Macro-diversity Reception in the Presence of Gamma Long-term Fading, Rayleigh Short-term Fading and Rician Co-channel Interference", *Annals of Telecommunications*, vol. 73, Issue 5–6, pp. 329–339, June 2018.

[12] Gradshteyn I.S. and Ryzhik I.M., *Tables of Integrals, Series and Products*, Academic Press, 2007.

[13] Savić M., Smilić M., Jakšić B., "*Analysis of Shannon Capacity for SC and MRC Diversity System in α−κ−μ Fading Channel*", Unuversity Thought, Publication in Natural Sciences, Vol.8, No.2, pp. 61–66, 2018.

[14] Alouini M.S. and Goldsmith A. J., "Capacity of Rayleigh Fading Channels Under Different Adaptive Transmission and Diversity-combining Techniques", *IEEE Trans. Veh. Technol.*, vol. 48, no. 4, pp. 1165–1181, 1999.

[15] Huang H. and Yuan C., "Ergodic Capacity of Composite Fading Channels in Cognitive Radios with Series Formula for Product of κ-μ and α-μ Fading Distributions", *IEICE Transactions on Communications*, Volume E103.B, Issue 4, Pages 458–466, 2020.

[16] Krstić D., Petrović N., Al-Azzoni I., "Model-driven Approach to Fading-aware Wireless Network Planning Leveraging Multiobjective Optimization and Deep Learning", *Mathematical Problems in Engineering*, vol. 2022, 4140522, pp. 1–23, 2022.

[17] Fung F., "Quantum Software: Quantum Machine Learning in Telecommunication", *Digitale Welt 6*, pp. 30–31, 2022.

[18] Petrović N., Vasić S., Milić D., Suljović S., Koničanin S., "GPU-supported Simulation for ABEP and QoS Analysis of a Combined Macro Diversity System in a Gamma-shadowed k-μ Fading Channel", *Facta Universitatis: Electronics and Energetics*, Vol. 34, No 1, pp. 89–104, March 2021.

[19] *Qiskit – Open-Source Quantum Development [online]*, available on: https://qiskit.org/, last accessed: 04/10/2022.

[20] Petrović N., Al-Azzoni I., Krstić D., Alqahtani A., "Base Station Anomaly Prediction Leveraging Model-driven Framework for Classification in Neo4j", *2022 International Conference on Broadband Communications for Next Generation Networks and Multimedia Applications (CoBCom)*, Graz, pp. 1–4, July 2022.

IoT technologies and applications

Recent Trends in Computational Sciences – Gururaj, Pooja & Flammini (Eds)
© 2024 The Author(s), ISBN 978-1-032-42685-3

Research insight of IoT (Internet of Things) fog optimal workload via fog offloading for delay minimisation using a fog collaboration model

Inzimam Ul Hassan & Inderdeep Kaur
Chandigarh University, India
ORCID ID: 0000-0001-7475-7039, 0009-0005-8345-0676

ABSTRACT: Due to the extensive expansion of IoT (Internet of Things) applications, the traditional centralised cloud computing paradigm these days encounter a variety of challenges, including latency, limited capacity, and network failure. To solve above mentioned problems, fog computing incorporate the cloud computing near to IoT devices. Fog analyses and store the data on local IoT devices rather than transmitting it to the cloud. The fog offers services that are more responsive and of better quality than the cloud. The greatest solution for allowing the IoT to provide efficient and safe services to a range of IoT consumers may thus be fog computing. In this research paper, we suggest an architecture with framework for fog computing to improve QoS through request offloading. The suggested solution uses a strategy of collaboration across FN (fog nodes) to enable data processing in a shared manner, thus satisfying QoS and serving the largest number of IoT requests. The suggested framework highlights important advantages of fog computing in the computing ecosystem and has the ability to achieve a sustainable network paradigm.

Keywords: IoT (Internet of Things), Offloading, Fog Computing, Cloud Computing, FaaS (Fog as a service)

1 INTRODUCTION

One of the developments receiving attention is the IoT (Internet of Things), It has the capacity to help our society in countless ways. Many of the things in our environment will soon be able to attach to the Internet and communicate with one another without the need for human involvement as the IoT (Internet of Things) advances [1]. The initial goal of the IoT (Internet of Things) was to eliminate the need for manual data entry, use a variety of sensors to collect data about the environment, and enable autonomous data processing and storage [3].

IoT allows for automatic data processing and storage while reducing the need for human data input efforts and using a variety of sensors to get environmental data [1]. There are many problems with IoT, including security, performance, reliability, and privacy. The Cloud of Things (CoT), an IoT integration solution that uses the cloud, offers a chance to solve these problems. CoT makes it simple for IoT data to flow while offering speedy integration and affordable installation for thorough data processing and deployment [2].

Numerous benefits for various IoT applications are brought about by the convergence of IoT and cloud computing. The design and creation of new IoT applications is a challenging undertaking, though, due to the increasing No. of IoT devices with diverse platforms [5]. Smart, internet-linked gadget that are connected to the internet enable connectivity to homes, workplaces, offices, and cars. However, as more people use and benefit from technology, these internet-connected devices are being overused in society [3]. As a result, uptime must be guaranteed and gateway consistency must grow to make the IoT a practical reality. IoT expands the use of the current internet and data centre infrastructure. The usage of systems, devices, and resources is optimised as a result of this combination of fog computing as well as cloud computing, which should be highlighted does not replace cloud computing. The development of fog computing resulted from efforts to deal with ongoing problems, incoming data action, and real-time processes [2]. Limitations on resources like processing power and bandwidth are also addressed. IoT apps use sensors and other IoT devices to generate enormous amounts of data. These huge data are then examined to come to judgments about different activities. All of this data demand an enormous amount of network bandwidth to send to the cloud [5]. Fog computing is used to address these problems.

The IoT (Internet of Things) uses fog computing to improve efficiency and performance while transferring less data in cloud for the purpose of analysis, process, and storage [8]. Internet of Things and fog computing work together to provide FaaS, in which a network provider deploys a number of fog nodes around its service area and then serves as an owner to several clients from various vertical industries. Every fog node offers nearby processing, networking, and storage resources. [9]. FaaS will enable new company models to provide their customers services. Contrary to clouds, which are typically run by large companies with the resources to afford the construction and operation of enormous data centres, Local and international enterprises will be able to setup and both manage public or private computing, storage, and control services at different scales to satisfy the needs of a range of clients thanks to FaaS. [10].

2   LITERATURE SURVEY

Agarwal *et al.* [17] proposed a three-layer design for clients, fog, and cloud and an algorithm to divide processing effort between fog and cloud layers. The algorithm determines if the fog node has enough capacity to complete tasks or send some to the cloud layer. It assumes each node in fog and cloud has a manager to oversee collaborations and performance, but doesn't fully show efficient performance of dispersed activities.

Aazam & Huh [27] proposed a single model for resource management in fog computing using resource prediction and allocation. It is a solid baseline for IoT and fog computing research and development and adaptable to cloud service providers' needs.

Skarlat *et al.* [26] proposed a framework for fog resource supply to reduce delays in delay sensitive computing by formalizing an enhancement solution. The framework was compared with previous models and found to be useful.

Bittencourt *et al.* [16] proposed a resource migration architecture for virtual machine (VM) migration across fog nodes (FN). The objective is to provide management tools for fog operation while maintaining QoS and accessibility of VMs during user movement. The VM includes user data and application components and migration is done carefully to avoid performance changes. However, the potential heavy load from multiple migrations to the FN is not addressed.

Abedin *et al.* [19] proposed an algorithm that promotes resource sharing among fogs. They define an efficiency measure for fog nodes to consider communication benefits when sharing resources and suggest a list of preferred fog node pairings based on this. A node sends a pairing request to its recommended partner and the decision to accept or deny is based on preferences and past approved requests. However, the algorithm has a limitation of only considering communication costs and not QoS like latency and bandwidth.

Gao *et al.* [20] proposed a hybrid data distribution paradigm using software-defined networks and DTN. The approach saves the FG approach by dividing into two planes, with fog servers as a data plane using DTN and cloud as a control plane for analysing content and managing data flows. The paradigm includes regular cloud and fog server data dissemination, as well as additional low-cost delay-tolerant data dissemination from cloud servers, mobile users, and fog servers.

Peter [28] discussed the real-time uses of fog computing, showing how it manages large amounts of IoT data, solves latency and congestion problems, and forms new business models and opportunities by handling the dispersed and dynamic nature of growing IoT infrastructures at the network edge.

Chiang and Zhang [29] reviewed the integration of fog and IoT, discussing challenges in IoT development and the need for new computing, storage and networking architecture. They examined the benefits and potential solutions offered by the fog architecture for IoT problems.

3 PROPOSED FOG SYSTEM

The (FN)fog nodes which are distributed at the network's edge are part of the fog computing infrastructure, which promises to give IoT applications a quick response time(RT) and low latency(LT). Generally, the distribution of fog nodes at the network's edge emphasises the need for new types of network management that regulate and improve fog performance, resulting in better resource consumption and management. We will describe a fog architecture and a methodology to control fog offloading for real-time IoT applications in this part.

3.1 *Optimal workload*

The proposed approach seeks to reduce latency through load distribution over the fog layer in order to enhance QoS (quality of service) for IoT applications. By unloading requests from a clogged fog to another fog or fogs that can handle the additional load, the request offloading methodology is the suggested method for balancing the load in the fog layer. The three main parts of the offloading strategy we developed are depicted in Figure 1.

First stage: The average waiting time for recently received requests is determined by regularly computing the queue size while accounting for the request type (heavy vs. light requests). When a request comes into fog, it often queues up till that fog is available to fulfil it. As a consequence, we can determine the overall delay by multiplying the time taken for processing the request by the time spent in the queue.

The second stage involves preserving the fog node's functionality. When the no. of requests in the queue exceeds the node's capacity, the fog node will signal. The processing power of each fog node is calculated in advance using the node capacity function. Offloading is necessary if the fog node is congested. Additionally, filter the size of queue to identify the load that has to be transferred from the overloaded node to a different node.

The decision to do unloading or not is taken at the third step. The basis for this is a time cost function, which is calculated as the interval of time between the process target and the network output. A list of the top prospective adjacent nodes that can manage the additional load and adhere to the IoT request's timeframe is the output of this step. The list is created using the location of the nodes (which is closer to the dense fog, the better for quicker offloading) and the current load on nearby nodes.

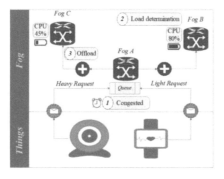

Figure 1.   Optimally distributed architecture.

267

Generally speaking, this proposed offloading approach offers solutions to the issues of what and where to offload. The additional IoT queries on a busy node. Offloading is also not necessary if all fog nodes have modest loads, and it is ineffective and if all fog nodes have significant loads when trying to reduce latency in an IoT network. When the load among the fog nodes varies significantly, then only it will be the effective offloading strategy.

3.2 *Simulation and evaluation*

This section describes the simulation of a proposed offloading approach for a healthcare monitoring system. The scenario involves a hospital using an IoT system to monitor patients with chronic diseases in real-time with low latency. The system includes a smart wearable, fog nodes distributed in hospital departments, and a cloud data center. The fog nodes collect realtime data from wearables and perform simple calculations, while the cloud handles data processing and storage. Multiple fog nodes are distributed for backup and load distribution. The suggested offloading strategy distributes the load across nodes to ensure accessibility and availability.

4 IMPLEMENTATION SETUP

A simulator was created using Python programming language to focus on delay minimization in fog collaboration model. The simulation was run on Google Collaboratory Notebook using Jupyter notebook environment. The offloading method, fog devices, and sensors were the fundamental parts of the framework. The fog device class described the hardware characteristics, connection to other fog devices and sensors, processing time, and queue offloading algorithm. The sensor class simulated IoT sensors with attributes such as data size, transmission rate, and start time. Different devices can be mimicked by adjusting these properties.

Table 1. Simulation parameters.

Parameters	Value
Sensor Nodes	1000
Fog Nodes	35
Simulation duration	45 mins
Counter step	100
Simulation Area	750*750(meter)
Architecture	Fog

5 RESULTS AND DISCUSSIONS

To assess the appropriateness of a fog architecture and the suggested offloading approach, the previously mentioned situation at a hospital was replicated. Considering how lightweight the application is (patient state assessment using both recently collected data and real-time data), According to Figure 2, every feasible combination between and (1 to 35) fog nodes and (1 to 1000) sensors was simulated. With each additional dimension you wish to compare (such as the quantity of sensor nodes and fog devices), the amount of information required grows rapidly. This is taken into consideration when programming the current system, which is very effective. The software only iterates through pertinent moments because a dictionary with all the pertinent timestamps has already been generated. Dynamically added information is added.

Figure 3's presentation of the findings demonstrates that queues are swiftly cleared when a message's processing time is low. When the transit time is longer than the process time, the offloading technique has no discernible effect and may even result in increased latency.

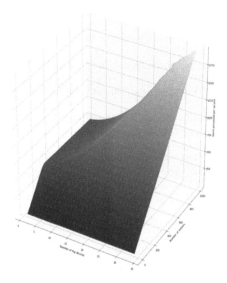 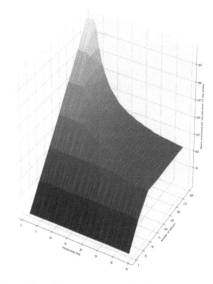

Figure 2.   Simulation result using fog and sensor.   Figure 3.   Request processing time.

These findings demonstrate that a fog node can only handle a finite amount of data, and that increasing the number of them results in a faster processing rate, which lowers latency and facilitates the creation of real-time applications. This argument is strengthened by the usage of an offloading algorithm because the load will be spread.

6   CONCLUSION AND FUTURE WORK

As an emerging network architecture in the IoT space, fog computing has a lot of promise to speed up time-sensitive IoT application processing. Fog also seeks to lessen the total stress placed on cloud computing. Fog, on the other hand, complements cloud rather than replacing it because it just extends the IoT network's edge computing and communication capabilities. Through request offloading, we address the problem of fog congestion in this paper. The outcome demonstrates that spreading the overload across numerous fog nodes considerably improves the performance of the fog layer. The suggested offloading technique emphasises the substantial advantages of fog in the computing ecosystem and has the ability to gain a sustainable network paradigm. In the future, we intend to optimise the offloading mechanism taking into account both the time cost and the power cost.

REFERENCES

[1]   Al-Khafajiy M., Baker T., Waraich A., Al-Jumeily D., and Hussain A., "Iot-fog Optimal Workload Via Fog Offloading," *Proc. – 11th IEEE/ACM Int. Conf. Util. Cloud Comput. Companion*, UCC Companion 2018, no. December, pp. 349–352, 2019, doi: 10.1109/UCC-Companion.2018.00081.

[2]   Mattern F. and Floerkemeier C., From the Internet of Computers to the Internet of Things, in *Lecture Notes in Computer Science (including subseries Lecture Notes in Artificial Intelligence and Lecture Notes in Bioinformatics)*, 2010, vol. 6462 LNCS, pp. 242259.

[3]   Vermesan O. and Peter Friess, *Internet of Things: Converging Technologies for Smart Environments and Integrated Ecosystems.* 2013.

[4]   Evans D., *The Internet of Things – How the Next Evolution of the Internet is Changing Everything*, CISCO white Pap., no. April, pp. 111, 2011.

[5]   Kai K., Cong W., and Tao L., Fog Computing for Vehicular Ad-hoc Networks: Paradigms, Scenarios, and Issues, *J. China Univ. Posts Telecommun.*, vol. 23, no. 2, 2016.

[6]   Bonomi F., Milito R., Zhu J., and Addepalli S., Fog Computing and Its Role in the Internet of Things, *Proc. first Ed. MCC Work. Mob. Cloud Comput.*, pp. 1316, 2012.

[7]   Brogi A. and Forti S., "QoS-Aware Deployment of IoT Applications Through the Fog," in *IEEE Internet of Things Journal*, vol. 4, no. 5, pp. 1185–1192, Oct. 2017.

[8]   Baker T., Mackay M., Shaheed A., Aldawsari B., "Security-oriented Cloud Platform for Soa-based Scada", *Cloud and Grid Computing (CCGrid), 2015 15th IEEE/ACM International Symposium on 2015 May 4* (pp. 961–970). IEEE.

[9]   Baker T., Mackay M., Randles M., Taleb-Bendiab A., "Intention-oriented Programming Support for Runtime Adaptive Autonomic Cloud-based Applications", *Computers & Electrical Engineering.* 2013.

[10]   Mukherjee M., Shu L. and Wang D., "Survey of Fog Computing: Fundamental, Network Applications, and Research Challenges," in *IEEE Communications Surveys & Tutorials.* doi: 10.1109/ COMST. 2018

[11]   Ottenwlder B., Koldehofe B., Rothermel K., and Ramachandran U., MigCEP: Operator Migration for Mobility Driven Distributed Complex Event Processing, in *Proceedings of the 7th ACM International Conference on Distributed Event-based Systems – DEBS* 13, 2013, p. 183.

[12]   Hong K. and Lillethun D., Mobile Fog: A Programming Model for Large-scale Applications on the Internet of Things, *Proc. Second ACM SIGCOMM Work. Mob. Cloud Comput.*, pp. 1520, 2013.

[13]   Bittencourt L. F., Lopes M. M., Petri I. and Rana O. F., "Towards Virtual Machine Migration in Fog Computing," *2015 10th International Conference on P2P, Parallel, Grid, Cloud and Internet Computing (3PGCIC)*, Krakow, pp. 1–8, 2015.

[14]   Swati Agarwal, Shashank Yadav, Arun Kumar Yadav, "An Efficient Architecture and Algorithm for Resource Provisioning in Fog Computing", International Journal of Information Engineering and Electronic Business(IJIEEB), Vol. 8, No. 1, pp. 48–61, 2016.

[15]   Kapsalis A., Kasnesis P., Venieris I. S., Kaklamani D. I., and Patrikakis C. Z., A Cooperative Fog Approach for Effective Workload Balancing, *IEEE Cloud Comput.*, vol. 4, no. 2, pp. 3645, Mar. 2017.

[16]   Abedin S. F., Alam M. G. R., Tran N. H., and Hong C. S., A Fog Based System Model for Cooperative IoT Node Pairing Using Matching Theory, in *Network Operations and Management Symposium (APNOMS), 2015 17th Asia-Pacific*, pp. 309314, 2015.

[17]   Gao L., Luan T. H., Yu S., Zhou W and Liu B., "FogRoute: DTN-Based Data Dissemination Model in Fog Computing," in *IEEE Internet of Things Journal*, vol. 4, no. 1, pp. 225–235, Feb. 2017. doi: 10.1109/JIOT, 2016.

[18]   Peter, N. FOG Computing and Its Real Time Applications. *Int. J. Emerg. Technol. Adv. Eng.* 2015, 5,266–269.

[19]   Chiang, M.; Zhang, T. Fog and IoT: An Overview of Research Opportunities. *IEEE Internet Things J.* 2016, 3, 854–864.

Organ procurement and transplantation network using IoT and blockchain

Benita Jose Chalissery
Department of MCA, St. Francis College
Department of MCA, New Horizon College of Engineering
ORCID ID: 0000-0002-3505-2068

V. Asha
Department of MCA, New Horizon College of Engineering
ORCID ID: 0000-0003-4803-099X

ABSTRACT: Human organ transplantation is a lifesaving miracle, but at the same time, costly in terms of money and risk involved. Healthier survival of the recipient, post-transplantation can be achieved by reducing the ischemia time and also by transplanting the deceased organ to a recipient who is at low risk of pathophysiological parameters at the time of allocation. These real-time parameters majorly influence the healthier survival span of the recipient. Also, the growing concern about corrupt practices involved in transplantation network, force the stakeholders to demand their right to know the reason for the change in allocation. Implementation of an organ procurement and transplantation network using IoT and Blockchain can be used to monitor the efficacy of the organ allocation system. This paper focuses on the solution to achieve real-time monitoring of the pathophysiological parameters and also to bring transparency in all the transactions which lead to an allocation which are not consistent with the organ allocation policy. Blockchain concepts are applied to bring more transparency to all the stakeholders during the various steps followed in the organ procurement and allocation process. The paper details the practical implementation of IoT and Blockchain technology in kidney procurement and transplantation.

Keywords: Human Organ Transplantation Network, Blockchain, Internet of Things, Survival Prognosis, Hyperledger Caliper

1   INTRODUCTION

Human Organ Transplantation is one of the most incredible achievements of medical science. The Organ Procurement and Transplantation System is a network of procurement and transplantation professionals including clinical transplant coordinators, transplant physicians, and surgeons along with the technological components which support organ procurement and transplantation. Organ shortage is a global issue and deceased organ donation is one of the major sustainable solutions [1]. The primary objective of the system is to make sure that available organs are not wasted and that the most appropriate recipient is transparently selected from the waiting list. The fairness of the organ transplantation and procurement network depends on meeting the urgency of the transplantation determined by the medical condition of the individual patient and also on the policy followed in the organ allocation process.

As seen from the studies [2], it is very clear that a matching system that takes into account all the below factors can select the most suitable recipient from the waiting list

- Age of the recipient
- Age of the donor

DOI: 10.1201/9781003363781-39

- Number of HLA mismatches
- BMI of the recipient at the time of organ transplantation
- Cold ischemia time for the kidney

New-age computer technology, like IoT and blockchain, based solutions has immense potential to streamline and automate organ procurement, accurate recipient identification, and also to reduce processing and transportation delays. Such a system is proposed in this research. This system takes into consideration the following work objectives

- To define an IoT and blockchain-based framework for Organ Procurement and Transplantation
- To achieve transparency for any transaction to all the participating entities.
- To decentralize organ matching algorithms using blockchain-based networks.
- To validate and authorize the transaction between any two entities of the network.
- To analyze the reason and track the changes to the priority list.
- Blockchain is suggested due to the following benefits/requirements as mentioned in this paper
- To achieve transparency for any transaction to all the participating entities.
- To decentralize organ matching algorithms (allocation) using blockchain-based networks.
- To validate and authorize the transaction between any two entities of the network.
- To analyze the reason and track the changes to the priority list

Blockchain meets these requirements as follows:

- Achieve transparency for any transaction to all the participating entities:
    Each participating entity in the blockchain-based system can trace the details of any transactions occurring in the system. Also, any alteration to the already executed transactions needs modification for the copies of blockchain blocks (ledger) owned by each entity of the system.
- Decentralize organ matching algorithms (allocation):
    Blockchain technology works on the decentralization principle. This means that no single party cannot influence or manipulate the network. Also, a common allocation policy can be followed for any organ across geographical boundaries.
- Validate and authorize the transaction between any two entities of the network:
    Both recipient and donor need to be registered in the system before any allocation/ transplantation of the organ. Distributed ledger copy concept followed in Blockchain will avoid organ trafficking and manipulation of the waiting list.
- Analyze the reason and track the changes to the priority list:
    Priority list change is considered a transaction in the proposed system. Each transaction needs verification through the mining concept of the blockchain. This ensures any illegal priority list change is detected and denied automatically by the system.

2 METHODS AND PROCEDURES

Blockchain enables the decentralized aggregation, ordering, timestamping, and archiving of all types of transactions involving information exchange. This ensures maximum transparency to all the stakeholders [3]. A system thus built with the Internet of Medical Things (IoMT) and Blockchain, can provide a transparent digital transplantation coordination system [4]. The blockchain-based system shall bring transparency by validating and logging every transaction happening in the system [5]. These transaction data (blocks in blockchain terminology) are interconnected with the previous transaction and provide a single point of truth.

Any change in the recipient waiting list requires consensus from the participants of the system such as registration centers, transplant hospitals, insurance agencies, regulatory bodies, etc. Similarly, the allocation of organs to a particular transplant center requires consensus from other entities. Consensus can be automated by acknowledgments from the participating entities, which get triggered when all the requirements are fulfilled for each entity. Organs once received by the transplant center need to be transplanted to the allocated recipient within the agreed timeline. Any deviation needs to be communicated by the transplant center to the blockchain system with proper justification. The lack of such a system may lead transplant centers to commit fraud by allocating organs to someone else who is not on top of the priority list.

All transactions which make a change in the priority and waiting list can be considered as blockchain transactions. Registration of a new recipient, the addition of a new Organ into the availability list, organ reallocation to a recipient who has a better survival prognosis, and successful organ transplantation are the 4 major transactions that will be broadcasted to all other nodes. The communication of the transaction data is by using the blocks in the Blockchain technology. As shown in Figure 1, Blockchain Blocks in the OPTB network, each block has a Block Header and Block Content. Block Header will have unique data of each block like Transaction or Block ID, Timestamp, Nonce, and Previous Block Hash. Block Content will be Transaction ID, Reason for the update (ENUM), Database identifier, Database Hash, Priority List status, and Hash of Priority list Database [5]. The database identified points to a database table that stores detailed information on each type of transaction.

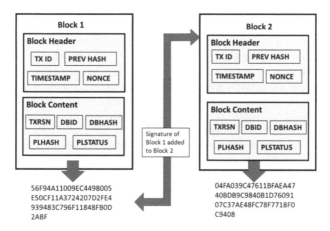

Figure 1.   Blockchain blocks in the OPTB network.

In this section, the logical view of the proposed Organ Procurement and Transplantation Blockchain (OPTB) framework is explained. Permissioned blockchain infrastructure known as Hyperledger Fabric is used for the proposed solution. Trusted Membership Service Providers are used to enroll stakeholders of the Blockchain network. The participants manage transactions and store data in different formats in the ledger (Blockchain block) in varying formats with the help of smart contracts. The transaction logic component maintains a log of all the transactions which has resulted in the current state of priority of each recipient in the waiting list and each stakeholder has a copy of the ledger of every network they belong to.

3   THE LOGICAL VIEW OF THE OPTB FRAMEWORK

The logical process flow of the system is depicted in Figure 2 Envisaged Process flow diagram. In a comparatively open network, privacy using channels is the key operational requirement. The logical entity channels ensure confidential transactions between a group of participants of a network.

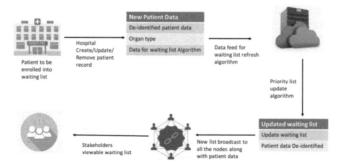

Figure 2.   Envisaged process flow diagram.

3.1   *Workflow of proposed system*

In the proposed blockchain-based OPTB network, the transaction happens in two phases-Registration phase and the Allocation phase. During the registration phase, patients who are suffering from end-stage disease and require an organ (kidney) will be registered using a system application, through any of the transplantation organizations. The applications of transplant organizations can access the ledger using smart contracts. The transactions generated will be endorsed by the agreed endorsement policies of the organizations, which are then written onto the ledger. The registration process automatically triggers the waiting list position determination algorithm. The algorithm runs in the central server OPTM. The OPTM returns the recipient_Id and the priority number of the new patient registered to the transplantation organization. The algorithm also updates the waiting list position of all the patients who are part of the waiting list on a need basis. The updated waiting list is shared across the transplantation organization along with the reason for the update. Ledger will be updated for each of these transactions. The priority number of the new registrant is determined by the priority factors taken into consideration and agreed upon by the organizations. The sequence diagram for a typical transaction of adding a patient to the waiting list of a transplantation center is depicted in Figure 3.

In the second phase, which is the allocation phase, when an organ is available from a brain-dead patient, the procurement organization registers the available organ details in the OPTB system, using a system application. The donor_Id is returned by the OPT manager.

The system will automatically call the allocation algorithm in the OPT manager. The process will return the recipient identified for organ allocation to all the transplantation organizations which have registered recipients. The transplantation centers then contact the identified recipient for the

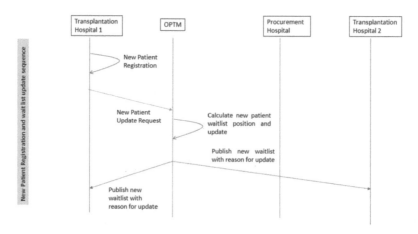

Figure 3.   Sequence diagram depicting patient registration.

possibility of immediate organ transplantation. In case the identified recipient is not available and not in a position for accepting the available organ, then the transplantation center updates the same in the system application. This will trigger another round of recipient identification process and identify the next best-suited recipient from the waiting list. This process continues till a recipient is available for accepting the organ. All these transactions will be logged in the ledger.

After the successful transplantation operation of the allocated patient, the transplantation center will update the status in the system application. This will remove the recipient from the waiting list. The algorithm at the OPT manager updates the waiting list and communicates the change to all the transplantation centers along with the reasons for the update. Transplantation organizations also can run a query to get the reason for the new priority number assigned to the recipients registered from a particular organization. This option can be utilized for patients who wanted to know their position and the reason for each position changes.

4 PERFORMANCE ANALYSIS OF THE DECENTRALIZED BLOCKCHAIN

For the performance analysis of the system, a private Blockchain network with 4 nodes is considered. The network is created using fabric docker images. The npm-based Caliper is used for Benchmarking [2]. Using Hyperledger a simple contract with functions to create a donor(createRecipient), modify or delete a donor(deleteDonor), and query the data (readWaitList) are created. These functions are executed several times in each round and the performance of the system is measured. The Caliper Benchmark Report for transactions is detailed in Figures 4, 5 and 6.

createRecipient	Fail	Send Rate (TPS)	Max Latency (s)	Min Latency (s)	Through put (TPS)
7204	0	246.6	0.2	0.01	246.5
7528	0	254.3	0.09	0	254.2
8796	0	297.1	0.05	0	297
9579	0	323.6	0.04	0	323.5
10222	0	345.3	0.03	0	345.2
11681	0	402.5	0.05	0.01	402.4
24754	0	416.7	0.04	0	416.6

Figure 4. Benchmark report for createRecipient transactions.

readWaitlist	Fail	Send Rate (TPS)	Max Latency (s)	Min Latency (s)	Avg Latency (s)	Throughput (TPS)
8054	0	272.1	0.08	0	0.01	272.1
8404	0	283.9	0.07	0	0.01	283.9
9072	0	306.4	0.06	0	0.01	306.4
9285	0	313.7	0.06	0	0.01	313.6
10493	0	359.3	0.07	0	0.01	359.2
11213	0	386.4	0.1	0	0.01	386.2
24463	0	411.8	0.03	0	0.01	411.7

Figure 5. Benchmark report for deleteRecipient transactions.

deleteRecipient	Fail	Send Rate (TPS)	Max Latency (s)	Min Latency (s)	Throughput (TPS)
7727	0	260.2	0.06	0	260.1
8426	0	288.2	0.3	0	288.2
9454	0	319.3	0.03	0	319.2
9765	0	329.8	0.03	0	329.7
10134	0	342.3	0.03	0	342.2
11795	0	406.5	0.04	0.01	406.4
24849	0	418.3	0.03	0	418.3

Figure 6.   Benchmark report for readWaitList transactions.

The latency more or less remained constant throughout the various input loads. Throughput per second (TPS) increased in direct proportion to the input load and then remained at a constant level. This indicates a stable system concerning donor creation, deleting, and reading transactions.

5 LIMITATIONS

Policymakers will frequently need to reevaluate the balance between fairness and usability when introducing new precision technology in organ allocation and organ transplantation. They will also need to consider how to create guidelines that represent our society's idea of a fair allocation system. Transparency of the system can be achieved for all transactions except those involving transactions between entities without registering into the system.

6 CONCLUSION

Transparency can be achieved for all the transactions initiated by every stakeholder in the transplantation system using Blockchain. The allocation policy based on the enhanced model including Blockchain not only optimizes the recipient selection but provides transparency for the whole process. This paper suggests a stable Blockchain based transparency improvement system as the way to improve the overall transparency of the organ transplantation system.

REFERENCES

[1]   Nallusamy, S., Shyamalapriya, Balaji, Ranjan, & Yogendran., "Organ Donation – Current Indian Scenario", *Journal of the Practice of Cardiovascular Sciences*, 4(3), 177, 2018.
[2]   Chalissery, B. J., Asha, V., & Sundaram, B. M., "More Accurate Organ Recipient Identification Using Survey Informatics of New Age Technologies", *Proceedings of the 3rd International Conference on Integrated Intelligent Computing Communication & Security (ICIIC 2021)*, 4, 6–13, 2021.
[3]   Vardhini, Dass, S. N., Sahana, & Chinnaiyan, R., "A Blockchain Based Electronic Medical Health Records Framework using Smart Contracts", *2021 International Conference on Computer Communication and Informatics*, ICCCI 2021.
[4]   Bai, C., & Sarkis, J. "A Supply Chain Transparency and Sustainability Technology Appraisal Model for Blockchain Technology", *International Journal of Production Research*, 58(7), 2142–2162, 2020. https://doi.org/10.1080/00207543.2019.1708989
[5]   Delgado-Mohatar, O., Tolosana, R., Fierrez, J., & Morales, A., "Blockchain in the Internet of Things: Architectures and Implementation". *Proceedings – 2020 IEEE 44th Annual Computers, Software, and Applications Conference*, COMPSAC 2020, 1072–1077,2020.

Recent Trends in Computational Sciences – Gururaj, Pooja & Flammini (Eds)
© 2024 The Author(s), ISBN 978-1-032-42685-3

Smart helmet for accident detection using IoT

H.T. Chethana
VVCE
ORCID ID: 0000-0003-0131-7395

S. Kunal & Sahil Jain
Department of Computer Science, Vidyavardhaka College of Engineering, Mysuru, India
ORCID ID: 0009-0003-1634-0309

ABSTRACT: Nowadays there is a drastic increase in the number of accidents occurring in two-wheeler than other vehicles. Even though many rules and regulations are issued by the government to avoid accidents but still many people will die every second due to accidents. The smart helmet is used for accident detection and wearing a smart helmet is much needed while riding the vehicle. An attempt is made in this research paper in which smart helmets can be used for accident detection using IOT. If the accident occurs, information such as the rider's current location, heart rate, the temperature of the rider, and consumption of alcohol by the rider is sent to the authorized person.

Keywords: Accidents, Alcohol Detection, Smart Helmet, Two Wheeler, Infrared Sensors

1 INTRODUCTION

The Internet of things (IoT) refers to a connection with the internet rather than people. It is found in many fields such as security systems, home automation, and medical and health care. IOT senses the data and provides the conclusion. An accident is an event that occurs unexpectedly. Sometimes these accidents may kill the person which may change the life of the own family. The number of accidents occurring in two-wheeler is more. More than 80,000 people died due to alcohol consumption and riding a two-wheeler without using a helmet. A smart helmet is much needed to provide safety and security to two-wheeler riders. The death rate can be reduced by wearing a helmet during riding a two-wheeler.

Sensors such as alcohol sensor, infrared sensor, pulse rate sensor, DHT11 sensor, and ADLX sensor are used in the smart helmet which sends the necessary information to the authorized person. If the rider rides a two-wheeler by consuming alcohol, the probability of an accident is higher. If a heart attack occurs during riding, this condition should be checked and informed to the authorized person. Suppose the rider rides a two-wheeler by wearing a helmet, an alcohol sensor is used to detect whether the rider has consumed the alcohol or not indicating through a buzzer, and the information is provided to the authorized person. Infrared sensors are used to detect whether any obstacle occurs while riding. The pulse rate

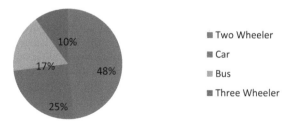

Figure 1.   Accidents rates through various modes.

DOI: 10.1201/9781003363781-40

sensor is used to detect the pulse of the rider and DHT11 sensors are used to detect the humidity and temperature [10]. By using the ADLX sensor, the accident can be detected and the message of the accident with the location is sent to the authorized person.

2 LITERATURE REVIEW

This section discusses a brief literature review on various methods implemented for accident detection using IOT. In India, the rate of accidents and death of two-wheeler riders is increasing drastically. To avoid these circumstances and to help the riders some applications have been developed.

Mohammed *et al.* [1] have proposed a system in which thermal screening is done through the helmet which avoids the spreading of covid-19. Jayasree *et al.* [2] have proposed a system that provides safety and security to construction workers during construction. Here accelerometer and gyroscope sensors are used for monitoring the workers.

Praveen M Dhulavvagol *et al.* [3] proposed a system for accident detection. Global positioning systems (GPS), Global systems for mobile communication (GSM), and the cloud are used to track the location of accidents. Jesudoss *et al.* [4] have proposed a system to detect whether a rider has consumed alcohol or not which helps in preventing accidents. In this system, the bike will not start provided the rider has consumed alcohol. If any accident occurs, then a notification will be sent to the authorized person. Sayan Tapadar *et al.* [5] have proposed a system with a vector mechanism to detect accidents and to check whether alcohol is consumed by the rider during an accident.

Jayasinghe *et al.* [6] have proposed a system using the smart helmet to detect the drowsiness of the rider. This system also detects the alcohol consumption of the rider thereby avoiding accidents. Keesari Shravya *et al.* [7] have proposed a system in which the detection of accidents, consumption of alcohol by the rider, and if the rider is falling during an accident can be detected by having various sensors installed in smart helmets.

Mohammed Atiqur Rahman *et al.* [8] have proposed a prototype with 3 modules a helmet circuit, an automobile circuit, and a mobile application. If an accident occurs, the mobile application is used to send notifications to the police station as well as to the authorized person. Rohith Ravindran *et al.* [9] have proposed a system that is used to detect whether the rider is under stress or not during riding a vehicle. Rithik Modi *et al.* [11] have proposed a system using IOT. This system detects whether the rider has to wear a helmet or not. If worn, then it checks the consumption of alcohol by the rider, and the GSM module is used for sending a message to the authorized person if any accident occurs.

According to a literature review, most of the works [12–14] are observed in the field of IOT for accident detection which will detect features such as alcohol consumption by the rider and the location of the accidents. But it is not observed that features such as detection of obstacles, consumption of alcohol by the rider, pulse rate, temperature, and humidity of the rider are not employed in the smart helmet for accident detection. Hence an attempt is made in this research work to detect accidents using smart helmets through IOT.

3 PROPOSED SYSTEM

The accident rates are increasing day by day, in order to reduce the rate of accidents and provide safety to the rider proposed system has been developed using IOT. In the proposed system, the various features of the rider can be detected by wearing a smart helmet using IOT for the rider. They are the consumption of alcohol by the rider, obstacle detection, temperature, and humidity of the rider. If the accident occurs, an emergency message is sent to the authorized person. The architecture of the proposed system using IOT is shown in Figure 2. It includes an IR sensor, alcohol sensor, temperature and humidity sensor, accelerometer sensor, and pulse rate sensor.

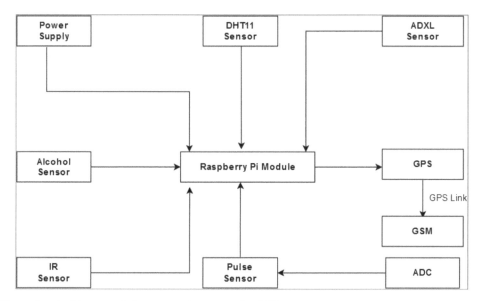

Figure 2.   Architecture of the proposed system using IOT.

The architecture of the proposed system consists of the following components. They are as follows

- **Battery:** A rechargeable battery of 12v is used in the proposed system to power the circuit and the proposed system needs a battery in real-time.
- **Raspberry PI:** It is a small PC that is integrated with the pulse rate sensors, GPS, and camera module. The emergency message will be sent automatically to defined contacts. With the help of a secure digital (SD) card, the proposed system can store all the information concerning it.
- **GPS Module:** It is a location tracker which is used to track the current location of the user. The information is used to know the exact address such as street, area, or nearby junction which helps in easily reaching the exact location. If GPS is not working, then longitude and latitude information is sent through mail or SMS.
- **GSM Module:** It is used for communication between the computer and the GPRS system. The sim card is inserted within the mobile phones. To receive and send SMS, the GSM sim card should be registered to the proposed system.
- **Buzzer:** Buzzer is the one that generates an alarm when there is a detection of alcohol by the rider.
- **Alcohol Sensor:** It is used to detect whether the rider has consumed alcohol or not. If consumed, then the amount of alcoholic substance by the rider is also detected using an alcohol sensor. When the consumption of alcohol by the rider is detected, the authorized person can make the rider safe.
- **IR Sensor:** It is used to detect obstacles. While riding the bike, if the rider encounters any obstacle then it can be detected and noticed.
- **ADXL (Accelerometer) Sensor:** Accelerometer is the device used to measure the speed. Using the ADXL sensor, accidents are detected when the angle of inclination changes towards gravity.
- **Pulse Rate Sensor:** It is used to check the pulse rate of the rider continuously. If it reaches irregular values, then the information is sent to the authorized one.

- **Temperature Sensor:** Due to various factors during an accident, the temperature of the rider changes abnormally and this can be recorded with the help of a temperature sensor.

Algorithm 1: Smart Helmet using IOT

Input: Smart Helmet with sensors
Output: Detection of alcohol, obstacle, speed, pulse rate, temperature, and humidity
Step 1: Start
Step 2: Wearing of helmet by the rider
Step 3: The temperature and humidity of the rider are detected.
Step 4: Alcohol is detected and the buzzer is ON
Step 5: Obstacle is detected
Step 6: Heart rate is detected
Step 7: If any accident occurs, the message is sent to the authorized people along with the location.
Step 8: Stop

---

4 EXPERIMENTAL RESULTS AND ANALYSIS

This section discusses the experimental results and their analysis, which involved two scenarios and various sensors including alcohol, infrared, pulse, temperature, and humidity sensors in the proposed system. In the first scenario, alcohol sensors detect whether the rider has consumed alcohol or not when wearing the helmet. IR sensors detect obstacles during riding, pulse sensors detect the rider's pulse rate, and temperature and humidity sensors detect the temperature and humidity of the rider. The second scenario involves sending an emergency message with the location to an authorized person in the event of an accident. The GSM and GPS modules enable the message to be sent to the authorized person. The proposed system was designed with IoT hardware components and continuously tested with software to obtain accurate results. Some modifications were made during the testing process, and individual and software modules were tested before being integrated with all the features. Results of the experiments for smart helmet detection using IoT can be found in Figures 3 to 14.

Figure 3.   Connecting device to the system.

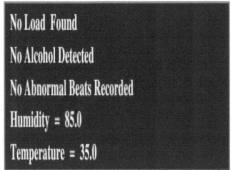

Figure 4.   View of output.

Figure 3 shows connecting the device to a system using a virtual network computing viewer with the password and user name and the device will run with default values.

Figure 4 shows the system showing the output with temperature and humidity, no consumption of alcohol by the rider, and no heartbeats as there is no pulse rate found with no obstacle detected. The humidity, temperature, and pulse rate are shown in Figures 5 and 6.

Figure 7 shows the exact location using the URL of the accident location. Figures 8 and 9 show the detection of obstacles (load) with the alcohol or not. If an accident occurs or if in case of any emergency, a message with the location is shown in Figure 10. The graph of varying temperature, humidity, and pulse rate is shown in Figures 11 to 14.

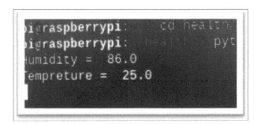

Figure 5.   View of humidity and temperature.   Figure 6.   View of pulse rate.

Figure 7.   View of exact location using URL.

Figure 8.   Detection of no load and alcohol.   Figure 9.   Detection of load but no alcohol.

EMERGENCY please visit the location visit location http://maps.google.com/maps?q=loc:13.0104383333,76.1025603333

EMERGENCY please visit the location visit location http://maps.google.com/maps?q=loc:13.0104503333,76.1025823333

Figure 10.   View of emergency message with location.

Figure 11.   Graph for varying temperature.

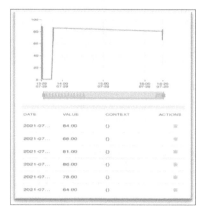

Figure 12.   Graph for varying pulse rate.

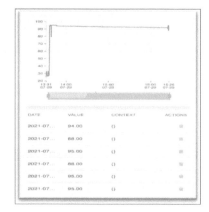

Figure 13.   Graph for varying humidity.

Figure 14.   Smart helmet using IOT.

5 CONCLUSION AND FUTURE WORK

A smart helmet that incorporates the Internet of Things (IoT) technology has been designed to increase the safety of two-wheeler riders. The system is programmed to send alert messages to authorized contacts, which helps monitor the safety of the rider. In addition, the helmet features a fault detector that can identify problems within the system and immediately sends alerts via messages. The proposed system is expected to improve rider safety by providing automatic detection of issues and sending help messages to parents or nearby police stations through IoT. This enhances rider security and ensures that prompt assistance can be provided if needed.

In the future, this technology could be expanded to incorporate an automatic system that includes a key to start the vehicle. This system would also be designed to detect the use of mobile phones while riding and restrict triple riding. These measures can help reduce the number of rider fatalities and improve overall rider safety.

REFERENCES

[1]   Mohammed M N, Halim Syamsudin, S Al-Zubaidi, Sairah A.K, Rusyaizila Ramli, Eddy Yusuf, "Noval Covid Detection and Diagnosis System Using IOT Based Smart Helmet", *International Journal of Psychosocial Rehabilitation*, Vol. 24, Issue 7, 2020.

[2]   Jayasree V., Nivetha Kumari M, "IOT Based Smart Helmet for Construction Workers", *IEEE 7th International Conference on Smart Structures and Systems*, ICSSS 2020.

[3]   Praveen Mhulavvagol, Ranjitha Shet, Prateeksha Nashipudi, Anand S Meti, and Renuka Ganige, "Smart Helmet with Cloud GPS GSM Technology for Alcohol and Accident Detection", Springer Nature, Singapore, 2018.

[4]   Jesudoss A, Vybhavi R, and Anusha B, "Design of Smart Helmet for Accident Avoidance", *International Conference on Communication and Signal Processing*, April 4–6, 2019, India.

[5]   Sayan Tapadar, Arnab Kumar Saha, Dr. Himadri Nath Saha, Shinjini Ray, and Robin Karlose, "*Accident and Alcohol Detection in Bluetooth Enabled Smart Helmet for Motorbikes*", IEEE, 2018.

[6]   Sanjana K, D S Jayasinghe, Udara S P, R Arachchige, "A Smart Helmet with Build Drowsiness and Alcohol Detection System", *Journal of Research Technology and Engineering*, Vol. 1, Issue 3, July 2020.

[7]   Keesari Shravya, Yamini Mandapati, Donuru Keerthi, Kothapu Harika, and Ranjan K. Senapati, "Smart Helmet for Safe Driving", *E3S Web Conference* 2019.

[8]   Mohammed Atiqur Rahman, Toufiq Ahmed, S M Ahsanuzzaman, Abid Ahsan and Ishman Rahman, "IOT Based Smart Helmet and Accident Identification System", *IEEE Region 10 Symposium*, 2020.

[9]   Rohith Ravindran, Hansini Vijayaraghavan and Mei-Yuan Huang, "*Smart Helmet for Safety in Mining Industries*", IEEE, 2018.

[10] K Suriyakrishnaan, R Arun Gandhi, R Babu, S Sakthivel and Saurabh Dev, "Smart Helmet in Coal Mining Using Arduino", *Turkish Journal of Computer and Mathematics Education*, 2021.

[11] Rithik Modi, Rohit Nair, Archit Gupta, Himanshu Sharma, and Yadvendra Bedi, "Smart Helmet Using IOT", *International Journal of Engineering, Business, and Management*, 2021.

[12] Mehata K M, Shankar S K, Karthikeyan N, Nandhinee K, and Robin Hedwig P, "IOT Based Safety and Health Monitoring for Construction Workers Helmet System with Data Log System", *International Conference on Innovations in Information and Communication Technology*, 2021.

[13] Divyasudha N, Arulmozhivarman P, Rajkumar E R, "*Analysis of Smart Helmets and Designing an IoT Based Smart Helmet: A Cost-effective Solution for Riders*", IEEE, 2022.

[14] Impana H C, Chethana H T, Hamsaveni M, "A Review on the Smart Helmet for Accident Detection Using IOT", *EAI Endorsed Transactions*, 2020.

Recent Trends in Computational Sciences – Gururaj, Pooja & Flammini (Eds)
© 2024 The Author(s), ISBN 978-1-032-42685-3

A survey on mission critical task placement and resource utilization methods in the IoT fog-cloud environment

Shifa Manihar & Ravindra Patel
University Institute of Technology, Gandhi Nagar, Bhopal (M.P.), India
ORCID ID: 0000-0002-4835-8058, 0009-0006-0731-5592

Sanjay Agrawal
National Institute of Technical Teachers' Training and Research (NITTTR), Shamla Hills, Bhopal, M.P, India
ORCID ID: 0000-0003-2963-6662

ABSTRACT: When n number of IoT applications arrive at servers, then rearranging and scheduling them in such a way to obtain the most optimal criteria is termed as Task Scheduling. The Optimal criterion may be the least delays, low energy consumption and cost, maximum resource utilization, etc. Because a fog has limited storage and processing capacity, these criteria become bottlenecks when these IoT applications reach the Fog scenario. As a result, under the Fog environment, all real-time applications cannot be scheduled. It is also necessary to allocate these resources in the most efficient manner feasible. As a result, it is recommended that mission-critical applications be scheduled on the fog and non-mission-critical applications be scheduled in the cloud. This paper surveys the various conventional methods as well as machine learning algorithms and meta-heuristic approaches adopted for placing the applications and allocating the resources to them in the fog-cloud scenarios proposed by different researchers.

Keywords: Cloud, Edge, Fog, Internet of Things, Latency, Machine Learning, Meta-heuristics, Resource allocation, Task scheduling

1 INTRODUCTION

Introducing Smartness into the IoT devices means to provide the devices (with unique identity and addresses) with the ability to amass and correlate the information independently relied upon the underlying communication protocol and also maintain the integrity of the information autonomously [1–3]. The cloud computing paradigm has pave the way for computing facility throughout the world, hence uninterrupted computation. But this cloud computing paradigm lags behind when it comes to several delays incurred with it. Transmission delays, network challenges such as congestion, and other factors contribute to these delays. Because of these limitations, it is unsuitable for delay-sensitive applications. As a result, we attempt to find such an arrangement and scenarios in which computation-intensive jobs may be scheduled in the cloud while latency-sensitive apps can be scheduled on a device close to the IoT device. Fog environments are merely a scaled-down version of the cloud, with restricted processing and storage capabilities. We can locate the Fog scenario's location between mobile IoT devices such as sensors, gadgets, and the cloud [4]. Fog computing is simply a network-edge extension of the cloud. This provides new benefits such as faster reaction times and improved security, integrity, and privacy [5]. The computation is done locally and the response is instantaneous. Not all jobs could be placed in the fog due to the bottlenecks associated with fog. As a result, research is

DOI: 10.1201/9781003363781-41

being conducted to identify mechanisms that can schedule applications between cloud and fog in order to reach ideal criteria such as low latency, energy consumption, cost, and so on. The problem statement here is how to divide jobs as efficiently as possible between the fog and cloud, as well as between several virtual machines on both ends [6]. In this article, mission critical tasks/applications refers to latency sensitive tasks/applications or energy saving applications, cost saving applications. Different authors used different parameter of the task and IoT devices to define this context. The context may include the location of the task and IoT device, storage and computational capacity of the Fog device, the task priority, delay sensitivity of the tasks, dependencies among tasks, etc. The rest of the paper is laid out as follows: Section II gives short description of the related works done in this field. Section III describes various task scheduling algorithms. Section IV gives a short discussion on the performance metrics of various task scheduling algorithms discussed in this literature. Section VI concludes this survey paper.

2 RELATED WORK

Potu Narayana *et al.* discussed conventional job schedulers in fog computing environment and enlisted all their advantages and limitations [7]. Xin Yang *et al.* has prepared an organized summary of the existing approaches for task scheduling, and their effect on fog computing [8]. The author has studied the future bottlenecks for the fog computing environments and how task scheduling can play an important role in solving them. Redowan Mahmud *et al.* have reviewed the present application management methodologies in the Fog environment and discussed them with respect to their architecture, placement and maintenance. The authors have also presented a complete classification and found out the research gaps in Fog-based application management [9]. Thang Le Duc *et al.* taken into consideration the environment of heterogeneous network and studied several works done in the field of resource provisioning using machine learning in the edge-cloud scenarios [10]. Essam H. Houssein *et al.* has presented a review on Meta heuristic based task placement algorithms in the cloud environment [11]. Rasha A. Al-Arasi *et al.* has reviewed concepts of resource scheduling and discussed task scheduling algorithm based on meta-heuristics working on different optimization criteria [12]. Khaled Matrouk *et al.* has also surveyed various modern task placement strategies in the fog environment [13].

3 TASK SCHEDULING METHODS

These methodologies are classified as: Conventional method; Meta-heuristic based methods; and Machine Learning based methods.

3.1 *Conventional solutions for task scheduling*

[14] established fog between IoT devices and the cloud, and assigned job priorities of high, medium, and low importance.Task classifications and virtual machine categorizations (TCVC) were used by the authors. They recorded the performance of these techniques using the MIN MAX algorithm. The simulator utilised was Cloudsim.

[15] proposed Petrel, a distributed and application-aware task scheduling system in their study. Petrel is utilised not just for load balancing, but also for ensuring an adaptable scheduling policy based on the task type. The basic motivation behind this is the computational offloading in the edge cloud environment. The cloudlets are properly interconnected to ease the computational offloading for the purpose of achieving low latency. They made use of the daemon cloud which incurs the least latency with the mobile devices in its vicinity. In the fog-cloud environment, [16] suggested a decentralised context aware 3-tier system for job scheduling. The response time of the jobs is used to determine the quality of service.

[17] took deadlines of the tasks as the priority and scheduled the tasks in the fog scenario. Both cloud and the fog layers had multiple data centres and the virtual machines. The fog

server manager (FSM) used to manage all these resources at each fog server. The author maintained three priorities queues (High, Medium, and Low) to accomplish task scheduling algorithm based on priorities.

In [18], the author used non pre-emptive task scheduling such that if some higher priority task arrives for execution during the execution of some lower task, then it will have to wait till it finishes its execution.

[19] in his work proposed a framework in which priority is assigned to tasks in such a way that minimum bandwidth is incurred. It took into account computational power need and the number of resources available at the Fog. Reults were recorded by placing tasks at Cloud only, Fog only and fog-cloud scenarios. This framework is known as RACE (Resource Aware Cost Efficient). RACE was compared with other baseline algorithms and showed better results.

In the fog environment, [20] used container instead of virtual machines in order to avoid overhead in the processing of the tasks.

On the fog cluster, [21] presented quality of service conscious task allocation techniques. The results showed that Throttled load balancing policy performed best irrespective of whatever scheduling algorithm taken.

In [22], the authors used the notion of context aware data replication to the different providers at the data replication level and investigated three scenarios: no replication, one replication, and full replication. Although replication adds overhead, it improves performance.

3.2 Meta heuristic based solutions for task scheduling

In the edge-cloud scenario, [23] used the disaster genetic algorithm to adopt a work scheduling algorithm. The task scheduling at the cloud is carried out based on the sensitivity factor on various virtual machines available at cloud.

[24] has applied nature-inspired Meta heuristic task scheduling based on the three architectures of the cloud fog environment, namely Ant Colony Optimization and Particle Swarm Optimization. The authors designed objective function based on the response time, and applied these algorithms to minimize it. For fog cloud environments, [25] has developed multiple resource sharing systems. Broker maintains the quality of service information of all available nodes. The QoS depends on the deadline, id application having strict time deadlines, then execute at the device nodes, if less strict deadlines then at fog nodes, if no deadlines then depending on the resources available at the fog or forwarded to the cloud. [26] has presented an application placement algorithm for dependent as well as independent tasks. The authors represented dependent tasks in the form of Directed Acyclic Graph (DAG) and applied weighted cost model to reduce the energy consumption and the response time of the applications in the fog and edge scenario. For fog cloud environments, [27] has developed multiple resource sharing systems.

In [28], in their work used cost aware task scheduling which is based on the genetic algorithm. The authors created master slave fog environment, where Master Fog node accomplishes the task of scheduling requests. The characteristics of the tasks are encoded as chromosomes and various operations are applied to maximize the fitness function (Minimize the cost function). The authors named this algorithm as Cost aware genetic based algorithm (CAG).

[29] has suggested a multi-stage greedy adjustment (MSGA) algorithm for task scheduling that takes into account task placement and network traffic.

[30] analysed a situation including five fog nodes with equally dispersed duties. In the fog cloud scenario, the author employed the NSGA II algorithm to optimise the delay and energy.

3.3 Machine learning based solutions for task scheduling

[31] a fog-based integrated classification method with task scheduling To do this, the authors used I-Apriori and Task Scheduling in Fog Computing.

[32] proposed a two-level neural network-based real-time task scheduling system. To reduce reaction time latency, it is recommended that dependent tasks be executed at a single

fog node. This not only reduced the makespan but also the communication cost that may incur if dependent tasks executed at different nodes. There exists no such model at the reinforcement layer i.e. no environmental parameters are set, the decision of task scheduling at different resources available at the fog is totally dependent on the experience or training with the objective to maximize the numerical reward. The policy selected based on given situation is evaluated using gradient decent policy selection method.

In the mobile fog environment, [33] employed an innovative adaptive job scheduling and server balancing technique. For task placement in the fog environment, [34] used History Analyzer module to keep track of all the task placements carried out in the history, and use this information to find out where to schedule the task in the fog. In a fog environment, [35] used a QCI-based neural network technique for load balancing and attaining minimal latency in a real-time scenario. According to [36], the author employed fuzzy rules based on predefined priorities, deadlines, and job size. Based on the priority, the work is assigned to the device, the dew layer, the fog layer, or the cloud layer. At these levels, the probabilistic neural network is utilised to balance load.

[37] proposed a reinforcement learning-based task scheduling system in which a cloud-based task scheduler keeps track of the task queue and feeds the status of all tasks to a reinforcement neural network, which subsequently sends the job to the cloud or the best fog node manager. [38] used an offline supervised training and an online unsupervised training technique, followed by a multiple regression model. [39] has presented a combined 'cloud-edge' aware task placement algorithm that makes use of the deep reinforcement learning exploration exploitation property.

4 DISCUSSION

All the task scheduling methods were based upon the context aware scheduling. The authors used to define this context in different ways considering different scheduling parameters. This can be depicted in the Figure 1.

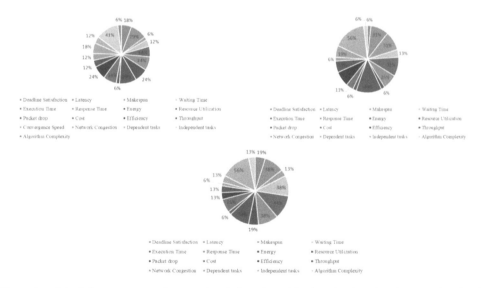

Figure 1. Scheduling parameters for meta heuristic and machine learning based and conventional task scheduling methods.

As can be observed from the scheduling metrics above, all types of algorithms considered response time, delay, makespan, etc. as their basic scheduling objective.

4.1 *Comparison between conventional, meta-heuristic and machine learning based solutions for task scheduling*

The running time of the Meta-heuristic based methods and machine learning based methods is higher when compared with the conventional methods of task scheduling. Meta heuristic methods incurs large running time because before giving an optimal solution, these algorithms have to converge. They try all possible solutions of the population so that they do not stuck into local optimum, hence giving the global optimum. Machine learning based solutions incur large running time because they are in need of training the data set first. Whereas there is no such requisites of the conventional methods. But as compared to conventional methods of task scheduling, the meta-heuristic based methods and machine learning based methods showed better performance in terms of response time, energy consumption, makespan, etc. Some authors used combination of these methods to form tradeoff between the running time and the other parameters. When compared to traditional methods, meta-heuristic and machine learning-based solutions were able to generate superior contexts on which task classification, scheduling, and allocation could be done.

5 CONCLUSION

The authors attempted to deploy various ways to make the fog layer intelligent with the goal of designing a context aware middleware between cloud and IoT devices. The authors considered various dependent and independent tasks, did workflow analysis as well and came up with different solutions. Several algorithms used combination of different Meta heuristic and machine learning algorithms both at the cloud and the fog level to get the most optimized solutions. Thus, several such combination of different streams can be done to improve the performance. We have made a comparative analysis as given in different tables (table 3, 4, 5) for tasks placement at both cloud or fog environments by taking into account the nature of the tasks, scheduling objectives and simulation environment attributes. It is analysed that if conventional task scheduling methods benefits in running time and resource allocation, the machine learning based and meta heuristic based task scheduling and resource allocation methods were able to develop better contexts for the most optimal solutions for mission critical applications. Thus, several authors used combination of these methods to fulfil the lacuna of each of these methods if used independently. In this paper, an extensive study of various categories of algorithms has been carried out which gives insight into the already existing solutions and also ushers researchers for carrying out such investigations to develop better solutions in the coming future.

REFERENCES

[1] Ahmed E., Yaqoob I., Gani A., Imran M., and Guizani M., "Internet-of Things-based Smart Environments: State of the Art, Taxonomy, and Open Research Challenges," *IEEE Wireless Commun.*, vol. 23, no. 5, pp. 10–16, Oct. 2016.
[2] Shafique K., Khawaja B. A., Sabir F., Qazi S. and Mustaqim M., "Internet of Things (IoT) for Next-Generation Smart Systems: A Review of Current Challenges, Future Trends and Prospects for Emerging 5G-IoT Scenarios," in *IEEE Access,* vol. 8, pp. 23022–23040, 2020, doi: 10.1109/ACCESS.2020.2970118.

[3]   *Co-Operation With the Working Group RFID of the ETP EPOSS, Internet of Things in 2020, Roadmap for the Future, Version 1.1, INFSO D.4 Networked Enterprise RFID INFSO G.2 Micro Nanosystems*, May 2008.

[4]   Gedeon, J., Jens Heuschkel, L. Wang and M. Mühlhäuser. *"Fog Computing: Current Research and Future Challenges."* (2018).

[5]   Dustdar S., Avasalcai C. and Murturi I., "Invited Paper: Edge and Fog Computing: Vision and Research Challenges," *2019 IEEE International Conference on Service-Oriented System Engineering (SOSE)*, San Francisco, CA, USA, 2019, pp. 96–9609, doi: 10.1109/SOSE.2019.00023.

[6]   Carvalho L. I., da Silva D. M. A. and Sofia R. C., "Leveraging Context-awareness to Better Support the IoT Cloud-Edge Continuum," *2020 Fifth International Conference on Fog and Mobile Edge Computing (FMEC)*, Paris, France, 2020, pp. 356–359, doi: 10.1109/FMEC49853.2020.9144760.

[7]   Narayana P., Parvataneni P. and Keerthi K., "A Research on Various Scheduling Strategies in Fog Computing Environment," *2020 International Conference on Emerging Trends in Information Technology and Engineering (ic-ETITE)*, Vellore, India, 2020, pp. 1–6, doi: 10.1109/ic-ETITE47903.2020.261.

[8]   Xin Yang, Nazanin Rahmani, "Task Scheduling Mechanisms in Fog Computing: Review, Trends, and Perspectives", *Kybernetes*, Vol. 50 No. 1, pp. 22–38, 2020. https://doi.org/10.1108/K-10-2019-0666.

[9]   Redowan Mahmud, Kotagiri Ramamohanarao, and Rajkumar Buyya, "Application Management in Fog Computing Environments: A Taxonomy, Review and Future Directions", *ACM Computing Surveys*, Vol. 53, No. 04, July 2020. https://doi.org/10.1145/3403955.

[10]   Thang Le Duc, Rafael García Leiva, Paolo Casari, Per-Olov Östberg, "Machine Learning Methods for Reliable Resource Provisioning in Edge-Cloud Computing: A Survey", *ACM Computing Surveys*, Vol. 52, No. 05, September 2019. https://doi.org/10.1145/3341145.

[11]   Essam H. Houssein, Ahmed G. Gad, Yaser M. Wazery, Ponnuthurai Nagaratnam Suganthan, "Task Scheduling in Cloud Computing Based on Meta-heuristics: *Review, Taxonomy, Open Challenges, and Future Trends, Swarm and Evolutionary Computation*, Volume 62, April 2021, 100841, ISSN 2210-6502, https://doi.org/10.1016/j.swevo.2021.100841.

[12]   Rasha A. Al-Arasi, Anwar Saif, "Task Scheduling in Cloud Computing Based on Metaheuristic Techniques: A Review Paper", *EAI Endorsed Transactions on Cloud Systems*, January 2020. doi: 10.4108/eai.13-7-2018.162829.

[13]   Khaled Matrouk, Kholoud Alatoun, "Scheduling Algorithms in Fog Computing: A Survey", *International Journal of Networked and Distributed Computing*, Vol. 9(1), pp. 59–74, January 2021. DOI:https://doi.org/10.2991/ijndc.k.210111.001.

[14]   Tahani Aladwani, "Scheduling IoT Healthcare Tasks in Fog Computing Based on their Importance", *Procedia Computer Science*, Volume 163, 2019, Pages 560–569, ISSN 1877-0509, https://doi.org/10.1016/j.procs.2019.12.138.

[15]   Lin L., P. Li, Xiong J. and Lin M., "Distributed and Application-Aware Task Scheduling in Edge-Clouds," *2018 14th International Conference on Mobile Ad-Hoc and Sensor Networks (MSN)*, Shenyang, China, 2018, pp. 165–170, doi: 10.1109/MSN.2018.000-1.

[16]   Minh-Quang Tran, Duy Tai Nguyen, Van An Le, Duc Hai Nguyen, Tran Vu Pham, "Task Placement on Fog Computing Made Efficient for IoT Application Provision", *Wireless Communications and Mobile Computing*, vol. 2019, Article ID 6215454, 17 pages, 2019. https://doi.org/10.1155/2019/6215454.

[17]   Choudhari, Tejaswini, "Prioritized Task Scheduling In Fog Computing" (2018). *Master's Projects*.581. DOI: https://doi.org/10.31979/etd.shqa-fdp6, https://scholarworks.sjsu.edu/etd_projects/581.

[18]   F. Fellir, A. El Attar, K. Nafil and L. Chung, "A Multi-Agent Based Model for Task Scheduling in Cloud-fog Computing Platform," *2020 IEEE International Conference on Informatics, IoT, and Enabling Technologies (ICIoT)*, Doha, Qatar, 2020, pp. 377–382, doi: 10.1109/ICIoT48696.2020.9089625.

[19]   Arshed J. U. and Ahmed M., "RACE: Resource Aware Cost-Efficient Scheduler for Cloud Fog Environment," in *IEEE Access*, doi: 10.1109/ACCESS.2021.3068817.

[20]   Yin L., Luo J. and Luo H., "Tasks Scheduling and Resource Allocation in Fog Computing Based on Containers for Smart Manufacturing," in *IEEE Transactions on Industrial Informatics*, vol. 14, no. 10, pp. 4712–4721, Oct. 2018, doi: 10.1109/TII.2018.2851241.

[21]   Elarbi Badidi, "QoS-Aware Placement of Tasks on a Fog Cluster in an Edge Computing Environment", *Journal of Ubiquitous Systems & Pervasive Networks*, Volume 13, No. 1 (2020) pp. 11–19. doi: 10.5383/JUSPN.13.01.002.

[22]   Breitbach M., Schäfer D., Edinger J. and Becker C., "Context-Aware Data and Task Placement in Edge Computing Environments," *2019 IEEE International Conference on Pervasive Computing and*

289

Communications (PerCom, Kyoto, Japan, 2019, pp. 1–10, doi: 10.1109/PERCOM.2019. 8767386.

[23] Shudong Wang, Yanqing Li, Shanchen Pang, Qinghua Lu, Shuyu Wang, Jianli Zhao, "A Task Scheduling Strategy in Edge-Cloud Collaborative Scenario Based on Deadline", *Scientific Programming*, vol. 2020, Article ID 3967847, 9 pages, 2020. https://doi.org/10.1155/2020/3967847.

[24] Hussein M. K. and Mousa M. H., "Efficient Task Offloading for IoT-Based Applications in Fog Computing Using Ant Colony Optimization," in *IEEE Access*, vol. 8, pp. 37191–37201, 2020, doi: 10.1109/ACCESS.2020.2975741.

[25] Qayyum T., Trabelsi Z., Malik A. W. and Hayawi K., "Multi-Level Resource Sharing Framework Using Collaborative Fog Environment for Smart Cities," in *IEEE Access*, vol. 9, pp. 21859–21869, 2021, doi: 10.1109/ACCESS.2021.3054420.

[26] Goudarzi M., Wu H., Palaniswami M. and Buyya R., "An Application Placement Technique for Concurrent IoT Applications in Edge and Fog Computing Environments," in *IEEE Transactions on Mobile Computing*, vol. 20, no. 4, pp. 1298–1311, 1 April 2021, doi: 10.1109/TMC.2020.2967041.

[27] Nikoui T. S., Balador A., Rahmani A. M. and Bakhshi Z., "Cost-Aware Task Scheduling in Fog-Cloud Environment," *2020 CSI/CPSSI International Symposium on Real-Time and Embedded Systems and Technologies (RTEST)*, Tehran, Iran, 2020, pp. 1–8, doi: 10.1109/RTEST49666.2020.9140118.

[28] Sahni Y., Cao J. and Yang L., "Data-Aware Task Allocation for Achieving Low Latency in Collaborative Edge Computing," *in IEEE Internet of Things Journal*, vol. 6, no. 2, pp. 3512–3524, April 2019, doi: 10.1109/JIOT.2018.2886757.

[29] Abbasi, M., Mohammadi Pasand, E. & Khosravi, M.R., "Workload Allocation in IoT-Fog-Cloud Architecture Using a Multi-Objective Genetic Algorithm" *J Grid Computing* 18, 43–56 (2020). https://doi.org/10.1007/s10723-020-09507-1.

[30] Lindong Liu, Deyu Qi, Naqin Zhou, Yilin Wu, "A Task Scheduling Algorithm Based on Classification Mining in Fog Computing Environment", *Wireless Communications and Mobile Computing*, vol. 2018, Article ID 2102348, 11 pages, 2018. https://doi.org/10.1155/2018/2102348.

[31] Mohammad Khalid Pandit, Roohie Naaz Mir, Mohammad Ahsan Chishti, "Adaptive Task Scheduling in IoT Using Reinforcement Learning", *International Journal of Intelligent Computing and Cybernetics*, Vol. 13 No. 3, pp. 261–282, 2020. https://doi.org/10.1108/IJICC-03-2020-0021.

[32] Xuejing Li, Yajuan Qin, Huachun Zhou, Du Chen, Shujie Yang, Zhewei Zhang, "An Intelligent Adaptive Algorithm for Servers Balancing and Tasks Scheduling over Mobile Fog Computing Networks", *Wireless Communications and Mobile Computing*, vol. 2020, Article ID 8863865, 16 pages, 2020. https://doi.org/10.1155/2020/8863865.

[33] Mostafaz N. "Resource Selection Service Based on Neural Network in Fog Environment", *Advances in Science, Technology and Engineering Systems Journal*, vol. 5, no. 1, pp. 408–417 (2020).

[34] Bhatia, M., Sood, S.K. & Kaur, S. Quantumized Approach of Load Scheduling in Fog Computing Environment for IoT Applications. *Computing* 102, 1097–1115 (2020). https://doi.org/10.1007/s00607-019-00786-5.

[35] Fatma M. Talaat, Shereen H. Ali, Ahmed I. Saleh, Hesham A. Ali, "Effective Load Balancing Strategy (ELBS) for Real-Time Fog Computing Environment Using Fuzzy and Probabilistic Neural Networks", *Journal of Network and Systems Management (IF 2.250)* Pub Date : 2019-02-06 , DOI: 10.1007/s10922-019-09490-3.

[36] He Li, Kaoru Ota, and Mianxiong Dong, "Deep Reinforcement Scheduling for Mobile Crowd Sensing in Fog Computing", *ACM Trans. Internet Technol.* 19, 2, Article 21 (April 2019), 18 pages. DOI: https://doi.org/10.1145/3234463.

[37] Kafle V. P. and Muktadir A. H. A., "Intelligent and Agile Control of Edge Resources for Latency-Sensitive IoT Services," *in IEEE Access*, vol. 8, pp. 207991–208002, 2020, doi: 10.1109/ACCESS.2020.3038439.

[38] Dong Y., Xu G., Zhang M. and Meng X., "A High-Efficient Joint 'Cloud-Edge' Aware Strategy for Task Deployment and Load Balancing," *in IEEE Access*, vol. 9, pp. 12791–12802, 2021, doi: 10.1109/ACCESS.2021.3051672.

[39] Panwar, Reena & Mallick, Bhawna. (2015). A Comparative Study of Load Balancing Algorithms in Cloud Computing. *International Journal of Computer Applications*. 117. 33–37. 10.5120/20890-3669

Recent Trends in Computational Sciences – Gururaj, Pooja & Flammini (Eds)
© 2024 The Author(s), ISBN 978-1-032-42685-3

Weather prediction using data mining techniques

B.H. Swathi, G.D. Nidhi, L.J. Sahana & M.N. Prathibha
Vidyavardhaka College of Engineering, Mysuru, India
ORCID ID: 0000-0001-6694-6075

ABSTRACT: Even with the recent century's enormous advancements in science and technology, forecasting the weather remains a very difficult undertaking. In the past ten years, data mining techniques and applications have advanced significantly. Several scholars have investigated the effective use of data mining technologies in predicting climate and weather change. The important characteristics that are essential for data mining methods to be integrated into forecasting weather models are discussed in the paper. This paper's major goal is to present a thorough comparative examination of the benefits, drawbacks, and outcomes of existing data mining approaches used to weather forecasting model. And also the comparative analysis is done based on the obtained results using different algorithms.

Keywords: data mining, weather prediction, naïve bayes, random forest, neural networks

1 INTRODUCTION

Weather forecasting is a critical area of research that has wide-ranging implications for agriculture, transportation, emergency management, and other industries. Accurate weather prediction can help to reduce the impact of severe weather events and enable more effective decision-making in a range of contexts. However, weather prediction is a complex and challenging problem due to the large amount of data involved, the inherent uncertainty of weather patterns, and the difficulty in accurately modelling these patterns [1].

In recent years, data mining techniques have emerged as a powerful tool for weather prediction, offering new ways to extract insights from large and complex datasets. Data mining involves the process of discovering patterns, relationships, and anomalies in data through the use of statistical and computational methods. By applying data mining techniques to weather data, researchers can identify key variables, predict future weather patterns, and develop models that can be used to improve the accuracy of weather forecasting [10,11].

In this paper, we provide a comprehensive review of various data mining techniques for weather prediction. We discuss the advantages and limitations of different techniques. We also discuss the challenges and future directions of the field, highlighting areas where further research is needed to advance the state-of-the-art in weather prediction using data mining. Overall, our goal is to provide a valuable resource for researchers and practitioners working in the field of weather prediction, and to contribute to ongoing efforts to improve the accuracy and effectiveness of weather forecasting using data mining techniques [7–9].

2 LITERATURE REVIEW

Weather prediction is a challenging task due to the complexity and dynamic nature of weather patterns. Data mining techniques have been used to improve the accuracy of weather prediction models by extracting useful information from large datasets. Here is a literature survey on weather prediction using data mining techniques:

DOI: 10.1201/9781003363781-42

"A survey of data mining techniques for weather forecasting" by Jha & Acharjya [2] (2012) – This survey provides an overview of various data mining techniques, such as artificial neural networks, decision trees, and fuzzy logic, that have been used for weather prediction. The authors also discuss the advantages and limitations of these techniques.

"Weather prediction using artificial neural networks: a literature review" by Liu & Gao [3] (2016) – This review focuses on the application of artificial neural networks (ANNs) for weather prediction. The authors discuss various ANN architectures and training algorithms that have been used in the literature, as well as the performance of these models.

"Weather forecasting using decision tree algorithm" by Kumar & Kumar (2018) [4] – This study proposes a decision tree algorithm for weather prediction. The authors compare the performance of their model with other data mining techniques, such as k-nearest neighbors and support vector machines, and show that their approach outperforms these methods.

"Weather forecasting using machine learning techniques: a review" by Ahmed & Tariq [5] (2020) – This review provides an overview of various machine learning techniques, such as regression analysis, support vector machines, and deep learning, that have been used for weather prediction. The authors also discuss the challenges and future directions of weather prediction using machine learning.

"Predicting hourly precipitation using machine learning: a review" by Liu and Sun [6] (2021) – This review focuses on the prediction of hourly precipitation using machine learning techniques. The authors discuss various models, such as random forest and convolutional neural networks, that have been used in the literature and evaluate their performance.

The Tables 1 and 2 shows the comparative analysis of the above literature survey by considering various parameters including strength and weakness of each paper.

Table 1. Comparative analysis of the literature review.

Study	Data Mining Techniques	Focus	Performance comparison	challenges
Jha and Acharjya (2012)	ANN, Decision Trees, Fuzzy Logic	Overview	N/A	Limitations of each technique
Liu and Gao (2016)	ANN	Overview of ANN architectures and training algorithm	N/A	N/A
Kumar and Kumar (2018)	Decision Trees, k-NN, SVM	Proposal of decision tree algorithm	Decision tree outperformed k-NN and SVM	N/A
Ahmed and Tariq (2020)	Regression analysis, SVM, deep learning	Overview of various machine learning techniques	N/A	Challenges in data quality and model interpretability
Liu and Sun (2021)	Random Forest, CNN	Overview of various models for hourly precipitation prediction	Comparison of model performance	N/A

Table 2. Comparison of the literature survey with their strengths and weakness.

Study	Strengths	Weaknesses
Jha and Acharjya (2012)	Provides an overview of various data mining techniques for weather forecasting	Does not include performance comparison or case studies
Liu and Gao (2016)	Focuses specifically on the application of artificial neural networks for weather prediction	Does not provide a comprehensive review of other data mining techniques
Kumar and Kumar (2018)	Proposes a decision tree algorithm for weather forecasting and provides a performance comparison with other data mining techniques	Only evaluates the proposed algorithm on one dataset
Ahmed and Tariq (2020)	Provides an overview of various machine learning techniques for weather forecasting and discusses the challenges and future directions of the field	Does not include a performance comparison or case studies
Liu and Sun (2021)	Focuses specifically on predicting hourly precipitation and provides a comprehensive review of various machine learning models	Does not include a discussion of the challenges and future directions of the field

3 SYSTEM ARCHITECTURE

Weather data can be collected from various sources, such as weather stations, satellites, and other sensors. The data may include information such as temperature, precipitation, wind speed, and humidity. The collected weather data is pre-processed to remove noise, fill missing values, and normalize the data. This step may also involve feature selection or extraction to identify the most relevant data for prediction. Various data mining techniques, such as artificial neural networks, decision trees, or support vector machines, may be used to develop models for weather prediction. The selection of a specific model may depend on the type and amount of data available, as well as the specific problem being addressed. The selected model is trained using historical weather data and validated using a separate set of data to assess its performance. The model may be refined or retrained based on the validation results. Once the model is trained, it can be used to predict future weather conditions based on input data. The predictions may be visualized using maps, graphs, or other visualizations to provide insights into the predicted weather patterns. The weather prediction system can be deployed for use in various applications, such as agriculture, transportation, or emergency management [12–15]. The system may also be updated or retrained periodically to improve its accuracy and performance. Same is shown in Figure 1.

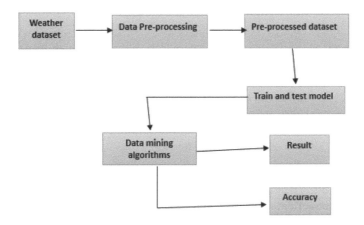

Figure 1. System architecture.

4 METHODOLOGY

To analyse the nature of the data, a descriptive statics technique is used. The data set contain seven parameters and 4576 records. For this data set the neural network, random forest and naïve bayes algorithms are applied to analyse the performance of each. In the following section the illustration of the methods are illustrated and in the next section the comparison of these techniques is discussed.

4.1 *Neural networks*

Neural networks can be used for weather prediction, and one common type of neural network used is the Multi-Layer Perceptron (MLP) model. The MLP model consists of multiple layers of interconnected nodes (or neurons), where each node in one layer is connected to all the nodes in the previous and next layers. The weights of the connections between the nodes are adjusted during training to minimize the error between the predicted output and the actual output.

Here's an example formula for predicting weather using an MLP neural network:

$$Y = f(W2 * f(W1 * X + b1) + b2) \tag{1}$$

Where:

Y is the predicted weather condition

X is the input weather data

W1 and W2 are the weight matrices for the connections between the input layer and the hidden layer, and between the hidden layer and the output layer, respectively. b1 and b2 are the bias terms for the hidden layer and output layer, respectively f is the activation function used in each neuron, such as the sigmoid or ReLU function.

The weights and biases of the neural network are learned during the training phase, where the network is trained on a subset of the historical weather data to minimize the error between the predicted output and the actual output. Once trained, the neural network can be used to predict the weather condition given a set of input weather data.

Note that the specific architecture and parameters of the neural network can vary depending on the specific application and data. Also, neural networks may require a large amount of training data and computational resources and may be more complex and difficult to interpret compared to other data mining techniques.

4.2 *Random forest*

Random Forest is another data mining technique that can be used for weather prediction. Random Forest works by constructing a multitude of decision trees at training time and outputting the class that is the mode of the classes (classification) or mean prediction (regression) of the individual trees. Here's an example formula for predicting weather using Random Forest:

$$Y = RF(X) \tag{2}$$

Where:

Y is the predicted weather condition

X is the input weather data

RF is the Random Forest model

The Random Forest model is trained on a subset of the historical weather data, where each tree in the forest is trained on a random subset of the input features and data samples. During prediction, the input weather data is passed through each decision tree in the forest, and the majority vote (in the case of classification) or mean prediction (in the case of regression) is taken as the final prediction.

Note that Random Forest is known for its robustness to noise and overfitting and can handle non-linear and high-dimensional data. However, it may not be as interpretable as other data mining techniques, and may require tuning of hyperparameters such as the number of trees in the forest and the size of the random feature and data subsets used for training.

4.3 *Naïve Bayes*

Naïve Bayes is a probabilistic data mining technique that can be used for weather prediction. The algorithm works by calculating the probability of each possible outcome (such as a specific weather condition) given the input weather data. The prediction is made based on the outcome with the highest probability.

Here's an example formula for calculating the probability of a specific weather condition given the input weather data using Naïve Bayes:

$$P(C|X) = P(X|C) * P(C)/P(X) \qquad (3)$$

Where:

P(C|X) is the probability of a specific weather condition (C) given the input weather data (X)

P(X|C) is the probability of the input weather data (X) given the specific weather condition (C)

P(C) is the prior probability of the specific weather condition (C)

P(X) is the marginal probability of the input weather data (X)

5 RESULT AND DISCUSSIONS

The Random Forest algorithm appears to have the highest reported accuracy, AUC, and F1 scores, which suggests that it may be the best performing algorithm for weather prediction. However, it also has the longest train time among the three algorithms, which may be a concern for applications where real-time predictions are required.

The Naive Bayes algorithm has the shortest train time among the three algorithms, which makes it a more attractive option for real-time weather prediction applications. However, its performance metrics, such as CA, F1, and Precision scores, are lower than those of the other algorithms, which may limit its usefulness for some applications.

The Neural Network algorithm appears to have a high AUC score, which suggests that it may perform well in predicting extreme weather events. However, its CA, F1, and Precision scores are lower than those of the Random Forest algorithm, which may limit its usefulness for some applications.

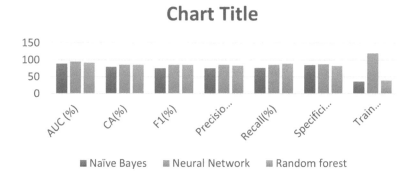

6 CONCLUSION

Weather prediction is an important area of research due to its impact on human activities and the environment. Data mining techniques have been used extensively in weather prediction, with algorithms such as Naive Bayes, Neural Networks, and Random Forest being commonly used. These algorithms work by analysing historical weather data to make predictions about future weather conditions.

The Random Forest algorithm appears to be the best performing algorithm for weather prediction, with high reported accuracy, AUC, and F1 scores in comparison to other algorithms. However, the performance of these algorithms may vary depending on the specific dataset and implementation. Overall, the use of data mining techniques for weather prediction has shown promising results, and continued research in this area may lead to improved accuracy and more reliable weather predictions.

REFERENCES

[1] Pushpa Mohan and Dr. Kiran KumariPatil: "Survey on Crop and Weather Forecasting Based on Agriculture Related Statistical Data", *International Journal of Innovative Research in Computer and Communication Engineering*, Volume 5, Issue 2

[2] Jha S. K. and Acharjya D. P., "A Survey of Data Mining Techniques for Weather Forecasting," *2012 International Conference on Computing*, Electronics and Electrical Technologies (ICCEET), 2012, pp. 108–113, doi: 10.1109/ICCEET.2012.6203897.

[3] Liu N. B. and Gao Y. H, "Weather Prediction Using Artificial Neural Networks: a Literature Review," *2016 IEEE International Conference on Mechatronics and Automation (ICMA)*, Harbin, 2016, pp. 655–660, doi: 10.1109/ICMA.2016.7558788.

[4] Kumar M. and Kumar N., "Weather Forecasting Using Decision Tree Algorithm," *2018 9th International Conference on Computing, Communication and Networking Technologies (ICCCNT)*, Bangalore, India, 2018, pp. 1–7, doi: 10.1109/ICCCNT.2018.8494024.

[5] Ahmed S. S. and Tariq R., "Weather Forecasting Using Machine Learning Techniques: A Review," *2020 2nd International Conference on Computer Science, Engineering and Applications (ICCSEA)*, Dubai, United Arab Emirates, 2020, pp. 1–7, doi: 10.1109/ICCSEA49056.2020.9140329.

[6] Liu C. and Y. Sun, "Predicting Hourly Precipitation Using Machine Learning: A Review," in *IEEE Access*, vol. 9, pp. 15520–15531, 2021, doi: 10.1109/ACCESS.2021.3050582.

[7] Rahman T. A., Nasiruzzaman A. B. M., and Saadat A. H. M., "Weather Prediction Using Data Mining Techniques: A Comprehensive Review," *International Journal of Computer Science and Information Security*, vol. 12, no. 8, pp. 23–31, 2014.

[8] Yarlagadda S. K. S. S., Raghuwanshi S. S., and Venkata V., "A Comparative Study of Data Mining Techniques for Weather Prediction," *International Journal of Computer Applications*, vol. 97, no. 12, pp. 18–25, 2014.

[9] Chatterjee S. and Paul, "Weather Prediction using KNN and SVM," *International Journal of Computer Applications*, vol. 97, no. 12, pp. 26–31, 2014.

[10] Alghamdi M. A. and Alzahrani A. H., "Weather Prediction Using Artificial Neural Networks: A Survey," *International Journal of Computer Science and Information Security*, vol. 16, no. 1, pp. 34–42, 2018.

[11] Shah S. S., Pathan S. S., and Zaidi T. A., "Prediction of Weather Using Machine Learning Techniques," in Proceedings of *2017 International Conference on Computing, Communication and Automation (ICCCA)*, pp. 1284–1288, 2017.

[12] Alharbi A. N. and Hassanien A. E., "A Hybrid Prediction Model for Weather Forecasting using Fuzzy Logic and Artificial Neural Networks," in Proceedings of *2015 International Conference on Computer, Communication, and Control Technology (I4CT)*, pp. 92–96, 2015.

[13] Shalaby S. A., El-Dahshan E. A, and Hassanien A. E., "Intelligent Weather Forecasting Using Fuzzy Logic and Neural Network," in Proceedings of *2014 IEEE 14th International Conference on Intelligent Systems Design and Applications (ISDA)*, pp. 510–515, 2014.

[14] Bhattacharyya S., Paul S., and Datta D., "A Comparative Study on Prediction of Weather Parameters Using Decision Trees," in Proceedings of *2016 International Conference on Advances in Computing, Communications and Informatics (ICACCI)*, pp. 1646–1651, 2016.

[15] Gupta N. and Soni M. K., "Performance Evaluation of Decision Tree and Artificial Neural Network in Weather Forecasting," in Proceedings of *2014 IEEE International Advance Computing Conference (IACC)*, pp. 699–703, 2014.

Blind spot monitoring for car

Aditya Ranjan & Ravi Shankar Pandey
Department of Computer Science and Engineering, Birla Institute of Technology, Mesra, Patna Campus India
ORCID ID: 0009-0004-2203-8349

ABSTRACT: Blind spot of a car is an area around the vehicle that cannot be seen while looking either forward or through side or rear view mirror. This paper introduces blind spot monitoring for vehicles using vehicle mounted cameras in order to reduce traffic accidents. We firstly describe how intelligent surveillance can be used to detect vehicle blind spots by using multiple cameras. We then describe the importance of blind spot detection including technical background. This paper provides an algorithm for vehicle detection in blind spot using Mean shift algorithm based on machine learning. In this approach, a target window is defined in an initial frame for a moving target in a video, and after that, the tracked object is separated from the back ground by processing the data within that window. After that alert message will be displayed on the screen.

Keywords: Mean Shift Algorithm, Image Processing, Blind Spot detection

1 INTRODUCTION

With the development of science and technology, distance sensing and tracking technology has been applied to many daily life facilities to promote life safety and convenience. For example, today's drivers are increasingly interested in safe driving, and automobile manufacturers are proposing many safety systems. In particular, collision warning technologies such as lane departure warning systems, vehicle rear support systems, and blind spot information systems are attracting attention. In particular, the blind spot information system greatly contributes to safe driving and is gradually gaining recognition from the market. Blind spots, as mentioned above, refer to areas that the driver cannot see through the rear view mirror. Blind spot information systems with vision solutions use cameras on both sides of the vehicle to capture blind spots and warn drivers to look out for oncoming vehicles when changing lanes. Therefore, a technique for estimating the distance of an approaching vehicle using a blind spot information system is required. Radar is another technology on the market. Radar uses ultrasonic ranging technology, infrared ranging radar, etc. to detect distance. Some vehicle radars can even detect the shape of objects. However, vehicle radars typically have a narrow field of view and blind spots. Blind spot coverage correlates with the number of vehicle radars installed. In addition, automotive radar has a limited detection range, making it difficult to detect objects moving over a wide range. Therefore, imaging technology for detecting moving objects overcomes the blind spot problem, and the driver can see images of all blind spots from the display device. According to the survey conducted for past 44 years, road accidents has increased at rapid speed. Road deaths and injuries are predicted to be the third leading contributor to the global burden of disease and injury [4]. There will be a death every 23 seconds, 43,640 monthly and 7,22,917 in a year in the world. Road deaths per thousand is highest in South East Asia in the world. The number of road traffic deaths continues to rise steadily [5]. According to the survey conducted in 2020, the total recorded 3,66,138 road accidents caused loss of 1,31,714 persons lives and injured

3,48,279 persons. Unfortunately, the worst affected age group in Road accidents is 18–45 years, which accounts for about 70 percent of total accidental deaths. When moving between lanes on the road, it is very easy for a driver to miss a car in his blind spot. This is the main motivation for the development of the Blind Spot monitoring system. To effectively reduce the number of accidents, multiple cameras should be installed and aimed at blind spots on the vehicle so that the algorithm can monitor all his events in real time. [1]. In this way, drivers are less likely to forget their blind spots leading to traffic accidents. A blind spot warning system (BSWS) based on direct vision was proposed in work described in [8]. In this paper, we will mainly focus our research on developing a blind spot monitoring system based on direct vision to detect whether there are any vehicles in the blind spots of a vehicle. from this if car will enter in the blind spot then it can be spotted according to the collected data and then warning will be displayed. The overall function of proposed work can be seen as shown in Figure 1.

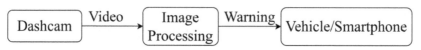

Figure 1.   Overall function of proposed work.

2   RELATED WORK

Recently, more and more car manufacturers, such as BMW has begun designing blind spot monitoring systems for its cars to prevent rear-end collisions ignorance of blind spots. There are many types of blind spot point warnings such as radar, ultrasonic and camera detection systems [6]. On an ultrasound basis BSDS have relatively low cost but detection range it tends to be relatively short, which requires more time for detection when using a car, its resolution is also relatively low [2]. Vision based system compared to other systems is more sensitive to the surrounding environment and the number of false detection (alarms) will decrease lateral resolution will increased [3]. Sensors and radars are always used in BSDS to improve detection accuracy. For example, Wong *et al.* [7] presented a system that used six ultrasonic sensors and three cameras mounted on a car to collect data about its surroundings. This system would predict if a vehicle is about to collide happens after processing the collected data. This system would predict whether a vehicle collision would happen after the collected data was processed. Visual cameras are mostly installed under the wing mirrors on both sides of the vehicle to detect moving object [12]. In 2012, a vision-based system [13] was proposed. This is a blind spot monitoring system that can detect possible vehicles by feature extraction from images captured by car-mounted cameras. This helps drivers spot potential blind spot hazards when changing lanes while avoiding vehicle collisions. Fernandez *et al.* [14] considered that blind spot monitoring could be supported using systems based on passive sensors, such as video cameras, or active sensors, including radar and laser sensors. In addition, blind spot monitoring can cover zones behind the camera that are 20 meters long and 4 meters wide on each side. In 2014, Tseng *et al.* [15] based on motion and static features he developed BSDS and divided the ground detection zone into his four main regions. Baek *et al.* [16] also performed vision-based object detection along the sides of vehicle blind spots and adapted the cascade classifier HOG for vehicle detection. Different from those existing work, in this paper, we propose an algorithm based on Machine learning, our contributions are:

(A)   data augmentation using the videos from four cameras,
(B)   Mean shift algorithm for blind spot analysis.
(C)   Outperformed results of blind spot monitoring.

3 PROPOSED METHOD

This algorithm works on the data collected from the video feed obtained from four cameras mounted on left side, right side and back of the car. it covers the four blind spots of a car located in right zone, left zone and back left zone, back right zone. It is quite difficult to observe the surrounding environment which may cause serious crashes [9]. In this method the cameras produce the target region for the algorithm then the algorithm using the computer vision covert the 2-D images obtained from the video feed in 1-D distance axis signal information which can be processed further using image processing. Now through the mean shift algorithm the danger region is identified. This region is represented as a polygon having a specified area. Now the algorithm keeps on detecting the vehicles in the close proximity of the car. if any vehicle comes in the area of the polygon the system alerts the driver regarding the potential threat of collision. The mean shift algorithm is actually a broader statistical concept related to clustering. The mean shift algorithm looks for locations in a data set where data points or clusters are concentrated. The algorithm places a kernel on each data point and sums them up into a Kernel Density Estimate (KDE). KDE has peaks and valleys corresponding to high and low data point densities. The algorithm takes a copy of the data points and shifts the points slightly to the nearest KDE peak in one iteration. The algorithm iterates over moving the point until it barely moves. This helps simplify the data and make the clusters more distinct. An image can be decomposed into the HSV color space, essentially one big data set. When I make KDE, each rainbow and blue color in the sky creates a spike. Subtle fading between colors is eliminated, making each color of the rainbow more distinct as the dots move repeatedly. As a result, each strip is more easily identifiable by a computer. This process is a form of image segmentation. The system keeps on closely monitoring the object until its out of the polygon. As soon as the object gets removed from the polygon, it no longer posses the threat of collision and thus the system removes the warning message. The Flowchart for the proposed work is shown in Figure 2.

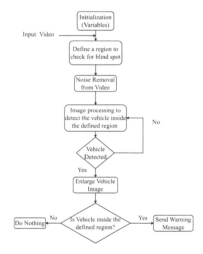

Figure 2.   Flowchart of proposed work.

4 RESULT

The aim of Blind spot monitoring for car system is to detect blind spot and alert the driver for the same. This is highly efficient and faster than any proposed methodology. This algorithm is used to develop a blind spot monitoring system for cars. To obtain the working model first Four cameras are mounted on front and rear of the car at each side. Figure 3 Depicts the positioning of the cameras mounted on the car and their coverage area.

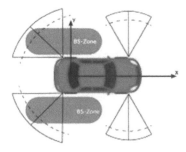

Figure 3. Positioning of cameras on the car.

These cameras constantly send the video feed to Python program working at the back end. The program performs the required image processing tasks on the obtained video feed and applies the proposed algorithm to monitor blind spots and produce the error messages to alert the driver against potential threat.

Figure 4 shows the defined boundary or region in which the vehicle detection occurs. After detecting blind spot from camera and algorithmic processing, if a car come in blind spot region then proposed algorithm will generate a warning message shown in Figure 5.

Figure 4. Blind spot for rear view camera.

Figure 5. Generate collision danger warning signal.

The experiment was conducted not only during the daytime, but also at night under different driving conditions. The system performs very well under all test conditions, as shown in Table 1 below.

Table 1. Results for blind spot monitoring system.

Test	Daytime	Night
Static	Pass	Pass
65km/h	Pass	Pass
overtake	Pass	Pass

5 CONCLUSION

This paper provided a method to detect and alert a driver when cars appear in their blind spots using videos from cameras installed in the car. Here Mean-shift algorithm is used to implement blind spot monitoring which is a highly efficient and fast algorithm as no any training data is required. It is directly processing the captured video frames. In this paper the algorithm proposed works in two parts one being the definition of target area and second, the application of mean shift algorithm for object tracking. The algorithm continuously ensures the safety of car as the driver is duly notified against potential dangers through warning messages and this serves its purpose.

REFERENCES

[1] W. Yan. *Introduction to Intelligent Surveillance: Surveillance Data Capture, Transmission, and Analytics.* Springer, 2017.

[2] R. Mahapatra, K. Kumar, G. Khurana, and R. Mahajan. Ultra Sonic Sensor Based Blindspot Accident Prevention System. In *IEEE International Conference on Advanced Computer Theory and Engineering*, pages 992–995, 2008.

[3] J. Alonso, E. Vidal, A. Rotter, and M. Muhlenberg. Lane-change Decision Aid System Based on Motion-Driven Vehicle Tracking. *IEEE Transactions on Vehicular Technology*, 57(5):27362746, 2008.

[4] Murray, C. J., & Lopez, A. D. (Eds.). (1996). *The Global Burden of Disease: A Comprehensive Assessment of Morality and Disability from Diseases, Injuries and Risk Factors in 1990 and Projected to 2020.* Boston: Harvard University Press.

[5] Jayaprakash G Hugar, Abu Waris, Mirza Muhammad Naseer, Muhammad Ajmal Khan, *Road Traffic Accident Research in India: A Scientometric Study from 1977 to 2020.*

[6] M. Ra, H. Jung, J. Suhr, and W. Kim. Part-based Vehicle Detection in Side-rectilinear Images for Blind-spot Detection. *Expert Systems with Applications*, 101:116–128, 2018.

[7] C. Wong and U. Qidwai. Intelligent Surround Sensing Using Fuzzy Inference System. In *IEEE Conference on Sensors*, volume 2, pages 1034–1037, 2005.

[8] M. Sotelo and J. Barriga. Blind Spot Detection Using Vision for Automotive Applications. *Journal of Zhejiang University – Science (A)*, 9(10):13691372, 1966.

[9] I. Arel, D. Rose, and T. Karnowski. Deep Machine Learning – A New Frontier in Artificial Intelligence Research. *IEEE Computational Intelligence*, 5(4):13–18, 2010.

[10] H. Lin, P. Liao and Y. Chang. (2018) Long-Distance Vehicle Detection Algorithm at Night for Driving Assistance. *2018 3rd IEEE International Conference on Intelligent Transportation Engineering*, Singapore. pp. 296–300.

[11] Comaniciu, V. Ramesh and Peter Meer, "Kernel-Based Object Tracking," *IEEE Transactions on Pattern Analysis and Machine Intelligence*, vol. 25, no. 5, may 2003.

[12] H. Jung, Y. Cho, and J. Kim. Integrated Side or Rear Safety System. *International Journal of Automotive Technology*, 11(4):541–553, 2010.

[13] S. Milos and L. Jan. *The New Approach of Evaluating Differential Signal of Airborne FMCW Radar-altimeter. Aerospace Science*, 17(1):1–6, 2012.

[14] C. Fernndez, D. Llorca, M. Sotelo, I. Daza, A. Helln, and S. lvarez. Real-time Vision-based Blindspot Warning System: Experiments with Motorcycles in Daytime/Nighttime Conditions. *International Journal of Automotive Technology*, 14(1):113–122, 2013.

[15] D. Tseng, C. Hsu, and W. Chen. *Blind-spot Vehicle Detection Using Motion and Static Features. International Journal of Machine Learning and Computing*, 4(6):516–521, 2014.

[16] J. Baek, E. Lee, M. Park, and D. Seo. Mono-camera Based Side Vehicle Detection for Blindspot Detection Systems. In *International Conference on Ubiquitous and Future Networks*, pages 147, 149, 2015.

Recent trends in image processing

Recent Trends in Computational Sciences – Gururaj, Pooja & Flammini (Eds)
© 2024 The Author(s), ISBN 978-1-032-42685-3

Literature review on industrial human activity and ergonomic risk analysis

Kavitha Jayaram, M.S. Brunda, Deepthi Rajakumar, M. Gautham & B.M. Bhuvan
Computer Science and Engineering, B.N.M.I.T, Bengaluru, Karnataka

ABSTRACT: Safety is one the most essential protocols and a moral responsibility that must be followed in any industry. Hazardous industries, such as fire departments, construction departments, and manufacturing units, require proper training to minimize dangerous accidents caused by heavy machinery. Some accidents that can occur are electrocution, falling from a height, or being hit by a moving vehicle. A statistical study conducted by the Bureau of Laboratory Statistics states that 31% of work-related injuries have caused Musculoskeletal Disorders. Since prevention is better than cure, we bring it to the limelight on the various works carried out for this purpose. In this article, a review of how AI models can be used for anomaly detection, feature recognition and rule-based human posture recognition is done along with the various methods to collect the data set for ergonomic analysis which can be used to broaden the scope of possible solutions to reduce industrial hazards.

1 INTRODUCTION

Work injuries are considered a major drawback in any industry that requires the construction, maintenance, or repair of structures, which reduces the efficiency and motivation of workers and causes a delay in production, which in turn results in poor feedback. Organizations that work on dangerous activities are found in huge numbers around the world and hence the probability of accidents and injuries are high. The different workplaces are designed and arranged such that their employees fit in based on the requirements. This process is known as ergonomics, which is a branch of science that deals with human limitations and abilities that can be used to collaborate with tools and environments. Industrial ergonomics is a subfield of this which deals with syncing the job requirements with the laborer's physical needs to perform it. An ergonomist analyzes the various tasks that laborers need to perform in an environment and predicts its effects on their health. This is essential for any workplace to reduce the possible risks that can occur. Identification of such ergonomically related problems was traditionally carried out using one to many sensors.

With the world's technology evolving towards artificial intelligence, the same problem can be addressed using computer vision, a sub-field of AI that deals with interpreting meaningful information from digital images or videos. At the Dartmouth Conference in New Hampshire in 1956, the scientific field of artificial intelligence was established [1]. Since then AI is being applied in varied sectors to minimize human tasks. Researchers in these fields are eager to come up with models that are robust and efficient. There are instances when AI in deployment have failed to make proper predictions. One such epic example would be the tragic diagnosis of COVID-19 and the Triage model, which was useless if not harmful due to the lack of good quality data [2].

Identifying the ergonomic situations in industrial sectors involves an analysis of the posture of individuals performing the various tasks. There is a need to know about the existing work carried out to solve this issue and identify the gap that can be filled effectively.

2 SURVEY OF POSTURE DETECTION MODELS

Weili Ding in his paper on Human posture recognition proposed a solution to identify various postures using the rule-learning algorithm. With the help of a Kinect sensor, human body Skeleton information is extracted. Figure 1(a) illustrates the joint points of the human body, which are very important for mobility. In addition, the RIPPER rule learning algorithm created a high-performance entity [3]. An overview of the steps followed for posture recognition is shown in Figure 1(b).

A method suggested by students of Amity University to detect postures from time-series data using Deep Learning. The knowledge of a long short-term model and fully connected convolutional network is discussed. Tri-axial accelerometers are attached to the back of an object using which nine different postures are identified. An average accuracy recorded by the model for classification is 99.91%. [4]

A solution is proposed in the IEEE Sensors Journal to use a hybrid model of machine learning and deep learning algorithms, resulting in an accuracy of more than 98%. ML algorithms are trained using cross-validation of 10-fold to find precision, recall, accuracy and f-measure. [5]

Students of Ewha Womans' University proposed a system which classifies the user's postures using deep learning and a specifically designed chair made of Arduino hardware. Pressure and ultrasonic sensors are used. Figure 3 shows how the dataset is collected through the sensors. [6]

A model implemented in the Philippines uses Motion capture data and RGB-D to assess a person's gait. The cameras are used to record the markers (39 retro-reflective) placed on the body parts of the subject. Compared to Mo-Cap data, RGB-D unfiltered has a 7.23% error rate, whereas filtered RGB-D has a 2.73%. [7]

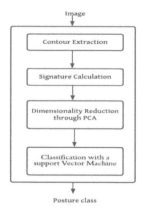

(a) Joints in Human Skeleton   (b)   Steps in Posture Recognition.

Figure 1.   Fosture recognition.

The two types of human identification algorithms are discussed in Guanjun Liu's article. One uses depth image algorithm, the other RGB models. The extensively utilised classification techniques in the field of posture recognition include the Support Vector Machine (SVM), Dynamic Time Warping (DTW), Hidden Markov Model (HMM) and Gaussian Mixture Model (GMM). Their final rate of posture recognition was 99.01 percent, which is higher than the majority of other posture recognitions. [8]

An evaluation of postures carried out by a university in China uses an intelligent assessment of the whole body with the help of convolutional pose machines. To calculate the reliability of the assessment, a motion capture system was used for comparison. An average root means squared error of 4.77% was found. Not only does this system calculate the working posture but also predicts risk levels for musculoskeletal disorders. [9]

A system developed by Varsha Bhosale, provides a method to correct yoga postures with posenet and KNN with an accuracy of 98.51%. Posenet is a deep learning algorithm that is used to identify key points within a body to train the model. Webcam is used for input video feed, which helps in acquiring real-time yoga postures. [10].

3   SURVEY ON AUTOMATION TOWARDS INDUSTRIAL SAFETY

JoonOh Seo *et al.* in their paper, have proposed a 2D image-based classification approach for the ergonomic assessment. The purpose of their model is to classify the different body postures from video or time lapsed images. The dataset consists of diverse postures created using virtual human modelling The classifier used is a machine learning algorithm, SVM (support vector machine). The results of the proposed system ranges from 78% to 88%. Demerits include lack of real-time validation and consideration of the complexity of postures from multiple viewpoints. [11]

Srimantha E.Mudiyanselage *et al.* in their paper, have discussed various machine learning models like decision trees, SVM, random forest tree and K nearest neighbor to classify the risks in postures maintained by industrial workers while performing material handling activities. The sEMG (electromyogram) sensors are used to detect harmful activities. Decision Tree algorithm outperformed the other models and a test accuracy of 99.05% was recorded. Demerits include lack of data since only one participant was used to collect training data, to be able to produce generalized results and use sophisticated algorithms with large data sets [12].

JuHyeong Ryu *et al.* in their paper have discussed an automated assessment tool which is used by systems that generate postures from motion caption wearable systems which are static. The rule-based a tests RULA, REBA, and OSWAS are performed for a bricklaying task. It was seen that higher artificial external forces can inflate the scores by rule-based tests; hence biomechanical analysis was carried out to prove that evaluation can be robust to distinguish various risks. Demerits include the scope of risk analysis for a single task, that is, bricklaying. [13]

Behnoosh Parsa *et al.* have proposed a multi task learning graph based approach to evaluate the different postures of humans. The paper describes GCN (graph convolutional network) model's effectiveness to extract spatial features from unstructured data and how the joint positions in skeletal representations can be mapped to the model's scores to solve ergonomic risk assessments. Demerits include that the performance of the model is limited to the observed activities labeled and kinematics such as velocity and acceleration, which are important to describe any injuries, are not considered. [14]

Manlio Massiris Fernández *et al.*, in their paper, have proposed a method that uses computer vision and machine learning to calculate the RULA scores (rapid upper limb assessment) from digital videos or snapshots. This score computation requires worker skeleton information. The neural network model proved to be working under acquisition conditions and for single or multiple work postures. Demerits seen are that anthropometric feature evaluations are not considered, which is useful for the prediction of the model in situations related to gender and somatotypes. [15]

Amir H. Behzadan *et al.* in their paper, have proposed an approach using machine learning with the help of mobile sensors to perform ergonomic risk analysis and productivity analysis. They have built a classifier model to record and identify the different activities like push, pull, inspect and load which occur in construction sites. The smart phone sensors are used to produce the activity time and frequencies which are used for the risk analysis. About 80% of accuracy was recorded for 7 out of 10 activities tested. The dataset consists of one worker and one inspector, this can be considered a demerit, hence for further enhancement of the model a greater number of workers would be needed. [16].

4   SUMMARY OF LITERATURE SURVEY

From the above survey, we observed that the on field worker's safety in Industry 4.0 has been an area of research recently. Various combinations of sensors and machine learning and deep learning models were tested to monitor the ergonomic risks of the worker.

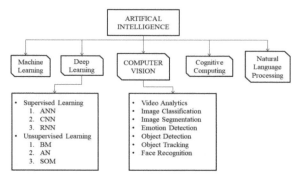

Figure 2.   Overview of AI concepts.

The Figure 2 displays the various concepts that come under the scope of AI, among which Deep learning with computer vision are largely explored fields, to solve many problems, which involve humans to see and analyze a situation, by replacing it with a computer. The complex problem of analyzing ergonomics of workers and making suitable predictions can be broken into two parts: one being continuous posture analysis in the surveillance video which is used for monitoring with the help of SVM or CNN (used in Pose Net model) and other being analyzing the risks involved with the help of classifier which classifies based on the risk calculated using the NOISH equation. The sequence flow is as shown in the Figure 3.

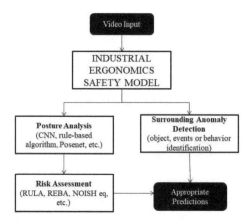

Figure 3.   Overview of steps involved in industrial ergonomics safety analysis.

5   CONCLUSION

Industrial worker's health is extremely crucial from both the industry as well as the worker's perspective. Companies take preventive measures to ensure ergonomics of employees by providing effective training for them to adapt to their job profiles. Even then, there can be situations where the employees consciously or subconsciously follow improper ergonomics which can jeopardize their health in the long run. This article provides a review of various research and work carried out on how to automate the monitoring of worker health and ergonomics. Finally, the paper has highlighted the method of approach for early detection of improper ergonomics of workers in various industries.

ACKNOWLEDGEMENT

We would like to thank everyone from Siemens Organization (STSPL, Bengaluru) for helping us with their valuable suggestions to complete this survey.

REFERENCES

[1] John McCarthy, Marvin L. Minsky, Nathaniel Rochesterand Claude E. Shannon, "A Proposal for the Dartmouth Summer Research Project on Artificial Intelligence," *AI Magazine* Volume 27 Number 4 (2006) (© AAAI)

[2] Epic AI Fails — A List of Failed Machine Learning Projects." *Analytics India Magazine*, 4 November 2022, https://analyticsindiamag.com/epicai-fails-a-list-of-failed-machine-learning-projects/. Accessed 8 December 2022

[3] Ding, W., Hu, B., Liu, H. *et al.* "Human Posture Recognition Based on Multiple Features and Rule Learning," *Int. J. Mach. Learn. Cyber.* 11, 2529–2540(2020). https://doi.org/10.1007/s13042-020-01138-y

[4] R. Gupta, D. Saini and S. Mishra, "Posture Detection Using Deep Learning for Time Series Data," *2020 Third International Conference on Smart Systems and Inventive Technology (ICSSIT)*, 2020, pp. 740–744, doi: 10.1109/ICSSIT48917.2020.9214223

[5] S. Liaqat, K. Dashtipour, K. Arshad, K. Assaleh and N. Ramzan, "A Hybrid Posture Detection Framework: Integrating Machine Learning and Deep Neural Networks," in *IEEE Sensors Journal*, vol. 21, no. 7, pp. 9515–9522, 1 April 1, 2021, doi: 10.1109/JSEN.2021.3055898.

[6] H. Cho, H.-J. Choi, C.-E. Lee and C.-W. Sir, "Sitting Posture Prediction and Correction System using Arduino-Based Chair and Deep Learning Model," 2019 *IEEE 12th Conference on ServiceOriented Computing and Applications (SOCA)*, pp. 98–102, doi:10.1109/SOCA.2019.00022.

[7] K. V. G. Castillo, N. K. T. Mendoza, C. A. S. Morales, A. D. A. Perez, J. Y. L. Unisa, and A. R. d. Cruz, "Denoising of Spatiotemporal Gait Waveforms from Motion-Sensing Depth Camera using Least Mean Square (LMS) Adaptive Filter," *IEEE 10th International Conference on Humanoid*, 2018, pp. 1–6, doi: 10.1109/HNICEM.2018.8666294

[8] Guanjun Liu, Liyuan Lin, Weibin Zhou, Rui Zhang, Hongyi Yin, Jingyu Chen, Hongfei Guo, "A Posture Recognition Method Applied to Smart Product Service," *Procedia CIRP*, Volume 83, 2019, Pages 425–428, ISSN 2212-8271, https://doi.org/10.1016/j.procir.2019.04.145.

[9] Z. Li, R. Zhang, C.H. Lee, and Y.C. Lee, "An Evaluation of Posture Recognition Based on Intelligent Rapid Entire Body Assessment System for Determining Musculoskeletal Disorders," *Sensors*, vol. 20, no. 16, p. 4414, Aug. 2020, doi: 10.3390/s20164414

[10] Bhosale, Varsha & Nandeshwar, Pranjal & Bale, Abhishek & Sankhe, Janmesh. (2022). "Yoga Pose Detection and Correction using Posenet and KNN."

[11] JoonOhSeo and SangHyunLee, "Automated Postural Ergonomic Risk Assessment Using Vision-based Posture Classification," *2021 published by ELSEVIER, Automation in Construction*, Vol.128 August 2021, 103725

[12] Srimantha E. Mudiyanselage, Phuong Hoang Dat Nguyen, Mohammad Sadra Rajabi and Reza Akhavian, "Automated Workers' Ergonomic Risk Assessment in Manual Material Handling Using sEMG Wearable Sensors and Machine Learning," *Electronics* 2021, vol 10, issue 20, doi: https://doi.org/10.3390/electronics10202558

[13] JuHyeong Ryu; Mohsen M. Diraneyya; Carl T. Haas, F. ASCE and Eihab Abdel-Rahman, "Analysis of the Limits of Automated Rule-Based Ergonomic Assessment in Brick-laying," 2021 *Journal of Construction Engineering and Management*, 147(2), 04020163, doi:10.1061/(asce)co.1943-7862.0001978

[14] B. Parsa and A. G. Banerjee, "A Multi-Task Learning Approach for Human Activity Segmentation and Ergonomics Risk Assessment," *2021 IEEE Winter Conference on Applications of Computer Vision (WACV)*, 2021, pp. 2351–2361, doi: 10.1109/WACV48630.2021.00240

[15] Manlio MassirisFernandez, J. Alvaro Fern andez, Juan M. Bajo, Claudio A. Del-rieux, "Ergonomic Risk Assessment Based on Computer Vision and Machine learning," *Computers & Industrial Engineering*, Volume 149, 2020,106816, ISSN 0360-8352, https://doi.org/10.1016/j.cie.2020.106816

[16] Nath, Nipun and Behzadan, Amir. (2017). Construction Productivity and Ergonomic Assessment Using Mobile Sensors and Machine Learning. 434–441. 10.1061/9780784480847.054

Recent Trends in Computational Sciences – Gururaj, Pooja & Flammini (Eds)
© 2024 The Author(s), ISBN 978-1-032-42685-3

Smart greenhouse: A bliss to the farmers of the cold arid region of Ladakh

S. Angchuk
Krishi Vigyan Kendra, Leh Ladakh

Aleem Ali
Department of Computer Science & Engineering, Uie, Chandigarh University, Punjab

ABSTRACT: Ladakh, the cold region of India is blessed with 300 cloud free days. The use of solar energy in a protected structure like a greenhouse is an ideal way to tap this natural energy to grow vegetables during the peak winter when the outside temperature drops to $-25°C$. But high temperature variation during day and night hampers the growth and yield of the vegetables in the greenhouse. Smart agriculture using IoT can be helpful to overcome these drawbacks. Using sensors of temperature controller, humidity, soil moisture will help in maintaining the temperature inside the greenhouse. The paper discusses the application of various IoT sensors in the greenhouse.

Keywords: Greenhouse, IoT, Smart Agriculture, Sensors, Cold Arid, Solar energy

1 INTRODUCTION

Greenhouse technology is the technique of growing crops by providing favourable environment growing conditions to the plants. It is used to guard crops from unfavourable weather conditions and to create the ideal environment for plant growth and maximum output, including the right amount of soil, light, temperature, humidity, carbon dioxide etc.Solar greenhouses offer immense scope and potential for farmers to utilize this technology in regions like Ladakh, which has 300 sunny cloud-free days each year [1]. It is difficult to visually monitor plant growth and manually turn on and off temperature controllers in a traditional hand-operated greenhouse. The ideal way to address these issues is by adopting an IoT-enabled Smart greenhouse. IoT agricultural sensors can provide farmers the crop production information, rainfall forecasting, soil nutrition, pest infestation, and other useful information. Agricultural sensors can be used to measure a variety of field factors, including temperature, humidity, soil moisture etc. Farmers can have an access of their field conditions from anywhere and interpret the data accordingly [2].

2 LITERATURE REVIEWED

Ravi Kishore Kodali *et al.* 2016 [3] "IoT based Smart Greenhouse". IEEE Region. 10th Humanitarian Technology Conference (R10-HTC). In this study, they developed a Smart Greenhouse model that enables farmers to carry out field work automatically without the need for intensive manual inspection. A water monitoring sensor, sensors that measure temperature and humidity, and a fogger are used to control temperature and air humidity.

Somnath *et al.* 2018 [4] "Smart Greenhouse using IoT and Cloud Computing", International Research Journal of Engineering and Technology. They have proposed the Smart Greenhouse farm employing IOT and cloud computing technology in this study. To operate the farm, several sensors like temperature, soil moisture, and sunshine are used. In addition to these sensors, an analogue to digital converter, a microcontroller and actuators

DOI: 10.1201/9781003363781-45

are used. When the defined climate variables cross the threshold, the sensors detect a change. Farmers can readily monitor the greenhouse farm once they have this information. The fan, water pump buzzer, and lights can all be conveniently monitored without any physical intervention. This system is being utilised to eliminate the system's issues by minimising human intervention to the greatest extent practicable.

Mohammad WoliUllah *et al.* 2018 [5], "Internet of Things Based Smart Greenhouse: Remote Monitoring and Automatic Control", International Conference on Electric and Intelligent Vehicles. They provide information regarding the system that monitors temperature and humidity, soil moisture, and takes action based on the data in this paper. It provides a database that will be useful for future analysis and reports. This device is ideal for deployment in locations such as the North Pole and cold climate countries where humans reside but plants do not grow owing to harsh winters.

RitikaSrivastava *et al.* 2020 [6] "A Research Paper on Smart Agriculture using IoT" International Research Journal of Engineering and Technology (IRJET). In this study, they have devised an automated method for smart agriculture which requires less time and resources than manual labour. This system makes advantage of Internet of Things technologies. The technology also detects water levels of the field and also the moisture content of the soil.

Shaweta Sachdewa & Aleem Ali, 2022 [7] "An Overview of Blockchain as a Security Tool in the Internet of Things", 2022.Blockchain – Principles and Application in IoT. In this paper they have discussed that the Internet of Things (IoT) can be linked to numerous smart devices in order to gather data from them and make them intelligent enough to make subsequent decisions. IoT prevents a risk to privacy and security by a feature called security by design – Block chain (BC) can assist to address with its security.

Shaweta Sachdewa & Aleem Ali, 2022 [8] "Rise of Telemedicine in Healthcare Systems Using Machine Learning: A Key Discussion". Internet of Healthcare Things – Machine Learning for Security and Privacy. Another important use of this technology is in Medical Department. Using the smart technology, medical service can be provide to the patients living in a far flung area by a specialized doctor from a metro cities through IoT.

Parveen, Nazia *et al.* 2021 [9] "IOT Based Automatic Vehicle Accident Alert System." IEEE 5th International Conference on Computing Communication and Automation. In this paper, they have shown the use of this technologies to locate the location of accident and the information of the location can be send through the GPS to the emergency offerings for assistance.

3 EXISTING AGRICULTURE PRACTICES

Like rest of state, agriculture is the main source of livelihood in Ladakh.Lack of information about IoT in agriculture with traditional manual methods and out-of-date equipment hinders production, limiting farming earnings in these region. Using traditional agricultural techniques, the farm requires more labour to achieve a good crop output. In order to achieve high crop yield, we need to protect the crop against natural calamities like high variation in day and night temperature in the greenhouse, high wind, weather etc. With the use of IoT, an innovative agriculture system is proposed for converting loss-making traditional farming into high crop yielding and profit-making in the greenhouse.

4 PROPOSED SYSTEM

The suggested system focuses on monitoring the greenhouse using sensors such as humidity, temperature, and soil moisture; Light Dependent Resistor is used to detect light intensity of the farm and Infraredsensor senses the humans and other animals by their body temperature and notifies the user via message format to their mobile device. These sensors serve as an interface to the Arduino-UNO processing module. The LCD is used to display the status of different sensors on the screen. When there is a change in temperature condition, the sensor detects and turns ON the DC and cools down the condition. The DC fan turns OFF as the temperature returns to normal. LDR (Light Dependent Resistor) is used to detect the light intensity in the farm. When the light intensity is less on the farm, the LDR senses the condition and turns ON the bulb. The bulb will switch OFF when the required level of light. The soil moisture sensor is used to sense the moisture level in soil (water level) when the water levels are reached low in the ground. The ground gets dry, and the sensor detects it, then turn ON the DC water pump. When floor gets moisturized, the DC water pump will turn OFF. The user can monitor these conditions in mobile phone with the help of Wi-Fi module through IOT mobile site [10].

5 SENSORS USED IN AGRICULTURAL AGRICULTURE FARMING

Sensors are useful components in IoT. Sensors are devices that collect and analyse data in order to do get an high yield of the crop. Smart farming, also known as precision agriculture, is a solution to many of the challenges that farmers confront. They can increase yield while using fewer resources such as seeds, fertilisers, and water. Sensors can assist you in analysing soil requirements based on sensor readings, and actuators can subsequently administer the equivalent amount of nutrients or water to the soil. It also helps in maintaining the optimum temperature inside a greenhouse. With the introduction of the Global Positioning System (GPS) and other sensors, it is now possible to simply generate a map of the field and assess or estimate crop yield in the field.

5.1 *Soil moisture sensor*

Soil moisture sensor helps in determining the volumetric water content of soil (Figure 1). It determines the soil moisture content of the soil using soil parameters such as resistivity, dielectric constant, and neutron interaction as well as external inputs such as soil type, temperature, and electrical conductivity. This device has two probes that are placed in the field, and when current flows through the probes, moisture% is calculated based on resistivity. Soil moisture measurement allows water to be applied just when needed, reducing wastage of water.

5.2 *Temperature sensor*

A temperature sensor is a device that detects whether an object is hot or cold. This sensor shows more accurate result than the thermistor which was used to measure the temperature earlier. As it has three terminals: input, output, and ground, the sensor is susceptible to overheating. Temperature sensors are available in a wide range of designs and sizes. The LM-35 IC, shown in Figure 2 is one type of temperature sensor.

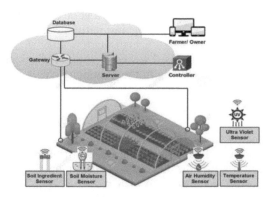

Figure 1.   Use of IoT in agriculture.

Figure 2.   Soil moisture sensor [11].

5.3   *Private infrared (PIR) sensor*

Anything that has a temperature higher than absolute zero emits heat energy in the form of radiation. Figure 3 depicts a PIR (private infrared) sensor capable of detecting infrared radiation emitted or reflected by an object. It is used to track the movements of humans, animals, and other objects. When an obstruction passes through the field, the temperature at the place rises above room temperature. The sensor converts it to voltage, which enables the detection.

Figure 3.   Water level sensor [11].

5.4   *Water level sensor*

A sensor for measuring the amount of water or other liquids is shown in Figure 4. It has a detecting probe capable of measuring the surface level of nearly any fluid, including water, salty water, and oils. This sensor is not easily broken, and can be connected to Aurdino with ease. It consist two buttons: one to record the lowest fluid level and the other one to record the highest fluid level. The level will be determined by the voltage.

Figure 4.   pH sensor [11].

5.5   *pH sensor*

Figure 5 shows a pH sensor which is used to determine the pH of a solution. The pH scale varies from 0 to 14, with 0-6 representing acidic, 7 neutral and 8-14 indicating non-acidic or basic. It computes the pH value based on the concentration of hydrogen ions measured by the pH electrode. The response time of this sensor exceeds 2 minutes. The temperature range

Figure 5. Temperature sensor [11].

Figure 6. PIR sensor [11].

is around 600 degrees Celsius with an input voltage of 5 volts and an output voltage of 414.12 volts.

5.6 *Temperature and humidity sensor*

A simple low-cost digital temperature and humidity sensor (DHT11) is shown in Figure 7. This sensor has two parts: a capacitive humidity sensor and a thermistor. The humidity sensor detects, measures and reports both the moisture and air temperature. The temperature ranges from 0°C to 50°C, while humidity levels range from 20% to 90%. Majority of these sensors are applied in the Internet of Things. There are many additional sensors besides these, but DHT11 is the most popular temperature sensor.

Figure 7. Temperature and humidity sensor [11].

6 MOBILE AS A TOOL

A smartphone can be used for a wide range of field applications, such as crop and soil observation by taking images of the crop and soil, determining the colour of the soil and leaves and detecting problems using GPS [12]. The user can view and control the field from any location using the android mobile application, commonly known as the android app.

7 CONCLUSION

IoT based applications have become very popular with the emerging technology in the last decade. The adoption of wired and wireless monitoring and management systems in greenhouse or agriculture is one of the best uses of IoT technology. By utilizing IoT in the greenhouse, one may enhance crop productivity, monitor weather conditions such as humidity, temperature and other parameters like moisture and dryness of the soil can be detected and modify the essentials required for the farm. We can detect pests and pedestrians in the field using an infrared sensor. Sensors and microcontrollers can be interfaced with each other via IoT and wireless communication between sensors. This can help the farmers to overcome the problems posed by the weather. Farmers can therefore monitor farm conditions using mobile devices or laptops. These systems can provide high crop yielding and output results. It can also reduce the labour work on the farm.

REFERENCES

[1] KunzangLamo, Parveen Kumar, D. Namgyal, S. Angchuk, Nasreen F Kacho. "Protected Cultivation: Indispensible for Cold Arid Ladakh", 2020. *International Journal of Advances in Agricultural Science and Technology*, Vol.7 Issue.1, pg. 75–80.

[2] Vivek Bhatnagar1, Gulbir Singh, Gautam Kumar, Rajeev Gupta, "Internet of Things in Smart Agriculture: Applications and Open Challenges", 2020. *International Journal of Students' Research in Technology & Management*. Vol 8, pp 11–17.

[3] Ravi Kishore Kodali, Vishal Jain and SumitKaragwal, "IoT Based Smart Greenhouse". 2016. *IEEE Region. 10th Humanitarian Technology Conference (R10-HTC)*.

[4] Somnath D. Bhagwat, AkashI. Hulloli, SurajB. Patil, Abulkalam. A. Khan,. A. S. Kamble, "Smart Greenhouse Using IoT and Cloud Computing", 2018. *International Research Journal of Engineering and Technology*.Vol 5, pp. 2330–2333

[5] Mohammad WoliUllah, Mohammad GolamMortuza, MdHumayanKabir, Zia Uddin Ahmed, Sovan Kumar DeySupta, Partho Das "Internet of Things Based Smart Greenhouse: Remote Monitoring and Automatic Control", 2018. *International Conference on Electric and Intelligent Vehicles*.

[6] RitikaSrivastava, Vandana Sharma, Vishal Jaiswal, Sumit Raj, "A Research Paper on Smart Agriculture Using IoT", 2020. *International Research Journal of Engineering and Technology (IRJET)*. Vol 7. pp. 2708–2710.

[7] ShawetaSachdewa, Aleem Ali, "*An Overview of Blockchain as a Security Tool in the Internet of Things*", 2022. Block chain – Principles and Application in IoT. Ist Edition. pp. 3–20

[8] ShawetaSachdewa, Aleem Ali, "Rise of Telemedicine in Healthcare Systems Using Machine Learning: A Key Discussion", 2022. *Internet of Healthcare Things: Machine Learning for Security and Privacy* (Spriner). 295–310

[9] Parveen, Nazia, Ashif Ali, and Aleem Ali, "IOT Based Automatic Vehicle Accident Alert System." 2020. *IEEE 5th International Conference on Computing Communication and Automation (ICCCA)*, pp. 330–333.

[10] CH Nishanthi, Dekonda Naveen, ChiramdasuSai Ram, KommineniDivya, Rachuri Ajay Kumar, *"Smart Farming using IoT"*

[11] G. Balakrishna, NageswaraRaoMoparthi, "Study Report on Indian Agriculture with IoT", 2020. *International Journal of Electrical and Computer Engineering (IJECE)*, Vol. 10, pp. 2322–2328.

[12] KamleshLakhwani, HemantGianey, NiketAgarwalandShashank Gupta. Development of IoT for SmartAgriculture a Review", 2019. *Advances in Intelligent Systems and Computing*, 841

Recent Trends in Computational Sciences – Gururaj, Pooja & Flammini (Eds)
© 2024 The Author(s), ISBN 978-1-032-42685-3

Real-time CAPTCHA using hand gesture recognition for highly secure websites

Lasya Priya, Beru Neha, B. Vivek Sai Chinna, Sai Sujay Reddy Balireddygari &
K.N. Divyaprabha
Department of Computer Science, PES University, Bangalore, India

ABSTRACT: Completely Automated Public Turing Tests to Tell Computers and Humans Apart (CAPTCHAs) are now almost a routine security measure which protect against unwanted and malicious bot programmes on the Internet. There are several security threats on websites and it would be a major risk to the nation if defense websites or other classified material were exposed. Several algorithms are in place to solve CAPTCHAs automatically. The true definition of CAPTCHA is that it must be able to determine that humans, not computers, are attempting to get into a password-protected account. Our work provides an efficient and highly secure alternative to classic CAPTCHA. The objective is to essentially device a two-step "bot-proof" authentication process with camera access as a requirement. According to studies, implementing Multi-factor authentication makes a specific account 99.9% less likely to be penetrated, and similarly building a two-level CAPTCHA would surely improve the security of the user being attacked. The first level involves a Text-CAPTCHA and in the second level, a new CAPTCHA technique that recognizes user hand gestures in real time aids in preventing the possibility of algorithm breaking by attackers. Hand gesture detection was first implemented using Support Vector Machine (SVM) and improvised with Convolution Neural Network (CNN). Finally, Mediapipe assisted and produced faster and accurate results in real time. The authentication will be granted to the website if the human performs the given task successfully in the second level.

Keywords: Hand Gesture Recognition, SVM, CNN, Mediapipe, Web Security

1 INTRODUCTION

This work proposes a simple, real-time and efficient two-step bot-proof process of authentication. The complete architecture of our work is as shown in Figure 1. The user needs to first login to the website and enter the credentials correctly. Text-based CAPTCHA [1] is used in the initial authentication phase to differentiate harmful bots from people. Background interference, noise lines, and geometric modifications are just a few of the resistance mechanisms that are added to CAPTCHA in order to increase security. To make it harder, the alphabets and numbers are distorted and rotated. The computer generates a string of six characters. The user must submit an input that matches the computer-generated string to advance to the second level. The user gets two attempts to enter the right answer in either case, failing to do which, the access to the website will be denied. However, all these protection systems have been dismantled thanks to the development of deep learning algorithms. The text-based CAPTCHA's security is frequently shielded by several mechanisms to hinder machine recognition [2]. Most prior attacks against text CAPTCHAs [1] have used a three-step strategy: pre-processing, segmentation and recognition. Thus, the introduction of second level hand gesture recognition.

DOI: 10.1201/9781003363781-46

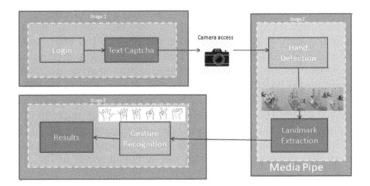

Figure 1.   High level design diagram.

The second level is a hand gesture recognition algorithm based on computer vision. The goal of gesture recognition study is to develop a system that can identify and manage applications that use human motion. Gesture recognition systems can be divided into two basic categories: vision-based and data-glove based [3]. Detection, tracking, and recognition are the three core components of most fully integrated hand interactive mechanisms that serve as the foundation of vision-based hand gesture recognition systems [4]. The scope of this work is that it needs camera access to pass the second step of CAPTCHA successfully and hence it is mainly for websites that require high security. A camera captures a live video stream, which is then used to create a snapshot via an interface. Only a few hand motions have been programmed into the system. After that, it is given a test gesture, which the system attempts to recognize. Several algorithms were utilized and evaluated in order to find the most accurate algorithm.

SVM [5] and CNN [6] algorithms were implemented by generating our own dataset of around 2000 images containing six gestures. Each gesture differs based on the number of fingers displayed by the user. When compared to SVM, CNN has a higher accuracy of around 98.3%. Finally, MediaPipe [7] was used for second level CAPTCHA, which relies on OpenCV [8] for video data handling. Each frame of the webcam video capture is used to run the MediaPipe Hands process function in the implementation. The outcomes give each hand detected a 3D landmark model [9] for each frame. Palm Detection model and Hand Land mark model are used which gives fast, accurate and real-time results

2   LITERATURE SURVEY

A comparison of various CAPTCHA algorithms for securing web pages is shown in [10].

1) Text CAPTCHA—After entering the CAPTCHA text, users can log in to the secure website, making the content easy to view. High character collapsing may be difficult for users to understand.
2) Image CAPTCHAs are easy for users to complete since, after verifying the image, they can quickly access a secure website. We may now finally access the secure website. Bright backgrounds are the only drawback to image-based CAPTCHA.
3) Audio CAPTCHA is extremely helpful for blind persons. The audio-based CAPTCHA presents difficulties for blind users when the network is unreliable or offline.

In [11] the proposed model, the CAPTCHA is displayed on screen leading to image acquisition, pre-processing of the input, matching the content with the database and finally displaying the result. Morphological Filtering filters out the noise from the binary image. There may be some parts of the hand showing 0 and background as 1. Edge detection gets the edge of the image to determine the shape Canny edge detection algorithm is used to eliminate the risk of multiple responses. 20 distinct people's hand gestures for training and testing the model, five individual gestures and a hundred photos are taken and kept in the database. The model has given an accuracy of 80% with the algorithms which has been mentioned.

The steps followed in [12] is to first take a frame of video and performing hand segmentation, tracking and recognition. Hand gesture segmentation analyses the photos and creates a Gaussian mixture model based on skin colors and additionally, it classifies hand movements using an AdaBoost classifier based on Haar features. The grayscale value change in the image is reflected by the Haar feature. The feature model is made up of the black and white rectangle portions. A total of 16000 hand gesture pictures are used in the experiment, which employs hand gestures ranging from 1 to 10 against a dynamic background. The test set then consists of 400 images from each of the 4000 categories of hand actions. Hand gesture identification has a 98.3 percent accuracy rate on average.

[7] Mediapipe offers a wide range of results within the holistic channel and it consists of three factors: pose, hand and face. The aggregate of 543 distinct landmarks is obtained using this holistic approach. The Mediapipe Python Module's Holistic Model for Mediapipe is what we are utilizing in which the function changes the image's BGR to RGB format before storing the result. This function returns the concatenated array of all the arrays containing the key point coordinates of the holistic model. The first step was data collection through our own data-set for the training and testing of deep learning model. The next step was to define an array of 10 gestures that would be used to train our recognition algorithm. Then collecting the videotape data for each gesture and use the extract keypoints function to collect the array of key points for each frame. The LSTM model is successfully operating with respect to the test dataset with an accuracy of 90%.

3   PROPOSED METHODOLOGY

The suggested approach considers the difficulties that earlier models experienced and works to reduce them. We created a system that does not trade off performance for efficiency. Our initial approach involved hand detection, followed by segmentation of the palm and fingers, and then hand gesture identification. The separation of the hands from the background and the non-uniform background were two of the most significant problems encountered. Later, SVM and CNN algorithms were used to overcome this drawback. We created our own dataset using OpenCV which was used for the above algorithms mentioned. CNN produced better results compared to SVM but having the need to use a dataset and store data was inefficient and did not produce real-time results. Using Google's Mediapipe solution and the OpenCV package, which creates the essential algorithms for analyzing the images and real-time movement recognition, we were able to solve that issue. It is used to recognize hand gestures in real time as well as to recognize landmarks on the hand. For the detection and recognition of hand gestures, several methods were employed.

3.1   *Hand detection*

The hand can be distinguished from other moving objects by its color. The skin's color is determined using the HSV model. The hue, saturation, and value (HSV) values for the skin tone are 315, 94, and 37, respectively.

Fingers and the palm can be segmented with ease using the palm mask. A matrix of 0s and 1s will be present. Once the palm point and wrist point have been established, an arrow can be drawn from the palm point to the middle point of the wrist line at the base of the hand. An algorithm for labeling is used to label the finger areas. Only the sections of sufficient size are considered fingers and are retained. The smallest bounding box to enclose each remaining region is a finger.

The centers of the fingers and the palm point are aligned. After that, the distances between these lines and the wrist line are calculated. The thumb will be seen in the hand image if the angle is less than fifty degrees. The palm line is initially explored to locate and identify the other fingers. The lines at the wrist and the palm are parallel. Once the fingers have been detected and identified, a straightforward rule classifier may be utilized to recognize the hand motion.

3.2 *Support Vector Machine (SVM)*

One of the supervised classification techniques is SVM. To make it function, a hyperplane is made with the greatest possible margin between it and the data, dividing each class of data. SVM is capable of handling learning tasks with a high number of features. That is the reason why we decided to use SVM as one of the algorithms in our data. When the sample can be separated linearly, SVM is the best classification plane that has been suggested. The best classification surface is created by extending the best classification line into high-dimensional space. Because SVM has numerous distinct advantages in tackling small sample, nonlinear, and high dimensional pattern recognition problems, it is employed in gesture recognition.

3.3 *Convolution Neural Network (CNN)*

CNN is a Deep Learning algorithm that can take in an input image, rank various items within the image, and distinguish between them. CNN is more widely used in the field of recognition and produces better results than other techniques. The images in the dataset first go through pre-processing. To facilitate predictions, images from the dataset are transformed to a suitable color palette.

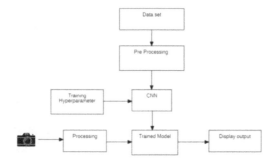

Figure 2.   Workflow of implementation using CNN.

Then, the images are loaded and ready for the algorithm training. For improved outcomes, the hyperparameters are tuned. Training and test sets of images are created. High precision is attained after several execution epochs. On providing the camera access, the results are displayed by processing and using the trained CNN model as shown in Figure 2.

The pseudo-code and general implementation of the proposed work is as follows:

//User needs to access a highly secure website
//User first fills in login info and the presented text captcha If (user login details or text captcha response) not right:
Re-enter respective details
If (level 1 passed within 2 attempts):
//Proceed to level 2
User MUST provide camera access
If (user replicates gesture image within 2 attempts):
//Provided access to his/her account Else:
//Denied access after level 2 captcha Else:
//Denied access after level 1 captcha Exit

3.4 MediaPipe

It is an open-source framework for creating machine learning processing pipelines for time-series data, including audio and video. 21 hand landmarks were detected and tracked, and the results are shown in Figure 1 for each x, y, and z axis. When a hand is found, it is immediately tracked more frequently for improved accuracy and temporal resolution.

Figure 3 displays the outcomes of the detection and tracking of each of the 21 x, y, and z hand landmarks. With the help of the coordinates generated by the hand landmarks model, we can compare the relative positions of different parts of each finger to detect whether it is folded or open. It is used to accurately show the number of fingers open in real-time speed. OpenCV's landmark model to detect and count the number of fingers a user is trying to show. This is done with the help of the coordinates in the landmark model. It is used to detect and show which fingers are open.

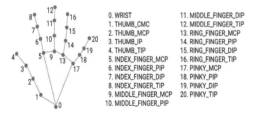

Figure 3.   Hand landmark model.

The experimental results illustrate that the number of fingers is accurately shown with real-time speed. Even the issue of the user unsteady is taken care of with the accuracy of the model. Other issues such as unstable video input and other disturbances in the background are taken care of by the model. Each frame of the webcam video capture is used to run the MediaPipe Hands process function in the implementation. The outcomes give each hand detected a 3D landmark model for each frame.

Gesture is recognized by the user by taking note of all the open fingers. Open fingers are recorded at every frame using finger coordinates obtained with the landmark model. These detected finger names are displayed in real time on the screen to aid the user further. The user then captures the webcam input after recreating the gesture at any time the user is

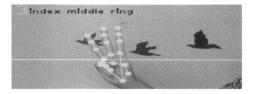

Figure 4.   Fingers open in question: ['index','middle'] fingers open in screencapture: ['index','middle','ring'].
WRONG ANSWER! TRY AGAIN.

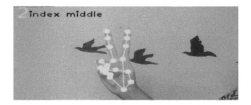

Figure 5.   Fingers open in question: ['index','middle'] fingers open in screencapture: ['index','middle'].
CORRECT ANSWER!

comfortable with, within the 20-second constraint. Using the above obtained information about the open fingers, we then verify whether the open fingers in the user's screen capture match the open fingers in the question. The set of detected open fingers is then compared to open fingers in the gesture. The user is supposed to recreate as shown in Figures 4 and 5, in order to check if the answer in correct or wrong and provide authentication to the website if correct.

4 RESULTS AND DISCUSSION

Six different hand gestures totaling 2450 photos were divided into training and testing data at the ration 80:20 for the SVM and CNN models, respectively. Images from the dataset are formatted to a suitable color palate to make predictions easier. On passing dataset through the SVM model, the overall accuracy was about 89.14% and for CNN model, accuracy of 99.362% is achieved after multiple epochs of execution as shown in Figure 6. In terms of accuracy and recall, the CNN model is 1% more accurate than the SVM model, although both model's rates of precision are equal. Mediapipe module offered by python helps in plotting landmarks on a detected hand. It detects outline of the user's hand and was seen to be accurate most of the time for the right hand. The model has an average accuracy of 95.7% overall. Since we altered it for only the right hand, it provides better accuracy. Since it gives real time gesture recognition, it proves to be a better alternative for SVM and CNN.

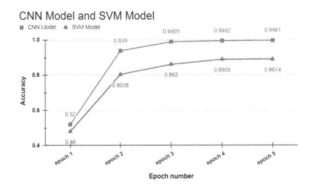

Figure 6.   Accuracy of SVM vs CNN plot.

We exclusively used real-world images for evaluation. Training using a big synthetic dataset lowers visual between frames in addition to improving quality. Based on this fact, we draw the conclusion that our real-world dataset can be increased for better generalization. We're aiming for real-time performance on extremely secure websites. MediaPipe Hands is a high-quality hand and finger tracking system. It offers a fantastic compromise between speed and quality.

5 CONCLUSIONS AND FUTURE WORK

Real-time vision-based hand gesture identification is one of the most challenging research issues in the field of human-computer interaction. An effective Hand Gesture Based CAPTCHA system was proposed and was implemented. In this method, a form containing a CAPTCHA image of a hand gesture from the database was displayed along with a message asking the user to copy the gesture. The user then made a gesture in front of the systems camera out of conscience, and the model checked to see if it matched the gesture in the image or not. As it is impossible for the robot to learn how to copy the identical move as displayed in the image on its own, using this approach would secure the identification of the user.

321

The experimental data demonstrates that our system will function better when the input image has good quality. It is safe and incredibly difficult to crack. This can also be expanded to a client-server-based architecture, where a CAPTCHA processing server is built up with an API to embed created CAPTCHA on different websites and carry out human or bot verification CAPTCHA.

REFERENCES

[1] Ping Wang, Haichang Gao; Ziyu, Zhongn Shi, and Yuan, Jiangping Hu *Simple and Easy: Transfer Learning-Based Attacks to Text CAPTCHA*, 2020.

[2] Jun Chen, Xiangyang Luo, Yanqing Guo, and Yi Zhang, Daofu Gong *A Survey on Breaking Technique of Text-Based CAPTCHA*. Hindawi, 2017.

[3] Lin Guo and Zongxing Lu, and Ligang Yao *Human-Machine Interaction Sensing Technology Based on Hand Gesture Recognition: A Review*, 2021.

[4] Rautaray S S and A. Vision-based Hand Gesture Recognition for Human Computer Interaction: A Survey. *Artif.Intell.Rev*, 43(1):1–54, 2015.

[5] Kai-Ping Feng. Fang Yuan Static Hand Gesture Recognition Based on HOG Characters and Support Vector Machines. *IMSNA*, 2013.

[6] Ming-Hsiang Hsien-I Lin† and Wei-Kai Hsu, Chen *Human Hand Gesture Recognition Using a Convolution Neural Network IEEE*, 2014.

[7] Sachin Agrawal, Agnishrota Chakraborty, and M, Rajalakshmi *Real-Time Hand Gesture Recognition System Using MediaPipe and LSTM*, 2022.

[8] Ruchi Manish Gurav and K Premanand. Kadbe Real time Finger Tracking and Contour Detection for Gesture Recognition using OpenCV. *ICIC*, 2015.

[9] Ahmad Khawaritzmi Abdallah, Dian Christy, Silpani, and Kaori Yoshida *Hand Gesture*. 2022.

[10] Shashank Awasthi, Arun Pratap Srivastava, and Swapnita Srivastava. Vipul Narayan." A Comparative Study of Various CAPTCHA Methods for Securing Web Pages. *ICACTM*, 2019.

[11] Pooja Panwar1, Monika1, Parveen Kumar1,2 and Ambalika Sharma."CHGR: Captcha Generation Using Hand Gesture Recognition. *IEEE*, 2018.

[12] Pranjali Manmode, Rupali Saha, and N Manisha. Amnerkar Real-Time Hand Gesture Recognition. *IJSRCSIT*, 2021.

[13] Jaya Prakash Sahoo and Samit Ari, Sarat Kumar Patra "*Hand Gesture Recognition using PCA based Deep CNN Reduced Features and SVM Classifier*, 2019.

[14] Okan Köpüklü1, Ahmet Gunduz1, Neslihan Kose2, Gerhard Rigoll. *Real-time Hand Gesture Detection and Classification Using Convolutional Neural Networks*, 2019.

[15] Gyutae Park, V K Chandrasegar, and Joonggun Park, Jinhwan Koh "*Increasing Accuracy of Hand Gesture Recognition Using Convolutional Neural Network*, 2022.

[16] Felix Zhan. Hand Gesture Recognition with Convolution Neural Networks. *IEEE 20th International Conference on Information Reuse and Integration for Data Science (IRI)*, 2019.

[17] Anju S R and Subusurendran. A Study on Different Hand Gesture Recognition Techniques. *International Journal of Engineering Research and Technology (IJERT)*, 03(04), 2014.

[18] Agrawal M, Ainapure R, Agrawal S, Bhosale S, and Desai S S. Models for Hand Gesture Recognition using Deep Learning. *IEEE 5th International Conference on Computing Communication and Automation*, pages 589–594, 2020.

[19] Andrea Tagliasacchi, Matthias Schroder, Anastasia Tkach, Sofien Bouaziz, Mario Botsch, and Mark Pauly. Robust Articulated-icp for Real-time Hand Tracking. *Computer Graphics Forum*, 34:101–114, 2015.

[20] Liuhao Ge, Hui Liang, Junsong Yuan, and Daniel Thalmann. Robust 3d Hand Pose estimation in Single Depth Images: From Single-view cnn to Multi-view cnns. *Proceedings of the IEEE Conference on Computer Vision and Pattern Recognition*, pages 3593–3601, 2016.

[21] Kumar B P and Manjunatha M B *Performance Analysis of KNN, SVM, and ANN Techniques for Gesture Recognition System*, 2016.

[22] Oudah M, Al-Naji A, and Chahl J. Hand Gesture Recognition Based on Computer Vision: A Review of Techniques. *Journal of Imaging*, 6(8):73–73, 2020.

Recent Trends in Computational Sciences – Gururaj, Pooja & Flammini (Eds)
© 2024 The Author(s), ISBN 978-1-032-42685-3

Histopathologic multi-organ cancer detection in lymph node tissues

C.S. Usha
Vidyavardhaka College of Engineering, Mysuru, India

Harshitha Suresh
Vidyavardhaka College of Engineering, Mysuru, India
ORCID ID: 0009-0009-8761-8279

ABSTRACT: In the current world, Image processing in the clinical field plays a crucial part in making the occupation of medicos unchallenging. The traditional methods of examining the reports physically are time-consuming and chances of missing out on smaller details are more. This can be replaced with modern technologies to avoid such problems. The uses are almost limitless, and with the popularity of Image Processing that is growing at a faster pace than many other fields, the usage of these Image processing algorithms is no where to cease in near future. One such field we are here trying to bring Machine Learning algorithms is the Detection of Metastases in Lymph Node Tissues. When a person is battling cancer, the tumor might break out from that particular region and rollout to other parts of the body including Lymph Nodes, as lymph fluid flows all over the body collecting wastes. This process of tumor growth in a secondary region in the body is called Metastasis. This project performs extracting of features from a tissue sample image and helps classify whether the given tissue sample has cancerous cells or not. This is a relevant problem in today's world as the number of cancer patients is increasing and faster solutions have to be formulated to comb at the disease as quickly as possible.

Keywords: Lymph node, Lymphoma, Convolutional neural network, Immuno toxic

1 INTRODUCTION

A lymph node is a small, bean-shaped organ that produces and stores blood cells which helps to fight diseases and infections. The lymph nodes become aggravated or broadened because of different diseases which range from inconsequential throat contamination to perilous malignant growth. The most widely recognized reasons for inflammation of lymph nodes are they become enlarged because of a disease, like a typical virus. In some cases, lymph node swelling is because of a hidden condition. The lymph nodes are present throughout the body and are more accumulated near the trunk region. Whenever lymph node swelling stays and is encircled by different symptoms, for example, fever, night sweats, or weight reduction, with no undeniable contamination, the time has come to see a specialist for testing and assessment.

IMAGE PROCESSING – Image processing is widely used in cancer detection for detecting cancer from images achieved through datasets. Diseases in various parts of the organs can be diagnosed using image processing. Image processing takes low-quality images as input and improves an image's quality as an output. Image processing includes the following: image enhancement, restoration, image acquisition, pre-processing along encoding and compression.

DEEP LEARNING – Deep Learning is a sub class of machine learning algorithms that utilizes ANN's. Nowadays, for the recognition and classification of plant leaf diseases, deep learning has been used. The different deep learning network architectures AlexNet, GoogleNet, VGG16Net, and QuocNet can be used for classification. Alexnet is a successful deep learning architecture and Google net is deeper than AlexNet.

DOI: 10.1201/9781003363781-47

CONVOLUTIONAL NEURAL NETWORK – Convolutional neural network is a multi-layer neural network used to extract features from input images. CNN does not require a lot of pre-processing. Convolutions and pooling can be used to fit an image into its basic features.

2   RELATED WORK

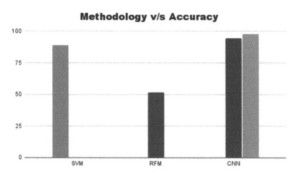

Figure 1.   Comparison of methodology with accuracy.

Irenaeus Rejani *et al.* [1] have purchased up a framework that spotlights on 2 distinct issues and its answer. One is to distinguish growths as dubious areas with extremely frail lighting and another is the manner by which to separate high lights that separate cancers.

In this paper, the proposed strategy incorporates the pictures separated with a Gaussian channel in light of standard deviation and network aspects like lines and sections. Then, at that point, the separated picture is utilized for contrast extending and afterward, the highlights are removed from the sectioned growth region. Then, at that point, the last stage is characterization utilizing the SVM classifier which gives a precision of 88.75%.

Mitsuru Futakuchi *et al.* [2] have proposed a 2-venture profound learning calculation that worked to manage the issue of false-positive prediction. A profound learning calculation became used to eliminate routinely misclassified non-cancerous regions.

The 2 models were accomplished contrastingly concerning lymphatic tissue forecast. Precision of model 01 which is of RFM and model 02 of CNN was 51.7% and 94.5%, separately.

The proposed technique is prepared and assessed utilizing a dataset from "Programmed Cancer Detection and Classification in Whole slide Lung Histopathology". All the more unequivocally, this dataset chose the initial 25 pictures to create a preparation set containing 124434 ordinary patches and 97588 cancer patches producing an exactness of 97%.

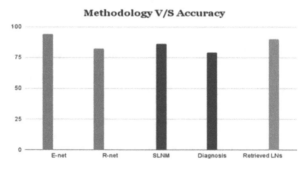

Figure 2.   Comparison between methodology and accuracy.

Pant *et al.* [4] have proposed two models, first model is a Efficient Net-based U-Net and second one is a ResNet-based U-Net model for pneumonia detection from chest X-Ray images, using a precautionary measure called Gaussian smoothing. The Efficient Net-B4 based U-Net resulted a accuracy of 94% and the Resent based U-Net led to 82% accuracy when these two were combined the ensemble of both models was about 90%.

Alexander R. Miller *et al.* [5] have proposed a mapping for rapid pathologic examination in patients receiving chemotherapy before surgery for breast carcinoma using sentinel lymph node classification. Sentinel Lymph Nodes Mapping dividing lymph nodes was 86% in 30 patients. Robert A. Ramirez *et al.* [6] have proposed Intrusion of intrapulmonary lymph nodes after routine pathologic examination of pulmonary cell degeneration. The pathologic nodal phase contributes to anticipation in patients with non-small cell lung cancer (NSCLC).

Additional Lymph Nodes were found to be 90% in 66 out of 73 patients and metastasis was 11% out of 56 of the 514 diagnosed Lymph Nodes out of 27% of all patients. It showed an unexpected metastasis of Lymph Node 12% in 6 of 50 non-node-negative patients. Metastatic satellite nodes were not detected in 3 different patients. The development of the pathologic phase was 11% in 8 out of 73 patients. SPE assembly decreases primarily due to experience, without adjusting the number of lymph nodes detected.

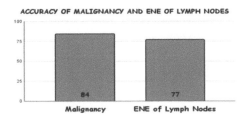

Figure 3.   Accuracy of Malignancy and ENE.

Tsung-Ying Ho *et al.* [7] modified a computer-based analysis method to assist in differentiating the images of lymph nodes in the victims of head and neck cancer. This model has been classified to be trained and tested on the histopathological image dataset using the Multi-layer perceptron neural network (MLP). Among the total of 6531 cancer patients with Gallbladder Cancer, the median wide variety of Lymph Nodes evaluated was 2; the simplest value of 21.1% i.e., (n = 1376) of victims had a presence of six or more Lymph Nodes evaluated. The median range of metastatic Lymph Nodes became zero. On multivariable analysis, assessment of lesser than four Lymph Nodes became associated with a better risk of dying whereas, patients who had 4 to 7 Lymph Nodes and greater than 7 Lymph Nodes which were evaluated had similar long-term mortality. Thus no difference was observed in the percentage of patients who had a small number of single metastatic Lymph nodes identified in each class of T primarily based on the total number of selected nodes.

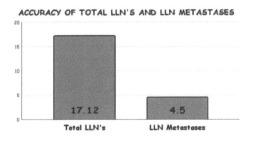

Figure 4.   Accuracy of LLN's and LLN metastases.

Jia *et al.* [8] examined the role of lingual lymph nodes (LLNs) within the recurrence of Squamous Cell Carcinoma (SCC) in the tongue and floor of the mouth. The cancer victims have been categorized into two groups:

1) The LNN group
2) The No-LLN group.

The segment and logical data varieties among the No-LLN foundation and the LLN bunch had been thought about. Factual investigations were accomplished utilizing the Pearson chi-square check. The average level of LLNs was 17.12% (19/111) and 5 patients (4.5%) showed LLN metastases. All patients with LLN metastases had neck lymph hub notoriety of the N2 group. Occurrence and metastases of Lingual Lymph Nodes were associated with neurotic groups of SCC of the tongue and lower lip.

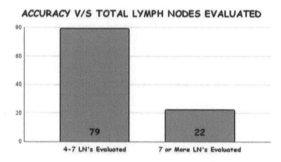

Figure 5.   Comparison of accuracy to total lymph nodes.

Diamantis I. Tsilimigras *et al.* [9] have performed examinations to find the insignificant number and the top quality assortment of lymph nodes to be inspected among victims with gallbladder cancer (GBC). A machine-based methodology was utilized in distinguishing the base assortment and scope of Lymph Nodes to survey comparative with long term impacts.

The Extra nodal Extension (ENE) is a lymph node neurotic element that has been shown to be clearly depressing in the oral cavity and various types of cancer. In this view, they demonstrated that they could use radiomic features separated by an open-source system and its addition from pre-programmed differences and upgraded T1 MRI images with a 5-layer neural organization to divide lymph nodes into three categories:

1) Benign Tumor
2) Malignant With ENE Tumor
3) Malignant without ENE Tumor

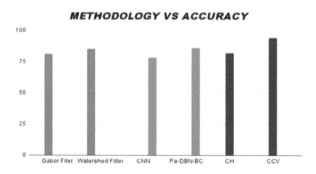

Figure 6.   Comparison of methodology to accuracy.

Mokhled S Altarawneh [10,11] had divided the image pre-processing stage into 3 steps i.e. Image enhancement, Segmentation, and Extraction, and has proposed a comparison study between a few methodologies used in each step and has mentioned which method is best suited for every step for analysis of histopathological lung cancer detection. This method offered auspicious effects evaluating with different techniques. Contingent upon well-known highlights, a regularity comparison was made. The identified capabilities for precise image contrast was pixel percentage and masks-labeling along with sturdy operation and high accuracy.

Jamaluddin and Fauzi *et al.* [12] Have proposed another new model which was based on CNN with the use of images with 12 convolutional layers with ReLU activation function and max-pooling to categorize slides with led to positive results of tumor and normal cases in lymph nodes tissue images using the feature color coherence vectors (CCV).

3 METHODOLOGY

Based on the literature survey the methodology has been depicted in Figure 7.

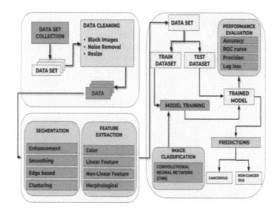

Figure 7. Methodology.

3.1 *Comparative analysis*

Table 1. Comparative analysis.

TITLE	OBJECTIVE	METHODOLOGY	RESULTS
EARLY DETECTION OFBREAST CANCER: USING SVM CLASSIFIER TECHNIQUE – Ireaneus Rejani 2018	To detect breast cancer using anSVM classifier.	Support Vector Machine (SVM)	In this paper, the mammogram was sifted with a Gaussian clear through basically founded on favored deviation and network aspects and afterward groupedutilizing the SVM classifier.
DETECTION OF LUNGCANCER LYMPH NODE METASTASES FROM WHOLE-SIDE HISTOPATHOLOGIC IMAGES USING A TWO-STEP DEEP LEARNING APPROACH – Mitsuru Futakuchi 2019	To detect tumorspresent in the lymph nodes of the lungs.	1) Random forest model (RFM) 2) ConvolutionNeural Network(CNN)	The 2 models wereachieved in a different way as for lymphoid follicle expectation. The exactness of model1 which is of RFMand model 2 of CNN was 51.7% and 94.5%. A well-fitting shape was shown in the model.

(continued)

Table 1.   Continued

TITLE	OBJECTIVE	METHODOLOGY	RESULTS
CNN-BASED METHOD FOR LUNG CANCER DETECTION IN WHOLE SLIDE HISTOPATHOLOGY IMAGES – Mladen Russo 2019	To detect lung cancer using CNN with whole slide images.	Convolution Neural Network	CNN model showed the possibility to recognize cellular breakdown in the lung cells from entire slide pictures, but an increase of classification accuracy has to be done with more effort.
PNEUMONIA DETECTION: AN EFFICIENT APPROACH USING DEEP LEARNING – Ayush Pant 2020	To develop a model that will detect Pneumonia in patients.	U-Net, Efficient Net-B4, ResNet-34, Gaussian blur, BCE and Dice Loss	Accuracy EfficientNet-B4 based U-Net – 94% Resent based U-Net – 82% Ensemble of Efficient Net-B4 based U-Net based U-Net and Resnet based U-Net – 90%
ANALYSIS OF SENTINEL LYMPH NODE MAPPING WITH IMMEDIATE PATHOLOGICAL REVIEW IN PATIENTS RECEIVING PREOPERATIVE CHEMOTHERAPY FOR BREAST CARCINOMA- Alexander R. Miller 2002	To distinguish cancers present in the breast lymph node carcinoma utilizing Sentinel lymph hub mapping(SLNM)	SLNM and axillary lymph node dissection Mapping is done with sulfur colloid in one patient's Lymphazurin color in certain patients.	SLNM effectively distinguished a sentinel lymph node 86%. The intraoperative pathologic analysis ended up being right at 79%.

4   CONCLUSION

After the study of various research papers, we came to a conclusion to propose the CNN (Convolutional Neural Network), an essentially based methodology to the detection of Histopathological multi-organ cancer detection in Lymph Node tissues. CNN architecture was used which indicates higher patch classification accuracy. Presented results suggest that CNNs have the potential for multi-organ cancer detection. This study proposes a CNN procedure that breaks down the dangerous growth tissue districts for the computerized recognition of this disease. Different CNN structures have been portrayed in this paper with appropriate differences. As the goal of the exploration is to diminish the time and space and to work on the precision of the proposed project when contrasted with past works, the size of the dataset is brought with the assessment time and percent of matching. Then, at that point, the hour of activity, dataset size, and matching rate are taken subsequent to lead tests considering edge identification-based CNN.

REFERENCES

[1]   Y. Irenaeus Anna Rejani *et al International Journal on Computer Science and Engineering* Vol.1(3), 2018, 127–130 Early Detection of breast cancer using SVM classifier technique.

[2]   Mitsuru Futakuchi, Andrey Bychkov, Tomoi Furukawa, Kiyoshi Kuroda, Junya Fukuoka. Department of Pathology, Kamed Medical Center, Kamogawa, Chiba, Japan, 18 September 2019. https://doi.org/10.1016/j.ajpath.2019.08.014

[3]   Mladen Russo FESB, University of Split, Split, Croatia. CNN-based Method for Lung Cancer Detection in Whole Slide Histopathology Images, 2019 4th International Conference on Smart and Sustainable Technologies(SpliTech)10.23919/SpliTech.2019.8783041

[4]   A. Pant, A. Jain, K. C. Nayak, D. Gandhi, and B. G. Prasad, "Pneumonia Detection: An Efficient Approach Using Deep Learning," 2020 11th International Conference on Computing, Communication and Networking Technologies (ICCCNT), 2020, pp. 1–6, doi:10.1109/ICCCNT49239.2020.9225543.

[5]   Miller, A.R., Thomason, V.E., Yeh, IT., *et al.* Analysis of sentinel lymph node mapping with immediate pathologic review in patients receiving preoperative chemotherapy for breast carcinoma. *Annals of Surgical Oncology* 9, 243–247 (2002). https://doi.org/10.1007/BF02573061

[6]   Robert A. Ramirez, Christopher G. Wang, Laura E. Miller, Courtney A. Adair, Allen Berry, *et al.*, "Incomplete Intrapulmonary Lymph Node Retrieval After Routine Pathologic Examination of Resected Lung Cancer", 2012 DOI: 10.1200/JCO.2011.39.2589 *Journal of Clinical Oncology* 30, no. 23 (August 10, 2012) 2823–2828.

[7]   Tsung-Ying Ho, Chun-Hung Chao, Shy-Chyi Chin, Shu-Hang Ng, Chung-Jan Kang, Ngan-Ming Tsang, "Classifying Neck Lymph Nodes of Head and Neck Squamous Cell Carcinoma in MRI Images with Radiomic Features", *Journal of Digital Imaging* 16 January 2020 DOI:10.1007/S10278-019-00309-W

[8]   Jun Jia, Meng-qi Jia, Hai-xiao Zou, "Lingual lymph nodes in patients withsquamous cell carcinoma of the tongue and the floor of the mouth", *Journal of the sciences and specialties of the head and the neck*, 26 July 2018 DOI: 10.1002/hed.25340

[9]   Diamantis I. Tsilimigras, J. Madison Hyer, Anghela Z. Paredes, Dimitrios Moris, Eliza W. Beal, Katiuscha Merath, Rittal Mehta, Aslam Ejaz, Jordan M. Cloyd, Timothy M. Pawlik, "The optimal number of lymph nodes toevaluate among patients undergoing surgery for gallbladder cancer: Correlating the number of nodesremoved with survival in 6531 patients", *Journal of Surgical Oncology*, 12 March 2019, DOI: 10.1002/hed.25340

[10] Mokhled S Altarawneh, "Lung Cancer Detection Using Image Processing Techniques", August 2012

[11] M F Jamaluddin, MFA Fauzi and F S Abas, "Tumor detection and whole slide classification of H&E lymph node images using convolutional neural network", 2017 IEEE International Conference on Signal and Image Processing Applications (IEEE ICSIPA 2017), Malaysia, September 12- 14, 2017

[12] Irum hirra, Mubashir Ahmad, Ayaz Hussain, M. Usman Ashraf, Iftikhar Ahmed Saeed, Syed Furqan Qadri, Ahmed M. Alghamd, and Ahmed S. Alfakeeh, "Breast Cancer Classification From Histopathological Images Using Patch-Based Deep Learning Modeling", February 2, 2021, DOI 10.1109/ACCESS.2021.3056516

Recent Trends in Computational Sciences – Gururaj, Pooja & Flammini (Eds)
© 2024 The Author(s), ISBN 978-1-032-42685-3

Literature survey on tracking template face mesh using stereo video

Kavitha Jayaram, M. Ameer Suhail, Aneesh Acharya, Manish K. Reddy &
Mohammed Raheel
BNMIT

ABSTRACT: The essence of facial motion capture is in employing cameras or laser scanners to electronically turn a user's facial motions into a digital repository. The capture may be in a 2D or two-dimensional form, if so, the process of capturing is frequently referred to as "expression tracking." A single camera system, along with a complementing piece of capture software, can be used to do this two-dimensional capture. This type of system faces more difficult to track and falls short of capturing three-dimensional actions like head rotation. Laser marker systems or multi-camera rigs are used to capture 3D data. These tracking systems can be marker or marker less. It might be difficult to recognize and follow the nuanced facial expressions that could be produced by quick changes in the lips and eyes. Since these motions frequently only measure a few millimeters, they necessitate even higher resolution, accuracy, and filtering methods than those typically utilized in full-body capture. The majority of techniques rely on intricate multi-camera arrangements, regulated illumination, or fiducial markers. This makes it impossible for them to be deployed in open spaces, outdoor environments, or in real-time on a set while capturing their digital personalities. The goal is to move facial motion capture from a studio and into a public setting, making it accessible to everyone.

1 INTRODUCTION

The majority of techniques in facial performance capture rely on intricate multi-camera configurations, calibrated illumination, or fiducial markers. This makes it impractical for them to be used in real-world environments such as outdoor scenes. The general flow of a basic face tracking model is shown in Figure 1 and used as a base for the face capture implementations as discussed in the literature survey. Figure 1 illustrates how the tracker handles noisy data after detecting an established face. This is accomplished via a Kalman filter, also referred to as the Linear Quadratic Estimation (LQE), it is essentially an algorithm that employs a series of measurements made over time and generates estimates of uncertainties that are typically more accurate than most estimates based solely on a single measurement. These filtered frames are then used to track the face and later if there is any drift in the tracking, the face positions are adjusted accordingly. This tracked data is then mapped onto a face mesh which will try to accurately depict all the facial movements as shown in the video input. If the number of frames processed has exceeded 500 frames, we can stop the capture.

2 SURVEY ON FACIAL CAPTURE AND TRACKING

The general approach to facial capture is to have a coarse-detail face mesh template be tracked in the first pass of a binocular stereo sequence, and then in the second pass, time-varying fine detail, like folds as well as wrinkles, is added to the monitored meshes.

Levi Valgaerts, Chenglei Wu *et al.*, in their paper, have proposed a method that expands and improves on existing methods for illumination estimates, shading-based refinement, and scene flow computation based on images. With the use of this field approach, it is feasible to take

DOI: 10.1201/9781003363781-48

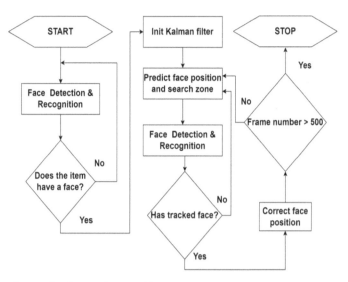

Figure 1.   Flowchart to implement face tracking.

facial performance capture outside of the studio and into a setting where anyone can use it. Strong shadows and low lighting are characteristics that call for a high-quality video source. The tracking step finds it difficult to interpret strongly moving shadows since they appear to be surface motion. It is suggested to add subtle time-varying information, such as wrinkles and folds, to the tracked meshes that were originally built by tracking a coarse-detailed facial template through a binocular stereo sequence. [1]

Yaming Wang, Xiangyang Peng et al., in their paper refer to a convolutional neural network that does both optimization and reconstruction, which entails the optimization of the camera matrix and the rebuilding of the three-dimensional structure along with the help of the two modules firstly an image optimisation module and secondly a depth reconstruction module. Although this technique is often quite helpful, there is an issue when enormous amounts of data must be computed, the recommended technique currently has the drawback of being sensitive to the matrix for giving a smooth output. The RONN is also incapable of handling data loss. [2]

Shridhar Ravikumar in his paper proposed a straightforward, light-weight, marker-less framework for capturing facial performance data that only accepts monocular video input, integrates active feature monitoring appearance models with 3D shape pre-constraints and leverages both to achieve an integrated objective function. The author uses a prebuilt ani-mation derived from accurate 3D tracking technologies to impose a constraint on the goal function to overcome the information loss and produce more logical outcomes and a Blendshape model adjustment technique to manage noise in predicting camera features and parameters while matching monitored 2D characteristics. Finally, the authors demonstrate that utilizing spatial limitations will result in more accurate results if the choice of Blendshapes has an impact. [3]

Luming Ma and Zhigang Deng based their paper on a monocular RGB video input where the research effort captures human-specific facial models with high-precision details and offers a unique method for real-time reconstruction of a high-resolution facial geometry. The shape-from-shading approach first reconstructs a rough face model from the input video, then recovers lighting, albedo texture, and displacement by matching the generated face with the RGB incoming video. Reconstruction of high resolution facial geometry and features in real time, is done by exploiting geometry-based techniques. The system operates auto-matically for a new user and doesn't need any offline preparation. Any faulty detections could result in incorrect subject identities or head postures. The system reconstructs broad head movements and facial emotions in real-time, as well as subtle wrinkles, lighting, and albedo. The refinement of the shadow surface from the human face shape is also reliably solved using the new hierarchical reconstruction techniques. [4]

From these papers, it can be inferred that a novel rigid stabilization method can be used via a dynamic rigidity approach to stabilize monocular real-time face tracking given a realistic facial performance data set is created using rigid pose and expression parameters that are based on ground truth. In this study, Chen Cao, Menglei Chai, and colleagues make the case that an effective face-tracking system requires a unique method for learning the hyper parameters of dynamic rigidity before the facial performance data set. [5]

Gaspard Zoss, Thabo Beeler *et al.*, developed an effective monocular real time face tracking that blends an expressiveness region based face model with the restrictions of sparse features and dense image processing techniques. Also,the system is required to solve for jaw pose which plays the main role in visual effects pipelines which is achieved by the system by learning a non-linear mapping from the skin deformation to the underlying jaw motion on a data set where the jaw poses have been acquired by direct observation, and then to re-target the mapping to new subjects. [6]

Another approach by Jiaman Li, Zhengfei Kuang et al., uses a self-supervised neural network to learn customized Blendshapes from a set of template expressions which uses a comprehensive library of over 4,000 facial scans which consists of varying pore level details, different expressions, and identities. [7]

The authors Bing-Fei Wu, Bo-Rui Chen et al., here explain the method which solve the problem of landmark shaking, with a simple solution consisting of a lightweight U-Net model with a dynamic optical flow (DOF) which can be used, to effectively capture the dynamic landmark detection. [8]

Shoou-I Yu Hyun Soo Park, Jae Shin Yoon et al., provide another novel approach to first circumvent by training a novel network that can directly generate a face model from a single 2D picture, we can eliminate the need for specific input data. The domain mismatch between the lab and uncontrolled settings is then addressed by self-supervised domain adaptation based on "consecutive frame texture consistency." [9]

Paulo Gotardo, Jeremy Riviere et al., describe a technique for acquiring dynamic features of face skin appearance. This study presents a time-varying model of appearances and surface details that may be inferred from multiview picture streams without the need for time-multiplexed lighting. However, due to the passive acquisition arrangement, the authors impose a few limits upon the appearing estimation by modeling the time varying diffused reflectance as a shift in albedo rather than any underlying scattering factors. [10]

3 SURVEY ON FACIAL ANIMATION

Chen Cao, Vasu Agrawal *et al.*, in their paper demonstrate how to use binocular video to produce real-time face animation on a 3D avatar. In order to account for fluctuations caused by light and other elements frequently found in nature, such as facial hair growth, cosmetics, and skin imperfections, this approach utilizes a highly accurate custom face model. There are two steps in this method, first, the approach is to learn Real-time inference from regression picture with room to avatar 3D forms and other textures by first solving the lighting model's parameters analytically on a brief video using the model's parameter pairs (rigid, non-rigid) and source photos. A real-time regression model is then improved using a fresh video. In the paper, the proposed methodology requires stable lighting conditions. It only models the features of the face for animation but not the following features such as hair, neck, or shoulder. [11]

Chenglei Wu, Takaaki Shiratori *et al.*, proposed an iterative learning framework for effective and precise tracking of face performance. This method alternates between the modeling step, which uses monitored meshes and contour maps to coach the deep learning-based statistical method, and the monitoring step, which takes the model's predictions of geometrical and texture from the measured images and optimizes the predicted geometry by reducing picture, geometrical, and facial feature errors. [12]

4 CONCLUSION

The essence of facial capture and animation lies in the ability to recreate human facial expressions and emotions through technology realistically. This has been made possible through advancements in computer graphics and machine learning algorithms, which enable the accurate capture and mapping of facial movements onto digital characters. This technology has revolutionized the entertainment industry by enabling the creation of lifelike and emotionally engaging characters in films, games, and virtual reality experiences.

The article focuses on reviewing research and work aimed at improving the efficiency and precision of the facial motion capture and animation process. Moreover, it highlights the approach taken toward achieving seamless facial capture and animation which intends to introduce a new degree of realism to the realm of digital entertainment by enabling the construction of realistic and emotionally compelling characters.

In conclusion, the field of facial capture and animation continues to evolve and has the potential to bring about new and exciting applications in the future.

REFERENCES

[1]   Levi Valgaerts, Chenglei Wu, Andrés Bruhn, Hans-Peter Seidel, and Christian Theobalt. "Lightweight Binocular Facial Performance Capture Under Uncontrolled Lighting", *ACM Trans. Graph. Volume 31, Issue 6, Article 187, November*, 2012.

[2]   Yaming Wang, Xiangyang Peng, Wenqing Huang, and Meiliang Wang, "A Convolutional Neural Network for Nonrigid Structure from Motion", *Hindawi International Journal of Digital Multimedia Broadcasting Volume 2022, Article ID 3582037*, 2022.

[3]   Shridhar Ravikumar, "Lightweight Markerless Monocular Face Capture With 3d Spatial Prior", arXiv preprint arXiv:1901. *05355, COMPUTER GRAPHICS Forum, January*, 2019.

[4]   Luming Ma and Zhigang Deng, "Real-time Hierarchical Facial Performance Capture", *Proceedings of the ACM SIGGRAPH Symposium on Interactive 3D Graphics and Games, Article No.: 11, May*, 2019.

[5]   Chen Cao, Menglei Chai, Oliver Woodford, and Linjie Luo, "Stabilized Real-time Face Tracking Via a Learned Dynamic Rigidity Prior", *Association for Computing Machinery, ACM Trans. Graph., Volume 37, Issue 6, Article No.: 233, December*, 2018.

[6]   Gaspard Zoss, Thabo Beeler, Markus Gross, and Derek Bradley, "Accurate Markerless Jaw Tracking for Facial Performance Capture", *Association for Computing Machinery, ACM Trans. Graph., Volume 38, Issue 4, Article No.: 50, July*, 2019.

[7]   Jiaman Li, Zhengfei Kuang, Yajie Zhao, Mingming He, Karl Bladin, and Hao Li, "Dynamic Facial Asset and Rig Generation from a Single Scan", *Association for Computing Machinery, ACM Trans. Graph., Volume 39, Issue 6, Article No.: 215, November*, 2020.

[8]   Bing-Fei Wu, Bo-Rui Chen, and Chun-Fei. Hsu, "Design of a Facial Landmark Detection System Using a Dynamic Optical Flow Approach", *IEEE Access*, vol. 9, 2021.

[9]   Shoou-I Yu Hyun Soo Park. Jae Shin Yoon, Takaaki Shiratori, "Self-supervised Adaptation of High-fidelity Face Models for Monocular Performance Tracking", *Proceedings of the IEEE/CVF Conference on Computer Vision and Pattern Recognition (CVPR)*, 2019.

[10] Paulo Gotardo, Jérémy Riviere, Derek Bradley, Abhijeet Ghosh, and Thabo Beeler, "Practical Dynamic Facial Appearance Modeling and Acquisition", *Association for Computing Machinery, ACM Trans. Graph., Volume 37, Issue 6, Article No.: 232, December*, 2018.

[11] Chen Cao, Vasu Agrawal, Fernando De la Torre, Lele Chen, Jason Saragih, Tomas Simon, and Yaser Sheikh, "Real-time 3d Neural Facial Animation from Binocular Video", *ACM Transactions on Graphics, Volume 40, Issue 4, Article No.: 87, August*, 2021.

[12] Chenglei Wu, Takaaki Shiratori, and Yaser Sheikh, "Deep Incremental Learning for Efficient High-fidelity Face Tracking", *Association for Computing Machinery, ACM Trans. Graph., Volume 37, Issue 6, Article No.: 234, December*, 2018.

Recent Trends in Computational Sciences – Gururaj, Pooja & Flammini (Eds)
© 2024 The Author(s), ISBN 978-1-032-42685-3

NFT minting and owner verification in an NFT E-commerce website

Shashwat & Ravi Shankar Pandey

*Department of Computer Science and Engineering, Birla Institute of Technology Mesra,
Patna Campus, India*

ABSTRACT: NFTs or non-fungible tokens are unique digital assets with identifiable characteristics. NFTs can take the form of digital art, GIFs, videos, collectibles, video game merchandise, music, and even tweets. Uniqueness of a digital asset encompasses more than just visual properties, but also factors such as item IDs and hash codes. This means that even though the two NFTs are visually indistinguishable, they have different IDs and codes. In the digital space, these elements are an integral part of an asset's identity as they contain important data about the object's history. The popularity of the Non-Fungible Token (NFT) has grown tremendously since 2020, becoming one of the most popular applications in the Fintech field. However there is still lack of platforms which provide all the functionalities regarding NFTs (including minting, buying and selling) at a single place. This research paper introduces an algorithm for creation of NFTs along with ensuring owner verification which can be used for minting the NFTs in an E-commerce website using the MOTOKO language for creation of canisters consisting the owner id for owner verification and the principal id which serves as a unique token id for the NFT.

Keywords: NFT, blockchain

1 INTRODUCTION

NFT stands for Non Fungible tokens. These are digital assets cryptographically recorded on the blockchain which serves as a mark of ownership as the blockchains are highly secure. The blockchain is an open, immutable, distributed public ledger that allows transactions to be processed in a decentralized manner without the need for a trusted third party [1]. The blockchain is extremely secure as the data stored in the block chain can not be tampered. The blockchain consists of a nonce which is basically a random number generated during mining of block, the data to be stored, the previous hash which is the hash value of the previous block and the current hash which is the hash value of the current block. The previous hash of the first block (called as the genesis block) is zero. The value of each current hash is generated using SHA 256 hashing algorithm which is highly secure. The current hash is formed by combining the data, nonce and the previous hash together. Every new block to be added is validated by comparing the previous hash of the new block with the current hash of the last block, this is how the linking occurs. Since the generation of hash involves the data as well, it ensures high security as if anyone tampers the data of the block, the value of the current hash will be changed which will disrupt the link. The advent of block chain had led to creation of various technologies based on block chain. Since the adoption of cryptocurrencies, the NFT market has grown exponentially over the past few months as it appears to be the most popular business application of blockchain technology [2]. As more and more people expect to enter networked digital environments like the Metaverse [3], it is clear that NFTs will play a major role in this Internet of Tomorrow [4] due to its uniqueness similar to physical items [5]. This next decade of human interaction on the Internet

DOI: 10.1201/9781003363781-49

could depend entirely on the NFTs. Now the question arises What are NFTs? NFTs have proven to be rare and unique digital assets that can be used to represent ownership [6]. They can be unique and rare works of art, collectible trading cards, and other assets whose rarity can increase their value [7,8]. These are digital assets, but they can also be used to represent physical assets. Some examples are Digital Country Certificates/Certificates of Competence. The biggest winners in the NFT space in the last few months are the Digital Artists who have sold over 2.5 billion dollars in art [9]. NFT was introduced by Ethereum [10] as an improvement proposal [11,12] to the Ethereum Request for Comments(ERC)-721 standard [6]. This allows anyone to implement a smart contract using the ERC-721 standard to create NFTs and track the tokens they generate ensuring the verification of the created token. The entire working of NFTs depends on the smart contracts. Smart contracts are basically the codes running on the block-chain. These can be like the simple if else statements that execute a certain transaction once the desired condition is met. NFTs are created through a process called minting, which converts images, videos, soundbites, and other digital files into crypto assets on the blockchain. When you create an NFT, you are essentially configuring the underlying smart contract code that determines the quality of your crypto assets. Several smart contract standards have been established to make it easier for NFTs to interact with applications. For example, there are several smart contract blockchains that use NFT creation tools such as TRON, EOS, and Tezos. Without a common standard for smart contracts and how to ultimately encode NFTs, NFTs created on different platforms may not be able to trade on the same NFT marketplace.

2 RELATED WORK

Currently, most blockchains are used in the financial sector, with an increasing number ofapplications in various fields.Traditional industries can consider blockchain and apply blockchain to their fields to improve the system. For example, a user's rating can be stored on her blockchain. At the same time, Emerging Industry can use blockchain to improve its performance. For example, ride-sharing startup, Arcade City [13] offers to open a market-place where riders connect directly with their drivers using blockchain technology. All these technologies use smart contracts to run programs on blockchain technology. A smart con-tract is a computerized transaction protocol that enforces the terms of the contract [14]. It has been proposed for a long time, and now the concept can be implemented using the blockchain technology. In blockchain, smart contracts are code snippets that miners can automatically execute. Smart contracts have transformative potential in many areas, including financial services and IoT. NFTs are "forged" (i.e. created) on the blockchain network. Data stored on the blockchain is continuously shared, replicated and synchronized between nodes in the network — individual computer systems or specialized hardware that communicate with each other and store and process information is. Central Authority or Intermediary. In the blockchain network, the participant uses the public key to encrypt her data and the private key to decrypt the data. Participants can also sign transactions with their private keys, and other users can verify these signatures with their corresponding public keys. There are types of blockchains. In most cases, there are several common features such as decentralization (i.e. no centralized authority), immutability (i.e. the blockchain record is immutable), pseudonymity (i.e. the real identity of the user is not directly exposed), etc. Typically the minting of NFTs is done on public, permissionless blockchains, which allow any node on the blockchain network to read and submit transactions. A smart contract will mint an NFT by executing lines of code that add the NFT data to the blockchain.

3 PROPOSED METHOD

The proposed algorithm provides the features for creation of NFTs and the owner verification. The algorithm is implemented through the smart contracts on the ICP blockchain. The existing methods are not user friendly as they are complex and difficult to be combined with a user interface which is easily understood by the general public without any technical knowledge of the working of the blockchain technology, as through these methods the users first have to create their crypto wallets which are used to hold the cryptocurrencies in the crypto marketspace. These wallets are linked with the blockchain (for example the etherium blockchain) on which they want to record their art pieces. Then the blocks are mined and NFTs are minted by cryptographically recording the digital assets on the chain of blocks. However the proposed method is extremely user friendly as in this the users are not required to have any prior technical knowledge of blockchain technology. The algorithm is implemented through the MOTOKO language which is used to create canisters that executes software programs on the ICP blockchain. In this method each NFT primarily contains three fields including the name of the NFT (a specific text entered by the user), the unique token id which is different for different NFTs and the Owner id which is basically principal id of the main canister and the NFT data. The minting of NFTs is done through smart contracts of ICP blockchain. Each class serves as a separate canister which represents an individual NFT. All these canisters are managed through a single canister (or the main canister) which links all the NFTs for a single user and the functions for displaying the NFT. This main canister is different for different users.

4 ALGORTITHM

- Create the main actor class which contains the NFT classes.
- To mint the NFT create a class for storing the NFT information.
- Pass the values of the name entered by the user for NFT, the principal id for themain canister (the actor class), the NFT data in the form of an array of natural numbers.
- Initialize the variables as:
  a) itemName: Name of NFT
  b) nftOwner: principal id of main class
  c) tokenId: principal id of the nft class

- Each time a new NFT is created a separate NFT class is formed.Create a function to return the NFT name.
- Create a function to return the NFT data.

5 RESULT

The proposed algorithm is incorporated with a frontend (Figure 1) to provide a user interactive platform for minting the NFTs and can be given a form of web application or an e-commerce website. The MOTOKO language is used to obtain the smart contracts on ICP block chain which are basically the computer programs that are tamperproof. The features of these smart contracts can be enumerated as:

- The smart contracts are general purpose i.e any possible computation can be performed through the blockchain.
- They are tamperproof as they run on blockchain thus the programs can not be tampered.
- They are autonomous as the network automatically executes the contracts and the user does not have to perform any action.
- They are decentralized that means they are not controlled by a particular individual or an organisation.

The MOTOKO program runs at the backend and interacts with the frontend through IDL, which stands for Interface description language. The idl just simply acts as a translator

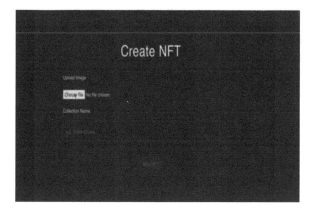

Figure 1.   The frontend for minting the NFTs.

between the MOTOKO backend and the javascript frontend. The frontend can be pretty simple as depicted in Figure 1. The user just needs to upload the NFT data which can be in the form of jpg, jpeg, png etc. by clicking on the choose file option.

As soon as the user provides the NFT data, the class name and clicks on the mint NFT button the NFT actor class is called which programmatically creates the NFT. The NFT created at the backend is displayed to user through the function,inside the NFT actor class which returns the NFT data Figure 2. Depicts the minted NFT.

Figure 2.   The minted NFT.

The use case for the NFT e-commerce website depicts its functional requirements (Figure 3). It has two actors the user (includes both buyer and seller). It covers the following functional aspects:

- Browse NFT
- View NFT
- Mint NFT
- Sell NFT The actors involved in the use case diagram are:
- User: This includes the buyers and sellers.
- Smart Contracts: The canisters consisting of programs to perform specific function. The actor user acts as the primary actor for the use cases: buy NFT, browse the NFTs, View NFT, Mint NFT and Sell NFT asset, and as a secondary actor for the use cases Display NFT, Payment, Upload File and Sell price.

337

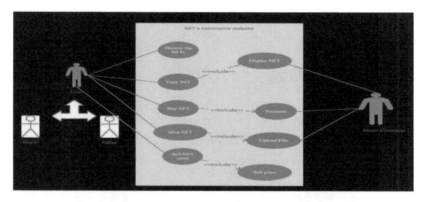

Figure 3.   The use case for the NFT e-commerce website.

6   CONCLUSION

This paper provides an algorithm for creation of NFTs and the owner verification which ensures enhanced verification and a simple method for minting NFTs. This can be seamlessly incorporated in an NFT E-commerce website which provides a single platform for minting, buying and selling NFTs. The algorithm proposed in this paper first creates an actor class which is a super class consisting multiple actors (the canisters containing the NFT) and then multiple actors are created for individual NFTs owned by a single user. Thus the algorithm provides owner verification through the unique principal id of the actor class and ensures the uniqueness of NFTs through the token id of each NFT which is different for different NFTs.

REFERENCES

[1]   Karthika, V. and Jaganathan, S. (2019) 'A Quick Synopsis of Blockchain Technology', *Int. J. Blockchains and Cryptocurrencies*, Vol. 1, No. 1, pp.54–66.

[2]   M. Dowling. (Apr. 29, 2021). "*Is Non-fungible Token Pricing Driven by Cryptocurrencies? — Elsevier Enhanced Reader*,"

[3]   Casey Newton. (Jul. 22, 2021). "Mark Zuckerberg is Betting Facebook's Future on the Metaverse – The Verge," *The Verge*.

[4]   Peter Allen Clark. (Nov. 15, 2021). "What is the Metaverse? here's Why it Matters — Time," *Time*

[5]   "*Non-fungible Tokens (NFT)*," ethereum.org,

[6]   "*ERC-721 Fon-fungible Token Standard*," ethereum.org,

[7]   R. Conti. (Apr. 29, 2021). "What You Need to Know About Non-fungible Tokens (NFTs)," *Forbes Advisor. Section: Investing*

[8]   J. Fairfield, "Tokenized: The Law of Non-fungible Tokens and Unique Digital Property," *Social Science Research Network*, J Rochester, NY, SSRN Scholarly Paper ID 3821102, Apr. 6, 2021.

[9]   (Jul. 5, 2021). "Off the Chain: NFT Market Surges to $2.5b So Far This Year," Aljazeera, 5(4):13–18, 2010.

[10]   D. G. Wood, "*Ethereum: A Secure Decentralised Generalised Transaction Ledger*," p. 39, 2014.

[11]   "*EIP-2309: ERC-721 Consecutive Transfer Extension*," Ethereum Improvement Proposals,

[12]   "*ERC*," Ethereum Improvement Proposals,

[13]   S. Solat and M. Potop-Butucaru, "*ZeroBlock: Timestamp-Free Prevention of BlockWithholding Attack in Bitcoin*," Sorbonne Universites, UPMC University of Paris 6, Technical Report, May 2016.

[14]   N. Szabo, "*The idea of Smart Contracts*," 1997.

Recent Trends in Computational Sciences – Gururaj, Pooja & Flammini (Eds)
© 2024 The Author(s), ISBN 978-1-032-42685-3

A smart health care ecosystem

V. Rohankumar, M.S. Sohan, K.N. Vasukipriya, B.B. Vishal & M.R. Pooja
Vidyavardhaka College of Engineering, Mysuru, India
ORCID ID: 0000-0001-6225-8486

ABSTRACT: A real-time health monitoring and alerting system is a system that continuously monitors a person's health and provides alerts in the event of a potential health issue. The system is able to gather data on a person's vital signs, such as heart rate, blood pressure, and body temperature, using sensors. This data is then analyzed in real-time to detect any abnormalities or potential health issues. If a potential problem is detected, the system can send an alert to the person or their healthcare provider, allowing for timely intervention and potentially preventing more serious health issues. The use of real-time health monitoring and the alerting system can help to improve patient outcomes and reduce the need for hospitalization.

Keywords: Wireless sensor network (WSN), Realtime monitoring, RF transmitter, Wearable, Healthcare

1   INTRODUCTION

An online real-time health monitoring system is a technology that allows individuals to monitor their health and wellbeing in real-time using the internet. This system can provide individuals with a convenient and efficient way to track their vital signs and other health measures, such as heart rate, blood pressure, and body temperature. By continuously monitoring these health indicators, individuals can detect potential health problems early and take action to address them before they become more serious. This can help individuals to maintain good health and prevent the onset of chronic diseases. Additionally, an online real-time health monitoring system can provide healthcare professionals with valuable data that can help them to provide better care and support for their patients. Overall, this technology has the potential to improve the health and wellbeing of individuals and communities around the world.

In addition to monitoring health indicators, an online real-time health monitoring system can also provide location tracing in real-time. This means that individuals can track their location and movements in real-time, allowing them to monitor their own safety and wellbeing. This can be particularly useful for individuals who are at a higher risk of falling or experiencing other health-related emergencies. By continuously tracking their location, individuals can quickly and Additionally, real-time location tracing can help healthcare professionals to monitor the movements and activities of their patients, allowing them to provide more personalized care and support. Overall, the combination of health monitoring and real-time location tracing can provide individuals and healthcare professionals with valuable tools for improving health and wellbeing.

An online real-time health monitoring system can also provide patients with the ability to communicate with their doctors and hospitals. This means that patients can easily and conveniently connect with their healthcare providers using the internet, allowing them to ask questions, receive guidance, and get support whenever they need it. This can be particularly useful for individuals who are managing chronic conditions or who need ongoing medical support. By providing patients with the ability to communicate with their healthcare providers, an online real-time health monitoring system can help to improve the quality of care and support that patients receive. Additionally, this feature can also help to reduce the burden on hospitals and other healthcare facilities, as patients can receive support and guidance remotely, without the need for in-person visits. Overall, the ability to communicate with doctors and hospitals can be a valuable addition to an online real-time health monitoring system

The working of an online real-time health monitoring system will vary depending on the specific design and features of the system. However, in general, the system will use sensors, wearable's, and other devices to collect data about an individual's health indicators and location. This data will be transmitted to the system in real-time, allowing it to be monitored and analyzed by the individual and their healthcare providers. The system can use this data to provide health monitoring and location tracking, as well as to alert healthcare professionals in emergency situations. Additionally, the system may include features that allow patients to communicate with their doctors and hospitals, providing support and guidance whenever needed. Overall, the working of an online real-time health monitoring system will involve the continuous collection and analysis of data, as well as the integration of various features and tools to support health monitoring and emergency response.

2 LITERATURE REVIEW

According to [1], The Real-time Online Activity and Mobility Monitor framework is designed to track health information, community mobility, and activity patterns. The goal of this framework is to develop a reliable and validated app for monitoring activity and mobility that can be used in research settings. The server for this framework is designed to be flexible, with a database model and view for displaying summary statistics corresponding to each set of features defined for the watch application. The main component of this framework is an application for the Samsung Gear S smartwatch, which collects data at a rate of 10 Hz and allows for variable construction. The web portal for this framework displays summary statistics of the data collected by the smartwatches in the field, and includes an archive module for accessing previously collected data and an administrative component where researchers can monitor the status of the smartwatches and validate the collected data in real-time.

According to [2], there are various helpline numbers available in each state, but the large number of them makes it difficult for commuters and citizens to quickly reach the desired helpline in an emergency. To address this issue, the proposed system is a mobile application that brings all of these helpline numbers together in a single place. It is assumed that in an emergency, people will likely have a smartphone with them or have access to one nearby. One potential question is why someone would need to create an account for an application that shows the same helpline numbers regardless of the user. Another scenario where the user might need to log out of their account is if they are switching to a new device. The proposed android application for emergency helpline services would be a valuable tool for the general public, as it would allow them to easily access helpline numbers in times of need without having to search for them.

According to [3], the proposed model for the automated medicine reminder is a medicine box that is equipped with a sensor that detects when the box is opened by the patient. When the box is opened, the light is reset and an alert is triggered. Regular medication containers can be converted into automated multi-pill dispensers for ease of use. The proposed model is designed using a microcontroller, which is used to track when the patient should take their pills. The system is also designed to monitor the health of elderly people at home and detect any abnormalities in their body condition. If an abnormality is detected, the system reminds the patient to take their medicine through the use of a buzzer and LCD display, and also alerts the patient's relatives or doctors about the abnormal condition. The advantage of this system is that it is easy to use and helps people who are busy with their work to remember to take their medicine.

According to [4], the indoor localization systems currently available on the market have various levels of accuracy, but a study titled precise indoor localization using smartphones presents an innovative system with an accuracy of up to 1.5 meters. This study also introduces a new method for statistical processing of RSSI data in smartphones, which outperforms existing deterministic techniques. The SOS Heart system plans to utilize this algorithm to provide precise localization of the user and transmit this information to the emergency department. It is evident that the combination of smartphones and artificial intelligence has the potential to improve the healthcare system. By advancing these technologies and involving clinicians and patients in research, it is possible to create smartphone applications that not only monitor and track health indicators but also detect or predict abnormalities and send data messages to physician offices. This study emphasizes the importance of mHealth applications in the healthcare system for monitoring and predicting health conditions.

According to [5], the use of the Internet of Things in healthcare can be helpful for tracking COVID-19 patients through a connected network. This technology involves large amounts of data, as well as cloud computing, to gather information from remote monitoring methods such as telecare and scheduled check-ups. The goal of this design is to lower mortality rates and hospitalizations by providing early notification of detected cases and directing patients to the nearest hospital. The system will also generate emergency alerts based on analysis of medical records. After testing positive for COVID-19, a patient may be given sensors to monitor their blood oxygen levels, heart rate, and body temperature from an isolation center. This data will be collected from the patient's body and sent to an IoT server via the cloud. Previous research and commentary suggests that smart monitoring and emergency alert systems are important for monitoring COVID-19 patients and can alert them in the event of an abnormal condition. One of the key benefits of using IoT in healthcare for COVID-19 patients is the ability to continuously monitor their condition remotely, which can help to prevent the spread of the virus by reducing the need for in-person visits.

According to [6], this paper presents a proposal for an internet of things (IoT) system for real-time electrocardiogram (ECG) monitoring that can function in both real-time and store-and-forward modes. Real-time mode can be challenging due to transmission delays and packet loss, which can affect the accuracy of the data, especially for real-time ECG signal transmission. The proposed system is intended for use in hospitals, allowing doctors or nurses to view the ECG signals of any patient at any time and place using a computer or smartphone, without physically going to the patient's ward. This system could potentially reduce the time and cost of travel for patients, especially those living in suburban or rural areas. In future work, it may be possible to incorporate an ECG self-interpretation algorithm into the system to detect abnormal ECG signals and trigger an alert. The system could also be expanded by adding additional health sensors to collect different health parameters.

According to [7], the Body Area Network (BAN) is made up of small, smart sensors that are placed on or near the human body and are used to measure various physiological parameters. These sensors, known as Wearable BAN (WBAN), transmit all of the data they collect to a wireless module called XBee, which uses radio frequency (RF) technology to transmit the data to a mobile device that serves as a monitor. The paper focuses on the design of a smart wearable healthcare system called Health Mate, which is based on WBAN and is distributed on the human body to transmit body parameters to a smartphone via IEEE 802.15.4 and to a cloud server for storage in a database. Doctors can access the data in real-time and remotely through the internet to monitor their patients. The Health Mate system uses a Fi module to transmit all of the sensed data to a mobile application, server, and website, and the transmission speed is fast due to the low delay in this process. The aim of this research was to develop the Health Mate system as a Smart Wearable System that allows patients to measure and monitor their vital body parameters and enables remote monitoring for doctors. In addition to this, the project achieved its goal of allowing doctors to remotely monitor their patients' body parameters through a website. Overall, most of the research objectives were achieved and the components were successfully tested, but the Health Mate system needs further refinement and additional features in order to be implemented in real-world applications.

According to [8], in this study, the healthcare system not only developed a passive care system that alerts caregivers when necessary, but also implemented a proactive care notification system to assist users in managing their health status. The system utilizes wearable devices, such as smart clothes and a healthy watch, to record users' physiological signs and behaviors. If the data values fall outside of the standard range, the care notification system will inform the caregiver and the user's family, who can then monitor the situation in real-time. In daily life mode, the system can continue to monitor physiological signals and unusual events over a long period of time, allowing caregivers to have a more comprehensive and systematic understanding of the resident's needs and provide more accurate and efficient care services to improve the quality of care. In addition to its primary function of monitoring and notification, the care notification system also includes various additional features, such as medication reminders, appointment scheduling, and health tracking, to help users manage their overall health and well-being.

According to [9], the application is designed to alert the user to an incoming alert message by playing a loud tone, even if the phone is in silent mode. This feature makes the device useful for alerting people to critical health conditions, as it is inexpensive and easy to use. The device includes two buttons that can be used to alert the user's relatives in case of an emergency. The user interface is simple and easy to use, allowing people to easily perform any necessary actions. Overall, the device is useful for alerting people to critical health conditions and is easy to use [10–12].

The ROAMM system is a technology that helps people to keep track of their health in real-time. It uses sensors and wearable devices to collect personal health data, and analyzes it to identify any potential health issues. If any issues are found, it sends notifications to both the individual and their healthcare provider. This system also includes an Android app that allows users to send emergency alerts, and can be customized and automated for ease of use. Additionally, it can monitor heart rate data from the iPhone's Health App and use sensors and an arduino controller to monitor patients and send data to an IoT server. The goal of this system is to help reduce mortality and hospital admissions by alerting healthcare professionals in emergency situations. The system also allows doctors to access ECG data in real-time via a cloud server remotely [13–15]. This system utilizes a wireless sensor network and wireless body area network to transmit data and has an API format that determines when to send alerts based on user-set thresholds. The patient is given a device with RF transmitter, buttons, Bluetooth transmitter and a microcontroller to use with the system. The detailed review about the smart healthcare ecosystem is mentioned in Table 1.

Table 1. Relevant studies on advantages and disadvantages of smart healthcare ecosystem.

Related Studies	Advantages	Disadvantages	Methodology
[1]	• Real-time monitoring of personal health metrics. • Early detection of potential health issues. • Improved communication and coordination between individuals and healthcare providers. • Enhanced overall health and wellbeing.	• Potential privacy concerns with the collection and sharing of personal health data. • Dependence on technology and potential technical issues. • Potential for false alarms or inaccurate readings.	• The ROAMM software infrastructure uses sensors and wearable devices to collect personal health data in real-time. • This data is then processed and analysed by the system to identify potential health issues. • If any potential issues are detected, the system sends alerts to both the individual and their healthcare provider.

(*continued*)

Table 1. Continued

Related Studies	Advantages	Disadvantages	Methodology
[2]	• Creating an Android app that allows for easy access to emergency helpline phone numbers for various emergency services. • In the event of an emergency, this system creates a digital profile of the user that can be sent to designated trusted contacts, allowing the responding team to better prepare for the situation.	• It relies on technology and may not work in the event of a technological failure or if the user does not have access to a device with the necessary sensors or internet connection. It also assumes that the user has designated trusted contacts who can be alerted in the event of an emergency.	• The Android app includes a feature that allows the user to send an alert message and their current location to specified contacts during emergency, either from the home module or a helpline pop-up with an "Alert" option.
[3]	• By setting reminders for medicine intake and alerting the user if a dose is missed, an IoT system can help improve adherence to prescribed medications. • Early warning for potential health issues.	• Implementing and maintaining an IoT system may be costly for both the user and the healthcare system. • An IoT system may be complex and require technical expertise to set up and use, which may not be accessible to all users.	• The system can be customized and automated using code, and utilizes a personal computer RDBMS product such as Oracle for the database.
[4]	• Smartphone-based heart rate monitoring system allows the user to easily track their heart rate at any time and place, as long as they have their phone with them. • Smartphone app can continuously monitor the user's heart rate and provide real-time data, allowing for more accurate and up-to-date monitoring.	• Smartphone-based heart rate monitoring system relies on the availability and functioning of a smartphone, which may not be reliable in all situations.	• The app receives a notification and collects heart rate data from the iPhone's Health App every time raw data is transferred, comparing it with previous entries to monitor the patient's health condition.
[5]	• An effective system for monitoring COVID-19 patients that includes smart monitoring and emergency alerts is crucial for ensuring the health and safety of these patients and can alert healthcare professionals if any abnormal situations arise.	• Particularly those with chronic diseases such as heart disease, that can alert hospitals in the event of abnormal blood pressure not caused by the coronavirus. This system is focused specifically on COVID-19 and its impact on the patient.	• System using sensors and an arduino controller monitors a patient in stage 1 of a disease and sends data to an IoT server. It hopes to reduce mortality and hospital admissions by alerting healthcare professionals.
[6]	• It is one of the solutions to solve health care issues in rural areas.	• Reducing jitter delay and eliminating noise signals is necessary to enhance the performance of the system.	• ECG data is collected and transmitted in real-time to a cloud server over the internet, allowing doctors to access and monitor the patient's ECG through their mobile cellular network in real-time.

(continued)

Table 1. Continued

Related Studies	Advantages	Disadvantages	Methodology
[7]	• This allows users to track their health in real-time and make any necessary adjustments to their life-style or treatment plan.	• The range of WBANs is typically limited to a few meters, which can be a disadvantage if the sensors need to communicate with devices that are farther away.	• The architecture of a wire-less sensor network (WSN) has been studied in pre-vious research, and this work is based on these findings and related litera-ture, including the use of wireless body area net-works (WBAN) to trans-mit data to a mobile application and remote server.
[8]	• Wearable devices gather and store raw data about an individual's health measurements, which is then incorporated into a personal health examina-tion report. The care noti-fication system tracks user's measurement values and proactively alerts them if these values fall outside of normal ranges.	• While most smart wear-able systems are designed to be accurate, there is still the possibility of errors in the data they collect.	• The healthcare system uti-lizes API format in the healthcare notification system, which determines whether to send alerts based on the alert thresh-olds set by each user for physiological value notifications.
[9]	• Cost effective • Easy to handle	• Some smart devices may not be compatible with certain Android versions or may not work properly on all devices.	• The patient is given a device with an RF trans-mitter, buttons, a blue-tooth transmitter, and a microcontroller. The device has two buttons that the patient is trained to use.

3   CONCLUSION

In conclusion, the use of online health monitoring systems that alert nearby individuals has the potential to greatly improve the overall health and wellbeing of communities. By allowing individuals to monitor their own health metrics and receive alerts when potential issues are detected, these systems can help to identify health problems earlier and allow for timely intervention. Additionally, by alerting nearby individuals, these systems can help to ensure that individuals receive the necessary assistance and support in the event of a health emergency. Overall, the use of online health monitoring systems has the potential to greatly enhance the health and safety of individuals and communities alike.

REFERENCES

[1]   Sanjay Nair, Matin Kheirkhahan, Anis Davoudi, Parisa Rashidi, Amal A. Wanigatung, Duane B. Corbett, Todd M. Manini and Sanjay Ranka, "ROAMM: A Software Infrastructure for Real-time

Monitoring of Personal Health", *2016 IEEE 18th International Conference on e-Health Networking, Applications and Services (Healthcom)*, DOI: 10.1109/HealthCom.2016.7749479

[2] Muthamilselvan S., Joshi C., Tca A. and Dutta A., "Android Application for Emergency Helpline Services," *2018 3rd International Conference on Communication and Electronics Systems (ICCES)*, 2018, pp. 1002–1006, doi: 10.1109/CESYS.2018.8723988.

[3] Ranjana P. and Alexander E., "Health Alert and Medicine Remainder using Internet of Things," *2018 IEEE International Conference on Computational Intelligence and Computing Research (ICCIC)*, 2018, pp. 1–4, doi: 10.1109/ICCIC.2018.8782349.

[4] Marinescu R. and Nedelcu A., "Smartphone Application for Heart Rate Monitoring," *2017 E-Health and Bioengineering Conference (EHB)*, 2017, pp. 141–144, doi: 10.1109/EHB.2017.7995381.

[5] Sabukunze I. D., Setyohadi D. B. and Sulistyoningsih M., "Designing an Iot Based Smart Monitoring and Emergency Alert System for Covid19 Patients," *2021 6th International Conference for Convergence in Technology (I2CT)*, 2021, pp. 1–5, doi: 10.1109/I2CT51068.2021.9418078.

[6] Yew H. T., Ng M. F., Ping S. Z., Chung S. K., Chekima A. and Dargham J. A., "IoT Based Real-Time Remote Patient Monitoring System," *2020 16th IEEE International Colloquium on Signal Processing & Its Applications (CSPA)*, 2020, pp. 176–179, doi: 10.1109/CSPA48992.2020.9068699.

[7] Omer R. M. D. and Al-Salihi N. K., "HealthMate: Smart Wearable System for Health Monitoring (SWSHM)," *2017 IEEE 14th International Conference on Networking, Sensing and Control (ICNSC)*, 2017, pp. 755–760, doi: 10.1109/ICNSC.2017.8000185.

[8] Huang P. C., Lin C. C., Wang Y. H. and Hsieh H. J., "Development of Health Care System Based on Wearable Devices," *2019 Prognostics and System Health Management Conference (PHM-Paris)*, 2019, pp. 249–252, doi: 10.1109/PHM-Paris.2019.00049.

[9] Prakash M., Gowshika U., Ravichandran T., "A Smart Device Integrated with an Android for Alerting a Person's Health Condition: *Internet of Things" 2016 Indian Journal of Science and Technology*, DOI: 10.17485/ijst/2016/v9i6/69545.

[10] Pushpalatha M. P. and Pooja M. R., "A Predictive Model for The Effective Prognosis of Asthma Using Asthma Severity Indicators," *2017 International Conference on Computer Communication and Informatics (ICCCI)*, Coimbatore, India, 2017, pp. 1–6, doi: 10.1109/ICCCI.2017.8117717.

[11] Pattanayak R. K., Kumar V. S., Raman K., Surya M. M., Pooja M. R. (2023). E-commerce Application with Analytics for Pharmaceutical Industry. In: Ranganathan G., Fernando X., Piramuthu S. (eds) *Soft Computing for Security Applications. Advances in Intelligent Systems and Computing*, vol 1428. Springer, Singapore. https://doi.org/10.1007/978-981-19-3590-9_22

[12] Pooja M R. (2022). A Predictive Model for The Early Prognosis and Characterization of Asthma. *Journal of Pharmaceutical Negative Results*, 13(4), 1–9. https://doi.org/10.47750/pnr.2022.13.04.001

[13] Salau A. O., Pooja M. R., Hasani N. F., Braide S. L. (2022). Model Based Risk Assessment to Evaluate Lung Functionality for Early Prognosis of Asthma using Neural Network Approach. *Mathematical Modelling of Engineering Problems*, Vol. 9, No. 4, pp. 1053–1060. https://doi.org/10.18280/mmep.090423

[14] Das S., Pooja M. R. and Anusha K. S., "IOT Based Case Study: Applications and Challenges," *2021 5th International Conference on Computing Methodologies and Communication (ICCMC)*, Erode, India, 2021, pp. 342–348, doi: 10.1109/ICCMC51019.2021.9418326

[15] Pushpalatha M. P. and Pooja M. R., "A Predictive Model for The Effective Prognosis of Asthma using Asthma Severity Indicators," *2017 International Conference on Computer Communication and Informatics (ICCCI)*, Coimbatore, India, 2017, pp. 1–6, doi: 10.1109/ICCCI.2017.8117717.

Author index